5th Global Conference on Materials Science and Engineering (CMSE 2016)

IOP Conference Series: Materials Science and Engineering
Volume 164

Taichung City, Taiwan
8 – 11 November 2016

Editor:

Anete Ashton

ISBN: 978-1-5108-3562-7
ISSN: 1757-8981

Printed from e-media with permission by:

Curran Associates, Inc.
57 Morehouse Lane
Red Hook, NY 12571

Some format issues inherent in the e-media version may also appear in this print version.

Printed by Curran Associates, Inc. (2017)

For permission requests, please contact the Institute of Physics
at the address below.

Institute of Physics
Dirac House, Temple Back
Bristol BS1 6BE UK

Phone: 44 1 17 929 7481
Fax: 44 1 17 920 0979

techtracking@iop.org

Additional copies of this publication are available from:

Curran Associates, Inc.
57 Morehouse Lane
Red Hook, NY 12571 USA
Phone: 845-758-0400
Fax: 845-758-2633
Email: curran@proceedings.com
Web: www.proceedings.com

Table of contents

Volume 164

5th Global Conference on Materials Science and Engineering
8–11 November 2016, Tunghai University, Taiwan

Accepted papers received: 15 December 2016
Published online: 23 January 2017

Preface

Papers

Preface

IOP Conference Series: Materials Science and Engineering is publishing a volume of conference proceedings that contains a selection of papers presented at the 5th Global Conference on Materials Science and Engineering (CMSE 2016).

CMSE 2016, organised by Wuhan Advance Materials Society and Taiwan Tunghai University, technically supported by Hubei University and Taiwan National Central University, was successfully held at Tunghai University in Taichung City, Taiwan during November 8th-11th, 2016.

CMSE is an annual event which has been started in 2012. This Conference aims to bring together leading academic scientists, researchers and scholars from interdisciplinary fields to exchange and share their experience and research results on all aspects of Materials Science and Engineering, and discuss the practical challenges encountered and the solutions adopted.

Located in the subtropical ocean, Taiwan offers comfortable weather, fluctuates during spring and winter, while in summer and autumn the weather is relatively stable, turns out to be an ideal meeting place for domestic and overseas participants of this annual international conference.

The conference program included keynote presentations, special sessions, oral and poster contributions. From several hundred submissions, 29 most promising and IOP mainstream-relevant contributions were included in this volume, which present original ideas or results of general significance supported by clear reasoning and compelling evidence, and methods, theories and practices relevant to the research. The authors state clearly the problems and the significance of their research to theory and practice.

The organisation of this conference and the preparation of proceedings volume would not be possible without the tremendous efforts and dedication of many individuals, especially of the CMSE Conference Secretary Ms. Liu Qin, who oversaw the organisation of the conference and the program, and a large team of reviewers with their timely submission of quality reports. We express our sincere thanks to the authors of all accepted papers for their high quality and fruitful contributions. We also express our gratitude to Editor Anete Ashton and all the Editorial Board members of IOP Publishing for their support in producing the proceedings of this event. We do hope that this volume will be beneficial for readers to their future research endeavours and careers.

Prof. Filippo Berto
Renowned Chair of Fatigue and Fracture
Norwegian University of Science and Technology, Norway

Dr. Alexander Khotsianovsky
Managing Editor of *Strength of Materials*
Pisarenko Institute of Problems of Strength of the National Academy of Sciences of Ukraine, Ukraine

IOP Conf. Series: Materials Science and Engineering **164** (2017) 011002 doi:10.1088/1757-899X/164/1/011002

Peer review statement

All papers published in this volume of *IOP Conference Series: Materials Science and Engineering* have been peer reviewed through processes administered by the proceedings Editors. Reviews were conducted by expert referees to the professional and scienti c standards expected of a proceedings journal published by IOP Publishing.

Numerical simulation and experimental research of the integrated high-power LED radiator

J H Xiang[1], C L Zhang, Z J Gan, C Zhou, C G Chen and S Chen

School of Mechanical and Electric Engineering, Guangzhou University, Guangzhou 510006, China

Email: xiangjh@gzhu.edu.cn

Abstract. The thermal management has become an urgent problem to be solved with the increasing power and the improving integration of the LED (light emitting diode) chip. In order to eliminate the contact resistance of the radiator, this paper presented an integrated high-power LED radiator based on phase-change heat transfer, which realized the seamless connection between the vapor chamber and the cooling fins. The radiator was optimized by combining the numerical simulation and the experimental research. The effects of the chamber diameter and the parameters of fin on the heat dissipation performance were analyzed. The numerical simulation results were compared with the measured values by experiment. The results showed that the fin thickness, the fin number, the fin height and the chamber diameter were the factors which affected the performance of radiator from primary to secondary.

Keywords: High-power LED radiator, Optimization, Experiment, Phase-change heat transfer.

1. Introduction

The thermal management issues, which hindered the further development of high-power LED, have increasingly attracted attention with the development of the high-power LED to the high integration and power. The heat flux density of high-power LED chips had reached as high as 200 W/cm^2 [1]. Poor heat dissipation leads to over-high temperatures of LED chips, which result in reduction of LED performances, or even burning down. The cooling of traditional high-power LED systems generally use solid radiators made of aluminum, copper or composite materials. Their large thermal diffusion leads to inefficient cooling, and has been difficult to meet the cooling demands of the LED with increasing power. On the other hand, the performance of phase-change heat transfer can be enhanced more than 10 times compared to conventional metal heat transfer [2], which has been widely applied in the heat dissipation in the fields of microelectronics and optoelectronics.

The application of solid aluminum radiators and phase-change heat transfer technology in the cooling of high-power LED systems has attracted many researchers' attention [3-4]. Compared with the phase-change radiators, the solid aluminum radiators have lower heat transfer power and larger thermal resistance, which are only suitable for low-power LED lights. However, the use of phase-change heat transfer technology becomes a trend in the high-power LED system cooling with the increase of the LED power [2]. Bladimir et al. [5] had simulated the cooling effects of a micro-jet and a micro-channel water-cooled plate on the LED using a 3D model, and calculated the thermal resistance of the influent with different flow rates. The results showed that the thermal resistance of the micro-channel water-cooled plate was lower than the micro-jet thermal resistance. Wang [6] established a radiator model of a LED spotlight in natural convection condition. By changing the fin spacing, the fin thickness and the LED substrate material, numerical simulations were performed with various placing angles. Cheng et al. [7] analyzed the integral multi-fin radiators on the PCB substrate under forced convection conditions using finite element analysis. The results showed that the junction temperature of the 1W LED arrays reached 123°C without fan and fins, while the junction temperature decreased to 82°C with forced convection cooling.

In this study, phase-change heat transfer and fin heat dissipation were combined in a radiator structure. In the phase-change chamber, the endothermic and exothermic processes of the working fluid during the cycle of vaporization-condensation realized efficient heat transfer, which result in an isothermal condition between the external fins at the top and the bottom of the radiator. With a certain heating power, the effects of the chamber diameter, the fin number, height and thickness on the heat transfer performance of the radiators were studied. The radiator was optimized using the approach combining the orthogonal tests and the numerical simulations, and the results were compared with the actual experiments.

2. Working principle and numerical modeling of the phase-change radiators

2.1. Working principle of the radiator

The working principle of the radiators is shown in Figure 1. The radiator is divided into two parts: the center section of the enhanced heat transfer and the periphery section of the cooling. When the LED heat the bottom of radiator, the working fluid (acetone) in the boiling structure is heated and evaporated. The steam rapidly flows and fills the whole chamber. When the steam contacts the condensing end, it rapidly condenses and releases the latent heat. Then, the condensed working fluid returns to the evaporation surface under the action of gravity. With this type of cycle, the heat continuously transfers from the evaporation surface to the condensation surface, and then dissipates into the atmosphere through the cooling fins. This structure can eliminate the contact resistance due to the seamless connection between the center section of the enhanced heat transfer and the surrounding section of the cooling.

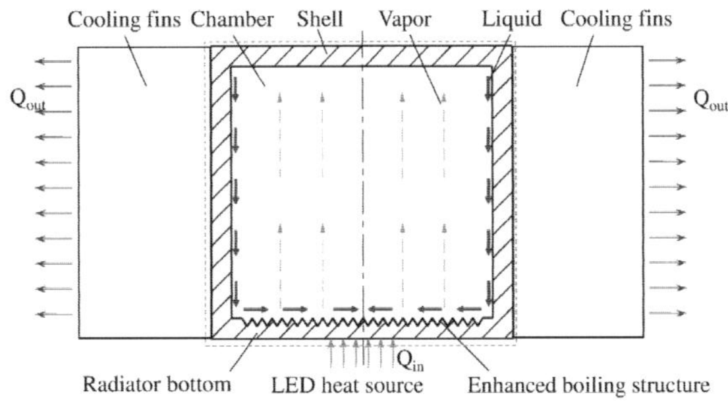

Figure 1. Working principle of the radiator.

2.2. Numerical modeling of the radiator cooling fins

According to the LED working requirements and the thermal design theory, the radiator was simplified before the simulations. The LED heat sources were simplified to the same geometric shape. The thermal resistance of the boiling structure was simplified as the thermal contact resistance between the hot end of the radiator and the phase-change chamber. The phase-change cavity was simplified as a solid cylinder with high thermal conductivity, and its axial heat transfer performance was evaluated as an equivalent thermal conductivity Y [8]:

$$Y = \frac{Q_{\text{in}}\,C}{(T_{\text{Z}} - T_{\text{L}})M} \tag{1}$$

where, Q_{in}-the heat source power, C-the effective length of the heat transfer, M-the sectional area, T_Z-the average temperature at the evaporating end, T_L-the average temperature at the condensation end.

The thermal conductivity of the phase-change cavity behaved as isotropic (Eq. 1). The temperature evaluation points in the cooling system located at the heat source, the hot end and the cold end of the phase-change cavity. After the property definition of the model, the setting of the initial and boundary conditions, the mesh generation and the solving, the average temperatures at the heat source and each evaluation point were exported.

Table 1. Factors and levels of orthogonal test.

Factors	Chamber diameter /A	Fin height /B (mm)	Fin thickness /C (mm)	Fin number /D (piece)
1#	55	160	3(4)	36
2#	65	155	2.5(3.5)	32
3#	70	145	1.5(2.5)	40
4#	60	150	2(3)	28

In order to improve the heat transfer performance of the radiator, the geometric parameters including the height, the thickness, the number, and the length of the fins in the phase-change radiator were optimized. The thicknesses of the inner and outer layer of the fin were represented as c(d). The test was four factors and four levels, and the interactions between factors were ignored. Factors and levels of orthogonal test are shown in Table 1. In order to facilitate the later analysis of variance, the orthogonal table with $L_{16}(4^5)$ type was selected. The model of the phase-change radiator is shown in

Figure 2.

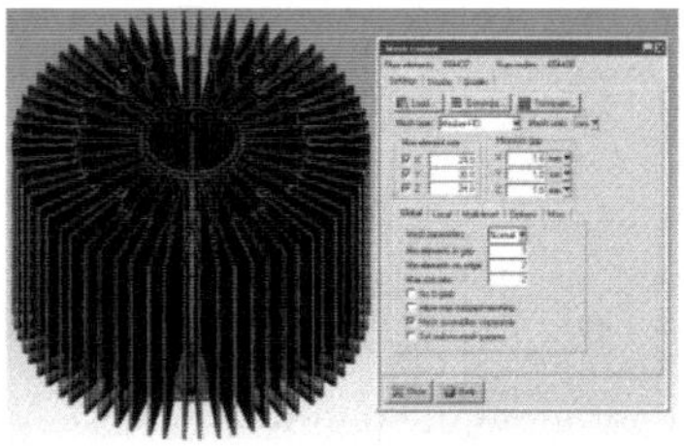

(a) Numerical model

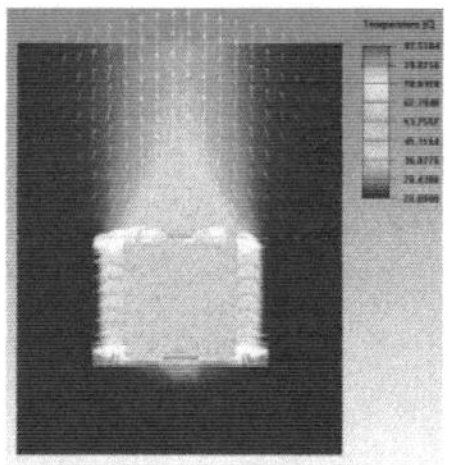

(b) slice image of the temperature and the airflow velocity

Figure 2. The numerical model and the temperature distribution of the radiator.

The calculation equation of the comprehensive index X, which take an overall evaluation of the results, is shown in Eq. 2.

$$X=\alpha_1 T_{hs}+\alpha_2 T_{ah}+\alpha_3(T_{hs}-T_{ah}) \quad (2)$$

where, T_{hs}-the central temperature in the heat source, T_{ah}-the average temperature of the heat source, α_1, α_2, and α_3 the proportion coefficients of the index significances. Comprehensively considering all factors, the values of α_1, α_2, and α_3 in this test were selected as 0.4, 0.3 and 0.3, respectively.

In the tests, the third column was randomly selected as the blank column, and the levels of each factor were filled in the orthogonal table. With temperature of 20°C, 16 models with various parameters were numerically simulated. The highest temperature and the average temperature of the heat source for each model were obtained, and the comprehensive index X was calculated.

3. Results and analysis

3.1. Intuitive analysis

In the intuitive analysis, the larger the range R is the greater the effect of the factor in this column on the index. Within the selected range of the level values, the significance sequence is C (fin thickness), D (fin number), B (fin height) and A (chamber diameter) from primary to secondary due to $R_C>R_D>R_B>R_A$. In the tests, the test index is corresponding to the cooling performance of the radiator. The cooling performance is reflected by the temperature level, therefore, it is better to have a smaller X value. The significance sequence of the factors on the index were both C, D, B, and A from primary to secondary form the variance analysis. The simulation parameters and the simulation results for the optimized scheme are showed in Table 2.

Table 2. The simulation results of optimized scheme.

A (mm)	B (mm)	C (mm)	D (piece)	T_{hs} (℃)	T_{ah} (℃)	X
60	155	1.5(2.5)	32	87.51	80.01	61.25

3.2. The influence of radiator height on the performance

To ensure junction temperature of chip $T_j≦85°C$, the verified heights were 55mm, 75mm, 95mm and 115mm in the orthogonal tests. Numerical simulations were conducted with various powers for the above verified heights. The values obtained from the simulations are shown in Figure 3.

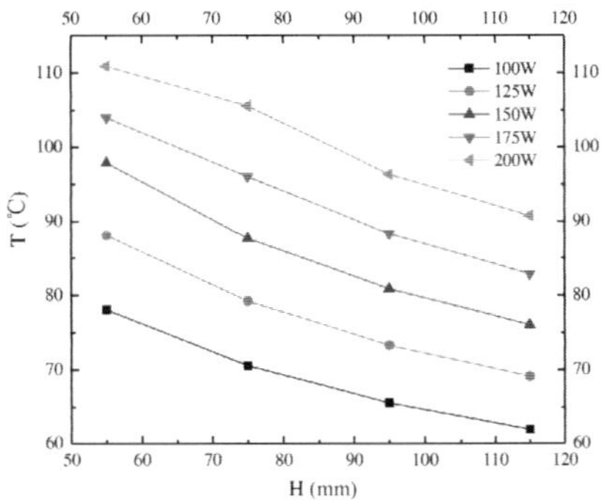

Figure 3. Simulated temperatures with various fin heights.

As can be seen from Figure 3, the height increase led to the temperature decrease with the same power, while the power increase resulted in the temperature increase with the same height. Regarding the cost reduction, the height value should be selected as low as possible in the test range with the premise of satisfying the cooling performance. Therefore, it was more appropriate to select a radiator height of 55, 75, 95 and 115 mm with a heating power of 100, 125, 150 and 200W, respectively.

3.3. Comparison between the experimental and the simulation results

With an ambient temperature of 20°C and heating powers of 50,100,125,150,175 and 200 W, the measured temperatures by thermocouples compared with the simulated temperatures at the corresponding points are shown in Figure 4.

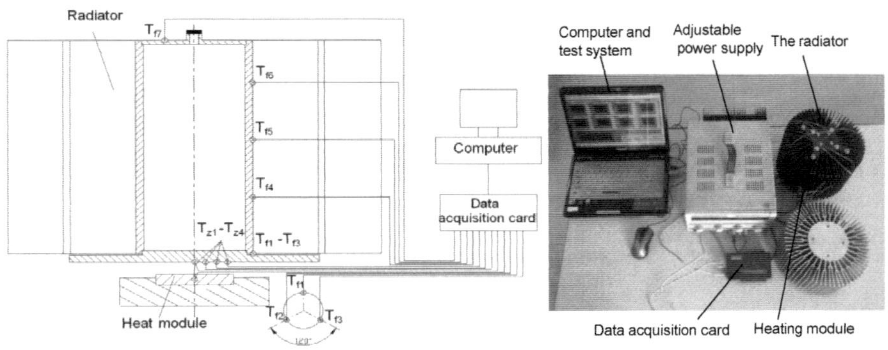

(a) Schematic diagram of thermocouple distribution (b) Testing process

Figure 4. Temperature testing system.

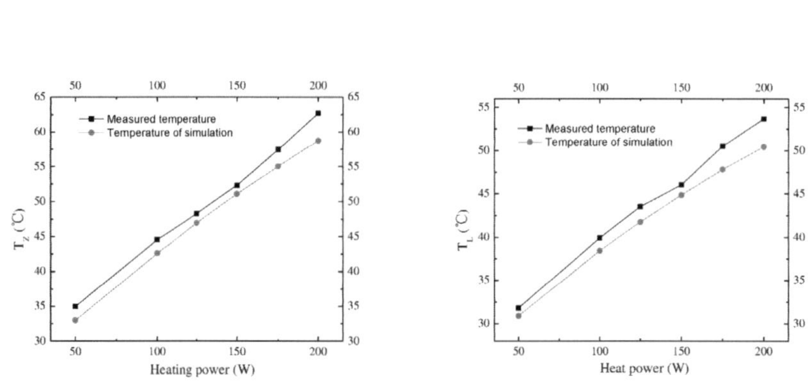

(a) Average temperature at the evaporating end (b) Average temperature at the condensation end

Figure 5. Comparison between the measured temperatures and the simulated temperatures.

From Figure 5, the trends of the measured temperatures and the simulated temperatures were consistent at the corresponding positions. The maximum error between the simulated and the measured temperature was 6°C. The relative errors were below 5% for most data, and were less than 10% for all the data. This was because the radiator model was simplified. Removing subtle geometric characteristics had some effect on the heat transfer process.

4. Conclusions

(1) The employment of the phase-change heat transfer technique led to an isothermal condition of the radiator fins. The radiator was optimized by using a combination of orthogonal tests and numerical simulations, resulting an improvement of the cooling efficiency.

(2) The significance sequence was the fin thickness, the fin number, the fin height and the chamber diameter from primary to secondary. The trend of the measured temperatures was consistent with the simulated temperatures at corresponding positions.

(3) With a certain heating power, the heat source temperature decreased with the increase of the radiator height. With a heating power of 100,125,150 and 200W, it was more appropriate to select a radiator height of 55, 75, 95 and 115mm, respectively.

Acknowledgements

This research were supported by the National Nature Science Foundation of China (51575115, 51275099); the project of Guangdong research program (Yq2013127, 2014A010105053); Project (201510010069) supported by Guangzhou research program; Guangzhou Key Laboratory for Monitoring and Control of Electromechanical Equipment (2060402).

References

[1] Wan Z M, Chen M, Liu W and Liu J 2010 Research on porous micro heat sink for thermal management of high power LED *Journal of Mechanical Engineering* **46**(8) 109-113

[2] Vasiliev L L 2008 Micro and miniature heat pipes-electronic component coolers *Applied Thermal Engineering* **28**(4) 266-273

[3] Kim L, Jong H C and Sun H J *et al* 2007 Thermal analysis of LED array system with heat pipe *Thermochimica Acta* **455**(1-2) 21-25

[4] Liu S, Yang J and Gan Z *et al* 2008 Structural optimization of a microjet based cooling system for high power LEDs *International Journal of Thermal Sciences* **47**(8) 1086-1095

[5] Bladimir R A, Feng B and Peterson G P 2013 Comparison and optimization of single-phase

liquid cooling devices for the heat dissipation of high-power LED arrays *Applied Thermal Engineering* **59**(1-2) 648-659

[6] Wang J C 2014 Thermal module design and analysis of a 230 W LED illumination lamp under three incline angles *Microelectronics Journal* **45**(4) 416-423

[7] Cheng H H, Huang D S and Lin M T 2012 Heat dissipation design and analysis of high power LED array using the finite element method *Microelectronics Reliability* **52**(5) 905-911

[8] Zhan D D and Qian J Y 2012 Study on effective conductivity of heat pipe embedded al plates *Electro-Mechanical Engineering* **04** 12-14

Effects of groove type on airflow speed and pressure during rotor spinning process

R H Yang[1], C Liu, Y Xue and W D Gao

Key Laboratory of Science & Technology for Eco-Textiles, Education Ministry, Jiangnan University, Wuxi 214122, Jiangsu, China

Email: yangrh@jiangnan.edu.cn

Abstract. Groove type is critical to the compactness of fibrous ring in groove and cohesion between fibers. The effect of groove type to high speed airflow during rotor spun yarn spinning process was investigated. Airflow speed and static pressure of G, T, U and S grooves of the 36 mm diameter rotor were studied by Fluent Software respectively. The results showed that under the same conditions, speeds in four slotted size were G>T>U>S within the range from 0° to 360° in groove. At 0° and 360° positions, the static pressures were G>S>U>T. While for the rest of angle position, the static pressures were S>U>T>G. Taking T slot as example, static pressures of the rotors were between -7330.80 Pa and -13719.63 Pa. High speed airflows were divided into two streams as soon as they enter into the inner wall of rotor (0° point), one clockwise and one reverse direction, which joined together at point of 180°. This phenomenon gives light to understand fiber strands stretch and twisting as yarn in rotor which can be used to optimize spinning parameters during spinning and design new rotor type.

Keywords: groove, numerical simulation, airflow field, speed, pressure.

1. Introduction

Rotor spinning is well known for its high output with wide raw materials [1-3]. During rotor spinning process, under the action of the centrifugal force of rotor rotation and fibers slip into the groove after they enter the slipped wall of the rotor by the high negative pressure, then the fibers gathered and twisted to form rotor spun yarn [4].

Coruh et al. studied the effect of the nozzle type as one of the most important parts of the open-end rotor spinning system on yarn quality and found that the nozzle type mostly affects yarn quality and yarn tenacity [5]. Roudbari et al. investigated effect of an increase in opening roller width on yarn quality including tenacity, strain at peak, work of rupture, evenness, imperfections, hairiness and fibre extent within the yarn structure and reported that an increase in fibre opening in lower level improves yarn quality [6]. Esfahani et al. investigated the influence of the navel and rotor type on the tenacity, elongation at break, mass irregularity, total number of imperfections, hairiness, and twist difference values of viscose rotor spun yarns, and found that samples showed a lower value of twist difference produced by a G-type rotor than a T-type rotor [7]. Groove type is critical to the compactness of fibrous ring in groove and cohesion between fibers [8]. There are mainly G, T, U and S types of groove.

In this paper, effects of groove types on airflow speed and pressure during rotor spinning process will be discussed by simulation rotor spinning process through a 3D model with ANSYS software which may favor the fiber strands stretch and twisting as yarn.

2. Models and experiments

The airflow during rotor spinning process obey mass conservation and momentum conservation in view of fluid mechanics [9-12].

Mass conservation equation

$$\frac{\partial(\rho u_k)}{\partial x_k} = 0 \tag{1}$$

Where u_k is the air velocity of x_k direction, and ρ is air density.

Momentum conservation equation

$$\frac{\partial(\rho u_i u_k)}{\partial x_k} = -\frac{\partial p}{\partial x_i} + \frac{1}{\mathrm{Re}}\frac{\partial \tau_{ij}}{\partial x_j} \tag{2}$$

Where ρ is air density, u_k is the air velocity of x_k direction, p is air pressure, Re is Reynolds number, and τ_{ij} is tensor of Newton fluid viscous stress.

$$\tau_{ij} = \mu\left(\frac{\partial u_i}{\partial x_j} + \frac{\partial u_j}{\partial x_i}\right) - \frac{2}{3}\mu\frac{\partial u_k}{\partial x_k}\delta_{ij} \tag{3}$$

Where μ is coefficient of dynamic viscosity, and δ_{ij} is the function of Komecker delta.

Standard k-ε turbulent model is applied to simulate the motion of air flow in rotor.

$$\frac{\partial(\rho k)}{\partial t} + \frac{\partial(\rho k u_i)}{\partial x_i} = \frac{\partial}{\partial x_j}\left[\left(\mu + \frac{\mu_t}{\sigma_t}\right)\frac{\partial k}{\partial x_j}\right] + G_k + G_h - \rho\varepsilon - Y_M + S_k \tag{4}$$

$$\frac{\partial(\rho\varepsilon)}{\partial t} + \frac{\partial(\rho\varepsilon u_i)}{\partial x_i} = \frac{\partial}{\partial x_j}\left[\left(\mu + \frac{\mu_t}{\sigma_\varepsilon}\right)\frac{\partial \varepsilon}{\partial x_j}\right] + C_{1\varepsilon}\frac{\varepsilon}{k}\left(G_k + C_{3\varepsilon}G_b\right) - C_{2\varepsilon}\rho\frac{\varepsilon^2}{k} + S_\varepsilon \tag{5}$$

Where G_k is the item caused by turbulent kinetic energy k which is generated by the average velocity gradient, G_b is the item caused by turbulent kinetic energy b which is generated by buoyancy, Y_M is on the behalf of pulsation expansion in the compressible turbulent flow, $C_{1\varepsilon}$, $C_{2\varepsilon}$ and $C_{3\varepsilon}$ are experimental constants, σ_k are σ_ε are Prandtl numblers according to turbulent energy k and dissipative energy ε separately, S_k and S_ε are source terms defined by users.

According to the recommended value by Launder et al. [13] and experimental verification, in this paper, model constants are determined as $C_{1\varepsilon}=1.42$, $C_{2\varepsilon}=1.68$, $C_{3\varepsilon}=0.09$, $\sigma_k=1.0$, $\sigma_\varepsilon=1.3$.

It's supposed that the airflow speed of inlet is 0.0054 m^3/s and the pressure of outlet is -8000Pa according to experiments while rotor wall is set as rotational moving wall with the speed of 120000r/min (Diameter 36 mm with G, T, U and S respectively). Angles of groove slot are 35°, 45°, 80° and 85° for G, T, U and S respectively. SIMPLE algorithm (Semi-Implicit Method for Pressure-Linked Equations) is used to solve the pressure and velocity coupled. Standard k-turbulent model is applied as the method of turbulent numerical simulation. As the development of turbulences is not sufficient, wall function method is used here. No slip boundary conditions are used in the wall. Geometric model of spin box was shown by Figure 1.

Rotor spinning process was recorded by olymbus i-speed3 in RF30C. Spinning unit is modified by clear plastic.

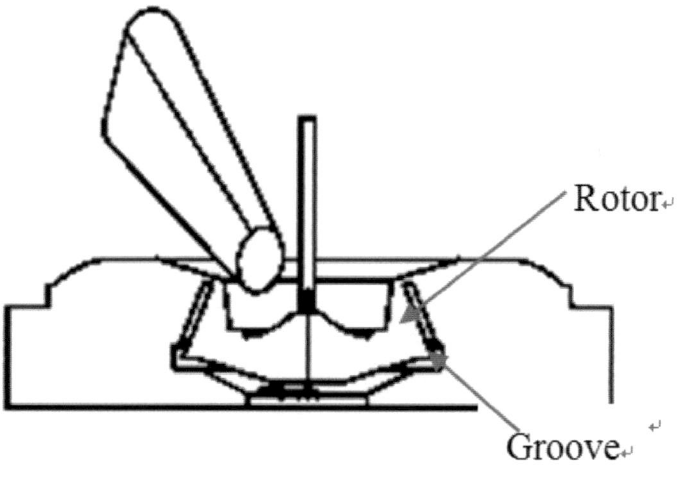

Figure 1. Geometric model of rotor spinning unit.

3. Results and discussion

Angle orders of four grooves type are $G<T<U<S$ (35°, 45°, 80° and 85° respectively). Figure 2-3 demonstrate that airflow speed orders are (S) $S_G>S_T>S_U>S_S$, pressure (P) $P_G<P_T<P_U<P_S$, hence the absolute value of negative pressure, $P_G>P_T>P_U>P_S$. It can be concluded that for grooves with small angle, airflow speeds achieve higher value and negative pressure are stronger. As experiments [4] demonstrated that yarns showed better quality when produced by rotors small angle grooves, it can be said that higher airflow speed and stronger negative pressure can improve yarn quality.

On the other side, short fibers are easier to combine and twist in groove with larger angle which is essential to yarn forming. For fibers which are soft and long, such as cotton, polyester, viscose yarns can be produced by G and T groove rotors which have smaller angles, while for fibers which are short and have high flexural rigidity which have larger angles, such as hemp, tough silk and wool, yarns can be produced by U and S groove rotors as experiments [4] have shown. And also thinner yarn can be produced by G and T which contents fewer fibers in cross section, while thicker yarn can be produced by U and S which contents more fibers in cross section.

The cross point of groove and the extension of fiber transport channel is set as 0° (also as 360°), following clockwise. High speed airflows were divided into two streams as soon as they enter into the inner wall of rotor (0° point), one clockwise and another reverse direction, which joined together at point of 180°, as demonstrated by Figure 3.

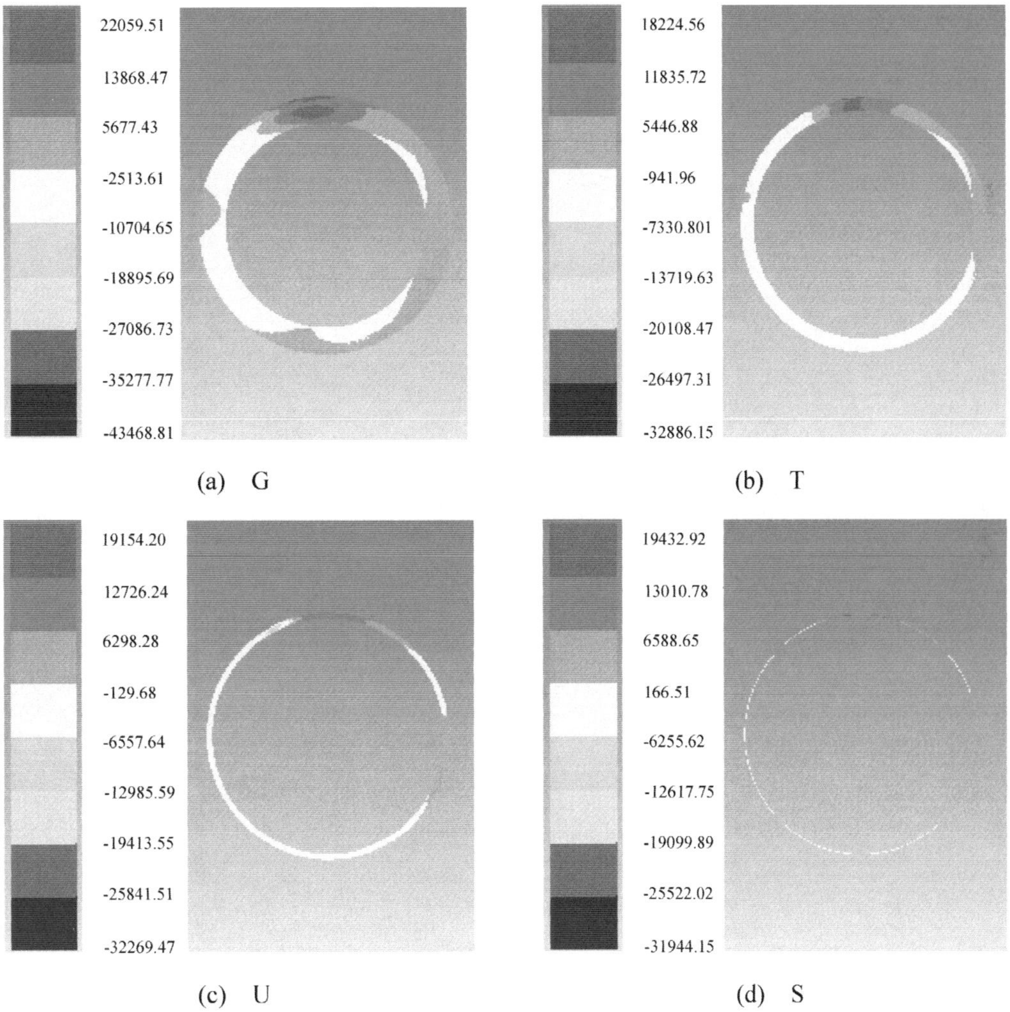

Figure 2. Static pressure (Pa) distribution of groove wall (a) G (b) T (c) U (d) S.

During rotor spinning process, the fibers enter the incline wall which is called the slipped wall inside the rotor and are circulated and piled up into rings like laminated layers, called 'fibrous ring' or 'yarn ring tail', which exerts a big doubling effect. When the piecing yarn enters the rotor, it will be thrown into the collecting groove and joined with the 'fibrous ring'. Then the delivery rollers deliver the yarn out and at the same time the rotor rotation twists the yarn tail. Twists were delivered from yarn tail to fibrous ring as figured out by Figure 4-5 from 90° to 0°. Compared with the reverse direction, speeds of clockwise airflow are faster and pressures are stronger, which can help fiber strands stretching and twisting as yarn which also be clarified by Figure 3.

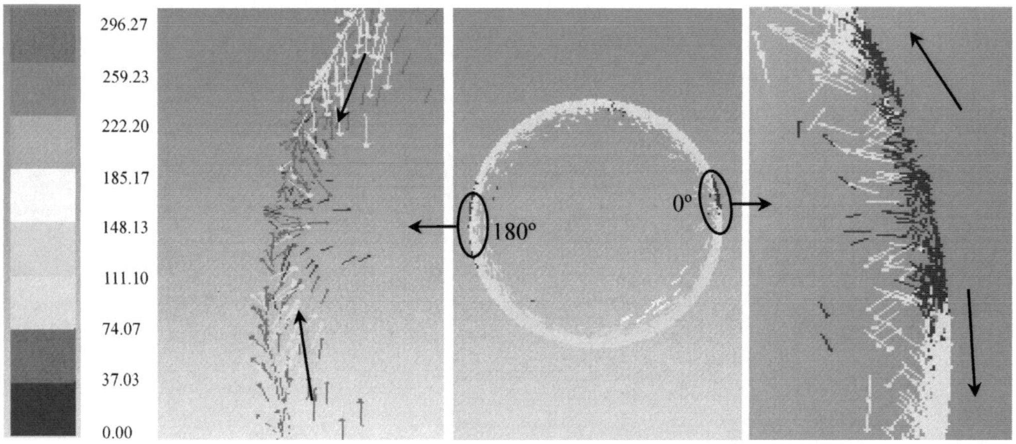

(a) G

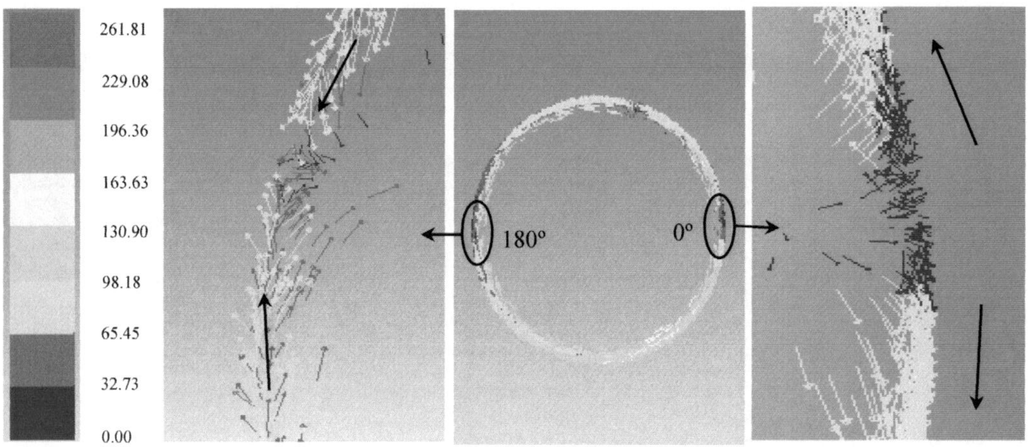

(b) T

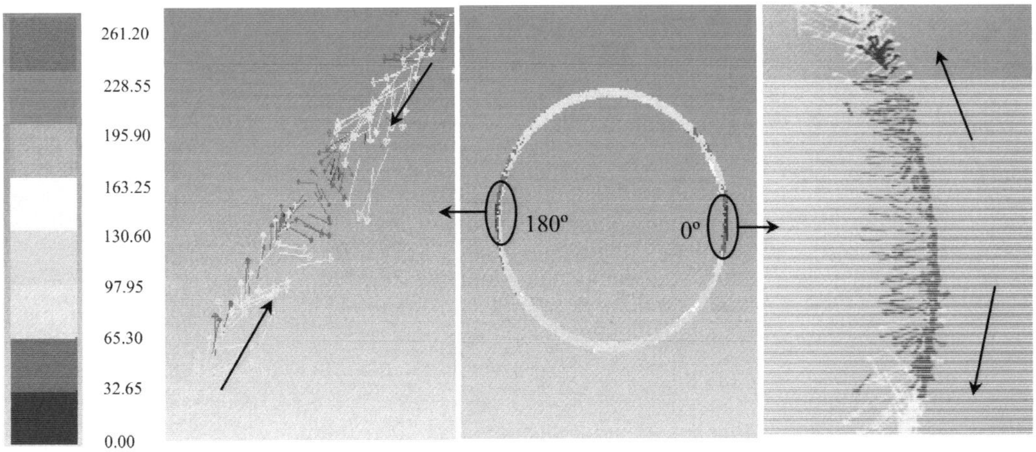

(c) U

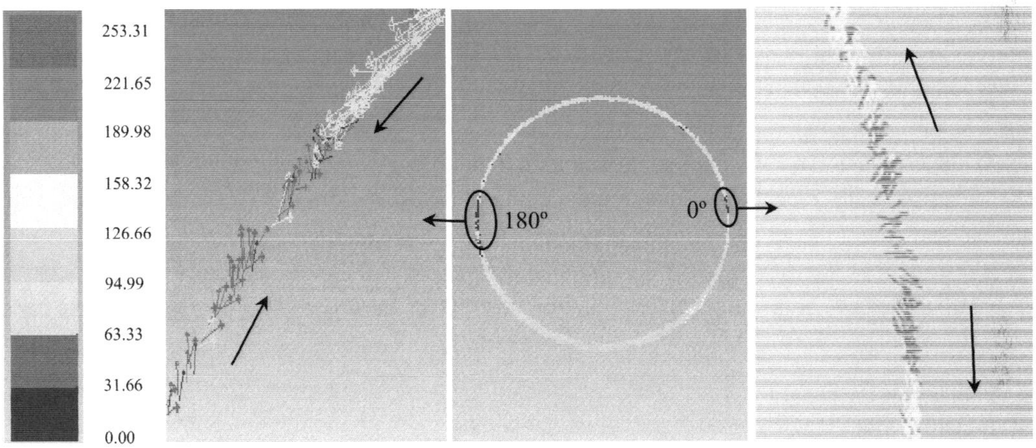

(d) S

Figure 3. Airflow speed (m^3/s) of groove wall (a) G (b) T (c) U (d) S.

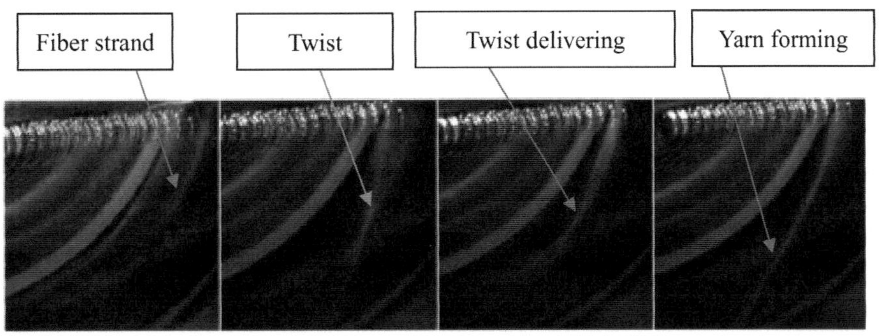

Figure 4. Twisting process of rotor spun yarn.

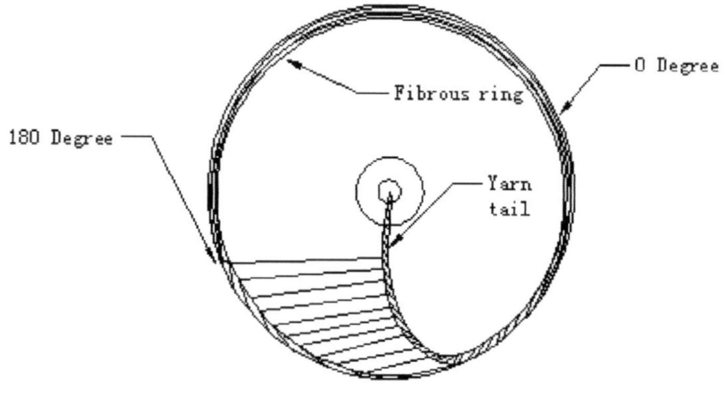

Figure 5. Fibrous rings and bridge fibers in rotor.

4. Conclusion

Airflow characteristics in rotors with G, T, U and S grooves of the 36 mm diameter during rotor spinning process were simulated and analyzed respectively. There are mainly two interesting and useful points. First, airflow speed (S) shows the order of $S_G>S_T>S_U>S_S$, pressure (P) $P_G<P_T<P_U<P_S$ (hence the absolute value of negative pressure, $P_G>P_T>P_U>P_S$). It can be concluded that, airflow speeds achieve higher value and negative pressures are stronger in grooves with small angle, which can enhance yarn quality especially thinner yarn. On the other side, short and rigid fibers are easier to combine and twist to form yarn in groove of larger angle. Second, high speed airflow were divided into two streams as soon as they enter into the inner wall of rotor (0° point), one clockwise and another reverse direction, which joined together at point of 180°. This phenomenon gives light to understand fiber strands stretch and twisting as yarn in rotor which can be used to optimize spinning parameters during spinning and design new rotor type.

Acknowledgement

This work was supported by Natural Science Foundation of Jiangsu Province of China No. BK20130148, the National Natural Science Foundation of China No. 51403085 and the Fundamental Research Funds for the Central Universities No. JUSRP51631A.

References

[1] Alamdar-Yazdi A and Heppler G R 2012 Abrasion behavior of yarns at right angle for ring and rotor spun yarn *Fibres & Textiles in Eastern Europe* **20**(6) 54-57

[2] Chattopadhyay R, Tyagi G K and Goyal A 2013 Studies of the hybrid effect in mechanical properties of tencel blended ring-, rotor- and air-jet spun yarns *Journal of the Textile Institute* **104**(3) 339-349

[3] Kuo C F J, Kao C Y and Wei H J 2011 Optimization of open-end rotor spinning frame parameter and estimation of relevant quality characteristics *Polymer-Plastics Technology and Engineering* **50**(9) 923-930

[4] Wang S Y and Yu X Y 2006 *New Textile Yarns* (Shanghai: Publication of Donghua University) p 93

[5] Coruh E and Celik N 2013 Influence of nozzle type on yarn quality in open-end rotor spinning *Fibres & Textiles in Eastern Europe* **21**(2) 38-42

[6] Roudbari B Y, Eskandarnejad S and Moghadam M B 2016 Effect of an increase in opening roller width on quality of rotor spun yarn *Journal of the Textile Institute* **107**(7) 864-872
[7] Esfahani R T and Shanbeh M 2014 Effect of navel and rotor type on physical and mechanical properties of viscose rotor spun yarns *Fibres & Textiles in Eastern Europe* **22**(3) 61-65
[8] Hasani H, Semnani D and Tabatabaei S 2010 Determining the optimum spinning conditions to produce the rotor yarns from cotton wastes *Industria Textila* **61**(6) 259-264
[9] Zeng Y C and Yu C W 2004 Numerical simulation of fiber motion in the nozzle of an air-jet spinning machine *Textile Research Journal* **74**(2) 117-122
[10] Pei Z G, Hu B Y, Diao C Y and Yu C W 2012 Investigation on the motion of different types of fibers in the vortex spinning nozzle *Polymer Engineering and Science* **52**(4) 856-867
[11] Zou Z Y, Cheng L D, Xi B J, Luo Y X and Liu Y C 2015 Investigation of fiber trajectory affected by some parameter variables in vortex spun yarn *Textile Research Journal* **85**(2) 180-187
[12] Li M L, Yu C W and Shang S S 2014 Effect of vortex tube structure on yarn quality in vortex spinning machine *Fibers and Polymers* **15**(8) 1786-1791
[13] Launder B E and Spalding D B 1972 *Lectures in Mathematical Models of Turbulence* (London: Academic Press) p 55

Experimental study on the influence of wire diameter on the internal flow behaviour of woven metal screens

Y B Peng, G Q Xu, X Luo, H W Li[1] and Y P Liu

National Key Laboratory of Science and Technology on Aero Engines Aero-Thermodynamics, The Collaborative Innovation Center for Advanced Aero-Engine of China, Beihang University, Beijing 100191, China

Email: 09620@buaa.edu.cn

Abstract. Sintered woven wire mesh structures are a classic porous medium. In this paper, we examined the internal flow behaviours of sintered metal wire mesh structures with 0.215mm diameter wires with different porosities. Following previous research results, the influence of woven mesh wire diameter on material penetrating quality was studied. The air that was applied by the gas source was used to investigate structure performance. The Reynolds numbers of the inlet changed from 8.2 to 66.1. The pressures and flow rate at the inlet and outlet were obtained to calculate the permeability and inertia coefficient of each specimen, as well as the friction factors. The experiment results showed that permeability increased and the inertia coefficient decreased as wire diameter increased. Moreover, structures with large wire diameters ($d_s = 0.215mm$) showed better penetrating quality at the same porosity levels. Increases in pressure drop kept pace with increases in diameter. The friction factor decreased as the Reynolds number increased, and tended to be constant.

1. Introduction

As technological advances improve thrust levels for jet engines, these engines need correspondingly better cooling techniques. A key problem associated with jet engine cooling lies in decreasing coolant consumption. The current composite cooling method on turbine blade offer suboptimal cooling for modern jet engines. Aero-engine engineers seek new and improved cooling methods. Transpiration cooling is a new cooling method that uses porous media to achieve higher cooling efficiency than traditional methods. Porous media effectively transfer heat because of the large heat exchange area within the unit volume [1]. Porous materials are widely used for a variety of applications including heat exchanger, solar energy collection, etc. [2-5]. To apply porous media to aero-engine cooling, it is indispensable to investigate the friction characteristics of porous structure.

Most research on flow and heat transfer behaviours focus on powder metal structures [6-8], which has been widely used in heat exchange, electronic equipment cooling, and other applications. For high centrifugal force environments such as the turbine blade in an aero engine, woven wire mesh structures have more application potential [9]; additionally, they are made of strong materials and they are relatively cheap to produce. Generally, wire mesh structures are woven. In this study, the flow resistance of Dutch weave metal screens were investigated. The Dutch woven matrix consists of machine direction and orthogonal cross-direction wires in each layer. Figure 1 shows the weaving pattern and close-up of a Dutch weave.

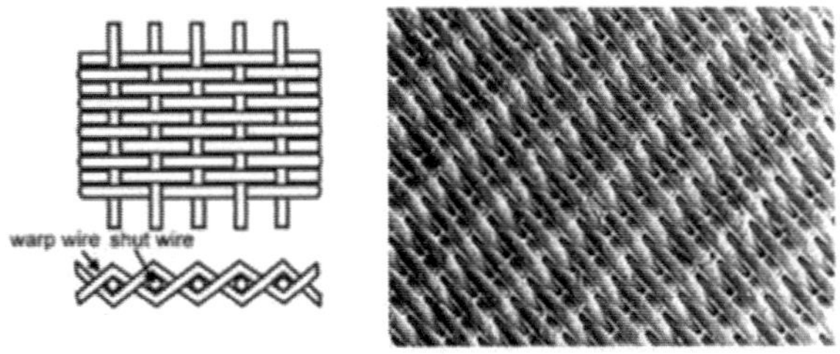

Figure 1. Weaving pattern and close-up of Dutch weave.

In 1968, Armour [10] studied the hydraulic resistance of several types of woven screens through experiments with a single screen layer. An expression to calculate the pressure drop was obtained. Xiong and Jiang [11] studied the influence of fluid compressibility on air flow resistance characteristics in Dutch weave wire mesh structures. Liu et al. [12] designed an experimental procedure to investigate the flow behaviour and heat transfer characteristics of Dutch weave metal screens with different porosities. They developed an empirical equation of friction characteristic for Dutch weave metal screens. Ma and Luo [13] set up the experiment to investigate the forced convection heat transfer of air in woven mesh porous structure with two layered media of different porosity. The experimental results showed that different porosity combinations significantly affected the flow and heat transfer parameters.

Despite the numerous works done to determine the pressure drop through woven-screen matrices, there have not been studies on the influence of Dutch woven metal mesh structural parameters on flow performance. In this study, a series of experimental tests was conducted to determine the influence of metal wire diameter (d_s) on the internal flow friction characteristic of Dutch weave metal screens, based on the research in Liu et al. [12]. In the previous reference study [12], the penetrating quality of sintered woven mesh with small diameter wires ($d_s = 0.14mm$) were experimentally studied to acquire the relationship between friction factor and Re in different porosities. To investigate the effects of wire diameter on woven mesh structure flow performance, we selected a large wire diameter ($d_s = 0.215mm$) test sample to obtain the flow resistance characteristic curve at different porosities. The woven mesh porosity was kept as a constant with the reference experiment [12].

2. Experimental system

2.1. Woven matrix and test section

Dutch weave screens are made of SAE 316 stainless steel wires with wire diameter in 0.215mm. The melting temperature of this type of stainless steel ranges from 1370°C~1398°C, which far exceeds the working temperature range of the current experiment. Thus, we can ignore material changes caused by temperatures in this experiment. The woven stainless steel wires suitably comprised the single-layer structure. Table 1 and Table 2 shows the parameters of the test pieces in the current experiment and reference [12], respectively. The averaged pore diameter values were measured by the manufacturer with a 0.5% measuring accuracy.

Table 1. Parameters in this experiment.

Text#	Wire diameter (mm)	Number of layers	Averaged pore diameter(μm)	Porosity
D1	0.215	15	154.7	52%
D2	0.215	18	126.6	46%
D3	0.215	22	76.4	35%

Table 2. Parameters in the experiment of Liu et al. [12].

Text#	Wire diameter (mm)	Number of layers	Averaged pore diameter(μm)	Porosity
S1	0.14	20	93.7	55%
S2	0.14	24	90.9	47%
S3	0.14	28	66.9	37%

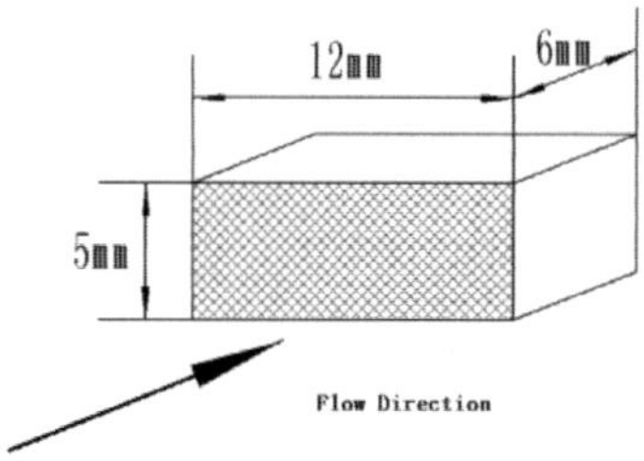

Figure 2. Schematic of the test sample.

As shown in Figure 2, the screen sample cross-section was 12mm × 5mm with 6mm thick in flow direction. The porosity in the present study is given by the following equation:

$$\varepsilon = 1 - \frac{M / V}{\rho_t} \tag{1}$$

where M, V, and ρ_t are the mass, volume, and density of the test piece, respectively.

The porous sample itself is installed into a test bench as shown in Figure 3. It is observed that the sample was fixed in the middle of the flow channel which was tightly attached by two plexiglass cover plates. Two copper blocks can stop test sample from moving normal to the flow direction.

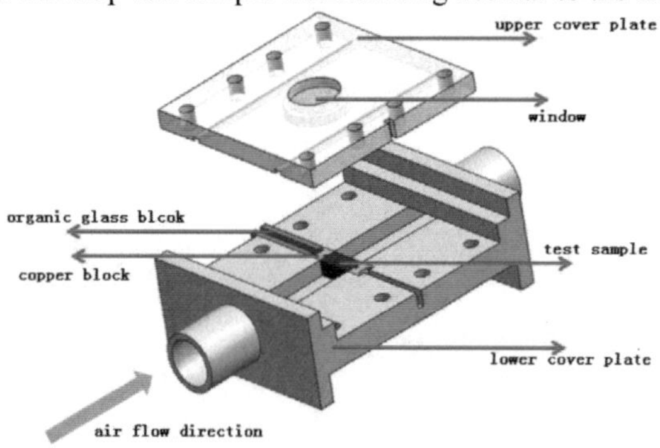

Figure 3. Test section.

2.2. Experimental apparatus

Originally constructed by Liu et al. [12], the experimental system used to investigate the flow and heat transfer performance of the sintered porous structure is shown in Figure 4. Compressed room temperature air was provided by an air compressor. The pressure and flow rate of the clean air which purified by filter were controlled through valve. Rosemount transducers (0.15% accuracy) were used for the pressure measurements which located before and after the test section. The K-type thermocouples with an accuracy of ±0.1 K were installed to measure the inlet and outlet gas temperatures. All the measurement instruments were connected to a data acquisition system with ADAM. The system was assumed to be at steady state when variations of the inlet and outlet fluid temperatures were all within ± 0.1K and the relative variations of the flow rate and pressure were all within $\pm 2\%$ for at least 10 min.

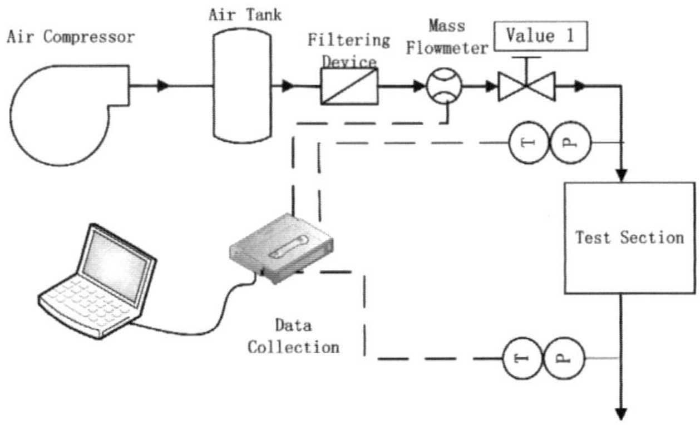

Figure 4. Experimental setup.

3. Experimental results and discussion

The modified Darcy equation [12] as shown in the following was used to calculated the pressure drop through a porous medium.

$$\frac{\Delta P}{\delta} = \frac{\mu}{K} u + R_f \rho u^2 \quad (2)$$

The value of permeability K and inertia coefficient R_f for each test sample should be evaluated in the experiment. δ is the test piece thickness, and u is the average velocity calculated by the following equation:

$$u = \frac{m}{\rho A_c} \quad (3)$$

where m is the air mass flow rate. A_c is the cross-section area of the channel.

In literature the pressure loss behaviour is commonly described with the dimensionless friction factor f.

In current study, the friction factor of the woven metal screen is defined as

$$f = \frac{\Delta P}{\frac{1}{2}\rho u^2} \cdot \frac{d_s}{\delta} \cdot \frac{\varepsilon^3}{1-\varepsilon} = \frac{2\Delta P \rho A_c^2}{m^2} \cdot \frac{d_s}{\delta} \cdot \frac{\varepsilon^3}{1-\varepsilon} \quad (4)$$

The characteristic length d_s in this equation is the wire diameter of the screen, and ε is the test piece porosity.

The Reynolds number is defined as

$$\mathrm{Re} = \frac{\rho u d_s}{\mu} \cdot (1-\varepsilon) = \frac{m d_s}{A_c \mu} \cdot (1-\varepsilon) \quad (5)$$

Following the uncertainty analysis in reference [12], the experimental uncertainty in the Reynolds number and friction factor are 6.1% and 9.3% of the presented data, respectively.

The value of permeability and the inertia coefficient for all the samples were determined from the experimental data are listed in Table 3. In this study, the experiments were repeated several times for ensuring the results reliable. For comparison, Table 4 shows the reference results [12].

Table 3. Properties of the investigated samples.

Text#	Wire diameter (mm)	Porosity	Permeability ($\times 10^{-10} m^2$)	Inertia coefficient ($\times 10^{-5} m^{-1}$)
D1	0.215	52%	1.45	0.181
D2	0.215	46%	1.19	0.261
D3	0.215	35%	0.29	1.377

Table 4. Properties of the investigated samples in Liu et al. [12].

Text#	Wire diameter (mm)	Porosity	Permeability ($\times 10^{-10} m^2$)	Inertia coefficient ($\times 10^{-5} m^{-1}$)
S1	0.14	55%	1.26	0.401
S2	0.14	47%	1.06	0.883
S3	0.14	37%	0.36	2.166

As shown in Tables 3 and 4, the variation trend of the permeability and inertia coefficient were similar on large and small diameters wires when the porosity changed. Permeability increased as porosity increased, while the inertia coefficient showed the opposite trend. Using these results, the relationship between pressure gradient and air velocity values could be obtained. Figure 5 shows the pressure gradient sharply increased with respect to the entrance velocity. The pressure gradient dropped with the porosity increase.

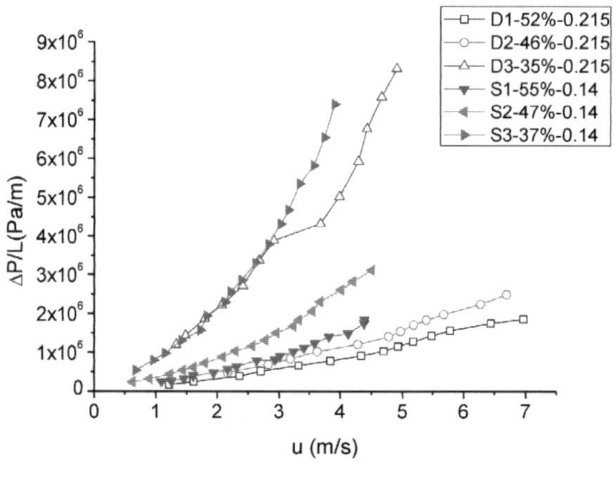

Figure 5. Penetrating quality curves for test pieces.

Test sample D1, D2, D3 $d_s = 0.215mm$;

Test sample S1, S2, S3 $d_s = 0.14mm$ from Liu et al. [12].

It can also be seen from Figure 5 that a larger wire diameter ($d_s = 0.215mm$) led to lower flow resistance and higher penetrating quality at the same porosity.

Figure 6 shows the pressure drop variations for small and large wire diameter samples. The pressure loss of each piece increased following the air flow rate increase. When under constant flow rate, decreasing wire diameter led to the fall of pressure. Note that the parameters from Tables 1 and 2 revealed that teat pieces with large diameter ($d_s = 0.215mm$) had higher averaged pore diameter when compared at the same porosity. The wire diameter decrease resulted in smaller averaged pore diameter. It means that more channels were formed in the unit area of flow section which generated lager area of fluid-solid interface. Thus the higher frictional dissipation was created by much greater contact area between air and inner wall of porous structure. This rule showed that a large diameter resulted in less effective contact area, thus decreasing the flow resistance.

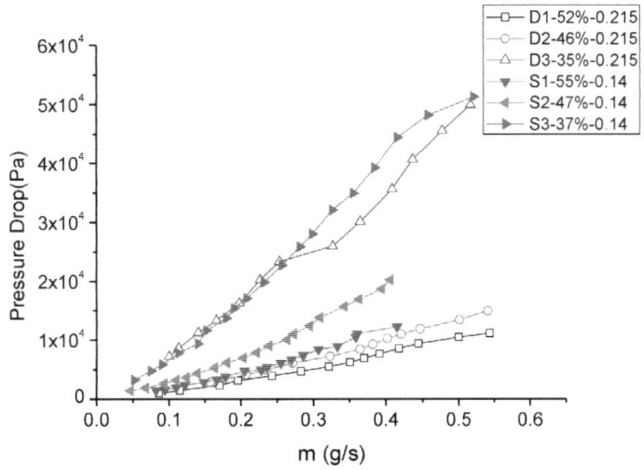

Figure 6. Pressure drop for samples in different diameters.

Test sample D1, D2, D3 $d_s = 0.215mm$;

Test sample S1, S2, S3$d_s = 0.14mm$ from Xiong et al. [11].

Figures 7 and 8 plot the friction factor variations at different Reynolds numbers for small and large diameter wires. With increasing Reynolds number Re, the friction factor f decreases significantly in the beginning and then the variety reduction. It is important to note that for small wire diameter samples as showed in Figure 7, the data for the test pieces showed a single curve within an error band. This indicates that an empirical equation about the friction factor in different porosities could be provided in small diameter wires.

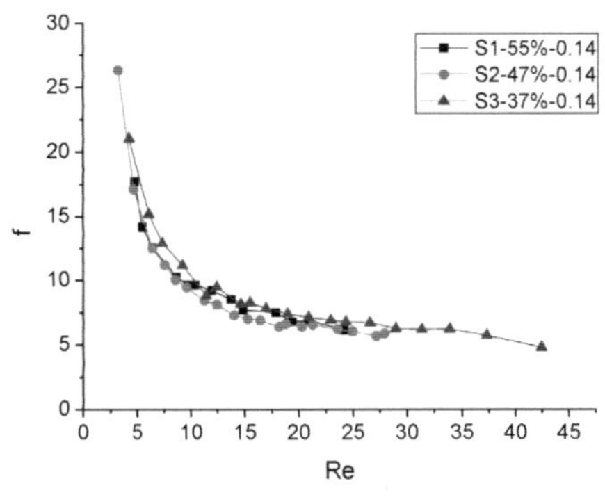

Figure 7. The friction factor for various Reynolds numbers at various porosities (55%, 47%, 37%).

Small wire diameter of test samples: $d_s = 0.14mm$.

However, there is a clear difference between the friction factor curves with large diameter wires, which mean the results cannot be represented by one empirical equation. Moreover, it can be seen from Figure 8 that the friction factor curves for 52% and 46% porosity were very close. The trend can also be seen in the pressure gradient curve from Figures 5 and 6. This phenomenon may be caused by comprehensive factors such as parameter definition and mesh geometry structure. Thus, it is necessary to further study this problem.

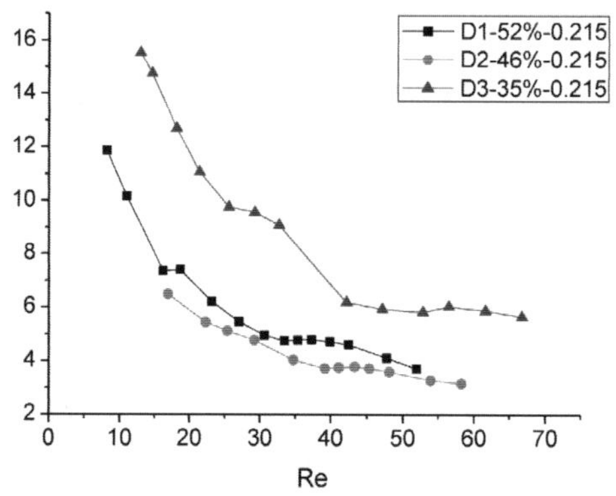

Figure 8. The friction factor for various Reynolds numbers at various porosities (52%, 46%, 35%). Large wire diameter of test samples: $d_s = 0.215mm$.

4. Conclusions

In this work, an experimental setup was established to measure the internal flow resistance performances of woven metal porous structure with the same 0.215mm in wire diameter. The influence of woven mesh wire diameter on the penetrating quality of the material was studied and compared with the research results in Liu et al. [12].

The permeability and inertia coefficient change tendency of woven mesh structures along with porosity in different wire diameters were similar. However, large wire diameter ($d_s = 0.215mm$) structures showed better penetrating quality at the same level of porosity. For the conditions studied here, the pressure drop increase with respect to the diameter. The friction factor decreased as the Reynolds number increased, and tended to be constant.

References

[1] Kays W M and London A L 1964 *Compact Heat Exchangers* (New York: McGraw-Hill)

[2] Antohe B V and Lage J L 1996 Numerical characterization of micro heat exchangers using experimentally tested porous aluminum layers *International Journal of Heat Fluid Flow* **17**(6) 594-603

[3] Kolb A, Winter E R F and Viskanta R 1999 Experimental studies on a solar collector with metal matrix absorber *Solar Energy* **65**(2) 91-98

[4] Tharur N S, Saini J S and Solanki S C 2003 Heat transfer and friction factor correlations for packed bed solar air heater for a low porosity system *Solar Energy* **74**(4) 319-329

[5] Richards P J and Tobinson M 1999 Wind loads on porous structure *Journal of Wind Engineering and Industrial Aerodynamics* **83**(1) 455-465

[6] Jiang P X, Li M, Lu T J, Yu L and Ren Z P 2004 Experimental research on convection heat transfer in sintered porous plate channels *International Journal of Heat and Mass Transfer* **47**(10) 2085-2096

[7] Jiang P X and Lu X C 2007 Numerical simulation and theoretical analysis of thermal boundary characteristics of convection heat transfer in porous media *International Journal of Heat and Fluid Flow* **28**(5) 1144-1156

[8] Xu R N, Huang Y L, Jiang P X and Wang B X 2012 Internal heat transfer coefficients in microporous media with rarefaction effects *Science China Technological Sciences* **55**(10) 2869-2876

[9] Polezhaev J 1997 The transpiration cooling for blades of high temperatures gas turbine *Energy Conversion and Management* **38**(10) 1123-1133

IOP Conf. Series: Materials Science and Engineering **164** (2017) 012003 doi:10.1088/1757-899X/164/1/012003

[10] Armour J C and Cannon J N 1968 Fluid flow through woven screens *AIChE Journal* **14**(3) 415-420

[11] Xiong Y B and Jiang P X 2011 Experimental study on Flow resistance characteristics of air flow through Dutch woven metal mesh *CSET Heat and Mass Transfer Academic Meeting* No.113567

[12] Liu Y, Xu G, Luo X, Li H W and Ma J D 2015 An experimental investigation on fluid flow and heat transfer characteristics of sintered woven wire mesh structures *Applied Thermal Engineering* **80** 118-126

[13] Ma J, Lv P, Luo X, Liu Y P, Li H W and Wen J 2016 Experimental investigation of flow and heat transfer characteristics in double-laminated sintered woven wire mesh *Applied Thermal Engineering* **95** 53-61

Advances in the numerical investigation of the immersion quenching process

D S Zhang[1,3], R Kopun[2], N Kosir[2] and W Edelbauer[1]

[1] AVL List GmBH, Graz, Austria
[2] AVL - Advance Simulation Technology d.o.o., Maribor, Slovenia

Email: dongsheng.zhang@avl.com

Abstract. A numerical investigation of the immersion quenching process is presented in this paper. Immersion quenching is recognized as one of the common ways to achieve the desirable microstructure, and to improve the mechanical properties after thermal treatment. Furthermore it is important to prevent distortion and cracking of the cast parts. Accurate prediction of all three boiling regimes and the heat transfer inside the structure during quenching are important to finally evaluate the residual stresses and deformations of thermally treated parts. Numerical details focus on the handling of the enthalpy with variable specific heat capacity in the solid. For two application cases, comparison between measured and simulated temperatures at different monitoring positions shows very good agreement. The study demonstrates the capability of the present model to overcome the numerical challenges occurring during immersion quenching and it is capable of predicting the complex physics with good accuracy.

Keywords: immersion quenching, numerical simulation, variable specific heat capacity, cooling curve.

1. Introduction

To satisfy ecological requirements, modern automotive industry is using various means to make vehicles become more efficient and more ecological. To further reduce fuel consumption and emissions, heavier metals have been increasingly replaced by those high strength-to-weight ratio and low-cost aluminium alloys [1]. Temperature sensitive microstructural defects can lead to early failure of working parts at the service temperature. Furthermore, Residual stress created during heat treatment can lead to stress corrosion cracking, fatigue cracking and distortion in components. Among many heat treatment techniques, optimized immersion quenching process has been extensively applied to improve the mechanical properties and to obtain the desirable microstructure, mechanical and metallurgical characteristics of final metallic alloy components during production. In addition, immersion quenching could produce a relatively even temperature distribution during the rapid cooling process, thus reduces the residual stress level and consequently lowers the risk of distortion and cracking of the cast parts [2].

In the immersion quenching process, rapid heat transfer happens immediately after the metal piece with a microstructure desirable temperature is dipped into the sub-cooled liquid pool. Water, oil or polymers are widely used as liquid quenchant [3]. In most cases, this microstructure dependent temperature is much higher than the normal Leidenfrost point, so film boiling, transition boiling and nucleate boiling are found during the immersion quenching cooling process [4]. Optimal quench uniformity is essential for distortion control and prevention of cracking problems. This demands the heat transfer between the quenchant and hot workpiece as uniform as possible throughout all three boiling regimes. Although experimental researches [5, 6] have showed invaluable insight into the fluid mechanics of the quenching process, they are generally incapable of continuously monitoring fluid flow in commercial quench tanks during heat treatment process. CFD modeling [7-12] is being extensively used to investigate fluid flow and heat transfer in quenching process and has clearly demonstrated the enormous prospect of promoting quenching performance with the objective of retrofitting for improved performance or designing new and innovative quench systems. Detailed description of the immersion quenching process and numerical aspects are described in the publications of Kopun et al. [9-10], Greif et al. [11] and Srinivasan et al. [12].

Previous works [9-13] focus mainly on constant thermal properties, while in many cases, thermal properties of the solid work pieces, i.e. specific heat capacity, C_p and thermal conductivity λ depend on temperature. The specific heat capacity, C_p is particularly a very critical parameter during the numerical simulation of the heat transfer. When the specific heat capacity C_p is constant over a temperature difference of ΔT, the static enthalpy change can be easily taken as $\Delta h = C_p\Delta T$; whereas when C_p is temperature dependent, the static enthalpy change is not $C_p\Delta T$. In this work, a correct treatment of the spatially and temporally changing C_p is developed and a more general expression for temperature dependence of thermal enthalpy change is presented.

The paper is organized as follows. Section 2 presents the mathematical model. Simulation set-up and details of two different industrial quenching processes are briefly introduced in section 3. The comparisons of the simulated results from two industrial quenching processes with the experimental data are presented and discussed in section 4. Conclusions and recommendations are given in section 5.

2. Mathematical model

In the framework of the Eulerian multi-fluid model, each phase is taken as interpenetrating continua coexisting in the flow domain with inter-phase transfer terms accounting for phase interactions. Two-sets of ensemble averaged balance equations for mass, momentum and energy [14] take the following form:

2.1. Continuity equation

$$\frac{\partial(\alpha\rho)_k}{\partial t} + \nabla\cdot\left(\alpha\rho\vec{\mathbf{V}}\right)_k = \sum_{l=1,l\neq k}^{N}\Gamma_{l\to k} \tag{1}$$

The compatibility condition is exerted on the continuity equation:

$$\sum_{k=1}^{N}\alpha_k = 1 \tag{2}$$

where the indices k refers to the phase (l for liquid and v for vapour). The volume fraction of each phase is denoted as α. ρ stands for density and $\vec{\mathbf{V}} = (u, v, w)$ is the velocity vector. The inter-phase mass change rate (in this particular case, boiling) is Γ_k.

2.2. Momentum equation

$$\frac{\partial\left(\alpha\rho\vec{\mathbf{V}}\right)_k}{\partial t} + \nabla\cdot\left(\alpha\rho\vec{\mathbf{V}}\vec{\mathbf{V}}\right)_k == -\alpha_k\nabla p + \nabla\cdot(\alpha\tau)_k + (\alpha\rho)_k\vec{\mathbf{g}} + \mathbf{M}_{lk} + \vec{\mathbf{V}}_i\Gamma_{l\to k} \tag{3}$$

where p is pressure; $\vec{g}$ is gravity vector, τ is stress tensor and interfacial velocity is denoted as $\vec{V}_i$. Interfacial momentum exchange term is calculated as

$$\mathbf{M}_{lk} = \mathbf{F}_D + \mathbf{F}_L + \mathbf{F}_{wl} \tag{4}$$

where the terms on the right hand side represents forces due to drag, F_D, wall lubrication, F_{wl} and lift F_L.

2.2.1. Drag force. The drag force of the liquid phase is given as

$$\mathbf{F}_D = \frac{3}{4}\alpha_v \rho_l \frac{C_D}{d_b}\left|\vec{\mathbf{V}}_v - \vec{\mathbf{V}}_l\right|\left(\vec{\mathbf{V}}_v - \vec{\mathbf{V}}_l\right) \tag{5}$$

where C_D stands for the drag coefficient, and d_b is the vapour bubble diameter.

2.2.2. Lift force. Lift force FL exerted on the liquid phase is described as:

$$\mathbf{F}_L = \alpha_v \rho_l C_L \left(\vec{\mathbf{V}}_v - \vec{\mathbf{V}}_l\right) \times \left(\nabla \times \vec{\mathbf{V}}_l\right) \tag{6}$$

where C_L is the lift coefficient.

2.2.3. Wall lubrication force. Wall lubrication force assures the experimentally found zero void condition near vertical walls and it is described by

$$\mathbf{F}_{WL} = \alpha_v \rho_l C_{WL} \left\{\left(\vec{\mathbf{V}}_v - \vec{\mathbf{V}}_l\right) - \left[\left(\vec{\mathbf{V}}_v - \vec{\mathbf{V}}_l\right)\cdot \vec{n}_w\right]\vec{n}_w\right\}^2 \vec{n}_w \tag{7}$$

where $\vec{n}_w$ is the unit wall normal vector pointing from the wall into the fluid, and C_{wl} stands for the wall lubrication coefficient.

For the above mentioned momentum interfacial forces, detailed models are referred to Kopun [9-10], Tomiyama [15] and Frank et al. [16].

2.3. Energy equation

In our previous works, vapour bubble condensation in fluid domain is assumed to be negligible, the phasic static enthalpy equation is then described by

$$\frac{\partial(\alpha\rho h)_k}{\partial t} + \nabla\cdot\left(\alpha\rho\vec{\mathbf{V}}h\right)_k = \nabla\cdot\left(\frac{\alpha\lambda_T}{c_p}\nabla h\right)_k + \frac{3}{4}\alpha_v\rho_l\frac{C_H}{db}\left(T_l - T_k\right) + \alpha_k\frac{\partial p}{\partial t}\alpha_k + S_h \tag{8}$$

Thermal conductivity is presented as λ_T, the inter-phase heat transfer coefficient between liquid phase and vapour phase is denoted as C_H. For simplicity, other terms are lumped into the source term S_h.

$$S_h = \nabla\cdot\left[\frac{\alpha\lambda_T}{c_p}\left(T - T_{ref}\right)\nabla C_p\right]_k + \left(\alpha\vec{\mathbf{V}}\right)_k\nabla p + \left(\alpha\tau : \nabla\vec{\mathbf{V}}\right)_k \tag{9}$$

2.4. Boiling model

Based on the previous works of Kopun [9, 10], the inter-phase mass change rate Γ_k due to boiling is evaluated as:

$$\Gamma_c = \frac{C_m C_{boil} h_{boil}\left(T_W - T_{sat}\right)}{H_{fg}}\frac{6\alpha_v}{d_b} \tag{10}$$

where C_m is the closure coefficient due to uncertainty, C_{boil} stands for boiling correction coefficient and and H_{fg} denotes the vaporization latent heat. h_{boil} is the boiling heat transfer coefficient. Clearly,

h_{boil} varies with boiling regime. h_{boil} in different boiling regime The following sub-sections describe the applied models for heat transfer coefficient.

2.4.1. Film boiling model. The Bromley's model [17] is employed to access the film boiling heat transfer coefficient h_{boil}:

$$h_{boil} = 0.62\left\{\frac{\lambda_v^3 \rho_v (\rho_l - \rho_v) g \left[H_{fg} + 0.4 C_{p,v} (T_w - T_{sat})\right]}{\mu_v (T_w - T_{sat}) d_b}\right\}^{0.25} \tag{11}$$

where vapor bubble diameter d_b stands for the length scale, the vapor thermal conductivity is denoted as λ_v and $C_{p,v}$ is the vapour specific heat.

2.4.2. Transition boiling mode. In transition boiling regime, heat transfer coefficient, h_{boil} is calculated as:

$$h_{boil} = \frac{q_{TB}}{T_w - T_{sat}} \tag{12}$$

where the transition boiling heat flux q_{TB} is given as follows

$$q_{TB} = \phi q_{CHF} + (1 - \phi) q_{MHF} \tag{13}$$

In Eq. (13), q_{CHF} denotes the Critical Heat Flux, and q_{MHF} stands for the Minimum Heat Flux. The critical heat flux is modeled as

$$q_{CHF} = 0.131 \rho_v H_{fg} \left\{\frac{\sigma g (\rho_l - \rho_v)}{\rho_v^2}\right\}^{0.25} \tag{14}$$

where σ is the liquid phase surface tension. The Minimum Heat Flux correlation for the quenching system in Eq. (13) is

$$q_{MHF} = 0.09 K_{MHF} \rho_v H_{fg} \left\{\frac{g(\rho_l - \rho_v)}{\rho_l + \rho_v}\right\}^{0.25} \left\{\frac{\sigma}{g(\rho_l - \rho_v)}\right\}^{0.25} \tag{15}$$

K_{MHF} serves as a model parameter and its typical value varies from 1 to 15. Further details about AVL quenching model can be obtained from AVL FIRE® Multi-fluid model solver theory guide [18] and/or previous works [9-13].

2.5. Variable specific heat capacity

Specific heat capacity, C_p is used to define enthalpy and to re-calculate temperature from enthalpy. When C_p varies with temperature, C_p becomes a critical parameter for the numerical simulation of the quenching process, in which temperature changes hundreds of Kelvin. In this case, the original definition of static enthalpy has to be used,

$$h = \int_0^T C_p(T) dT \tag{16}$$

During numerical simulation, the temperature dependent specific heat capacity is usually provided within a specified application range, $[T_0, T_u]$. T_0 is the lower bound where Cp(T) begins to be valid and T_u is the upper bound. Represented as polynomial function the specific heat capacity can be modeled by the following equation,

$$C_p(T) = a_0 + a_1 T + a_2 T^2 + a_3 T^3 \qquad T_0 \le T < T_u \tag{17}$$

For one aluminum work piece the polynomial coefficients are a_0 = -100.3, a_1 = 0.07751, a_2 = 0.0065, $a_3 = -4\times10^{-6}$, T_0 = 283.15 and T_u = 1100 K. In practice, T_0 is lower than the initial quenchant pool temperature and T_u is higher than the initial hot solid piece temperature. As from Eq. (16), enthalpy is defined from a reference temperature of 0 K, so specific heat capacity C_p is needed in the range of 0 K till highest temperature of the hot piece. Figure 1a shows how the specific heat capacity varies with temperature if Eq. (17) is applied in the range of $0<T<1200$. Clearly it is found in Figure 1a that negative C_p appears, and this will introduce errors in the simulation and spoils the simulation results. Hence, a special treatment for variable C_p is needed. Therefore Eq. (16) is re-written as

$$h = \begin{cases} \int_0^{T_0} S(T)dT & T < T_0 \\ \int_0^{T_0} S(T)dT + \int_{T_0}^{T} C_p(T)dT & T \geq T_0 \end{cases} \tag{18}$$

where S(T) is an unknown function. Since the solid temperature change during quenching takes place within T_0 and T_u, the numerical simulation focuses only on this interested temperature range of $T_0 < T < T_u$. Furthermore, only the difference pf the enthalpy is needed during the simulation, so, an exact expression of S(T) in Eq. (18) is not strictly demanded. To keep the simulation procedure simple, here we set it as constant

$$S(T) = C_p(T_0) \tag{19}$$

With the help of Eq. (18), temperature dependent $C_{p,new}$ in the hot solid part is defined as

$$C_{p,new}(T) = \begin{cases} C_p(T_0) & T < T_0 \\ C_p(T) & T_0 \leq T \leq T_u \end{cases} \tag{20}$$

Figure 1b displays variation of $C_{p,new}$ in the range of $0<T<1200$.

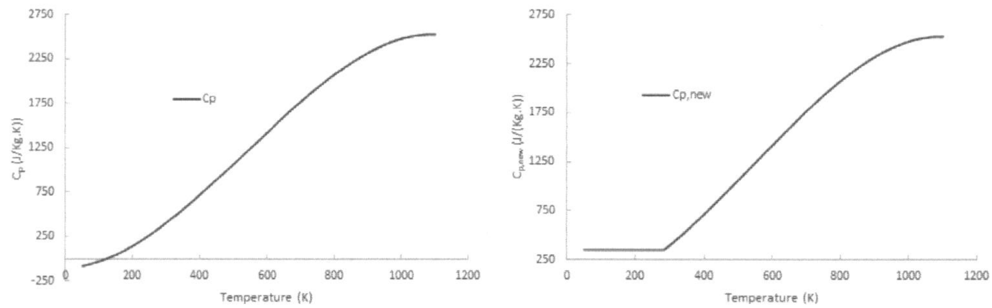

(a) Original (b) improved

Figure 1. Variation of heat capacity C_p with temperature.

2.6. *Thermal homogeneous assumption*

In previous works, the dispersed bubbly phase and the continuous phase are assumed to be in thermal equilibrium and a homogeneous enthalpy equation was solved. Rather to solve this homogeneous enthalpy equation, in this work, both enthalpy equations of bubble phase and continuous phase are solved. In order to achieve a thermal homogeneous field, a very high value is set as inter-phase heat transfer coefficient, C_H in Eq. (8).

3. Boundary condition and simulation set up

To evaluate the performance of the above mentioned simulation methodology, two immersion quenching processes, a simple step plate geometry and a real case aluminum cylinder head, are nummerically simulated here.

IOP Conf. Series: Materials Science and Engineering **164** (2017) 012004 doi:10.1088/1757-899X/164/1/012004

As its names says, the thickness of the step plate work piece varies along its length, and it is displayed in Figure 2a. Temperature history of three different positions, named as T_1, T_2 and T_3 in the middle of the test piece were measured in the experiments.

The aluminium cylinder head of the second test case and its 9 measuring points are illustrated in Figure 2b.

Figure 3a displays the polyhedral computational mesh and the boundary conditions of the step plate test case. The total computational domain is subdivided into 245830 grid cell, in which the solid step plate takes approximately 61000 cells. Fixed atmospheric pressure (1 atm) is applied at the outlet boundary on the top (light red plane), and a velocity boundary condition is exerted on the inlet boundary (at the bottom, light pink plane) to model the dipping process. The dipping velocity is used to determine the inlet velocity rising up from the initial water level (purple part) toward top. In this simulations, inlet velocity at the first 0.5 seconds takes a value of 0, which implies that the step plate is air cooled for the first 0.5 seconds, and then a constant velocity of 0.14 m/s lasts for 2 seconds, so that the final submerging depth has been reached; then inlet velocity is 0 again. The domain walls at the sides were treated adiabatic. Initial water pool temperature is 353.15K and step plate has an initial temperature of 774.15K.

Figure 3b schematically depicts the polyhedral computational mesh and the boundary conditions exerted in the cylinder head test case. The present computational domain consists of approximately 1.65 million cells in total, where the cylinder head has about 0.7 million cells. Fixed atmospheric pressure (1 atm) is applied at the outlet boundary on the top (red plane), and on the inlet section (at the bottom, pink plane), velocity is given. In this simulation, inlet velocity at the first 0.5 seconds is 0 (the aluminium cylinder head is first air cooled for 0.5 s), then a constant inlet velocity of 0.14 m/s lasts for 2.35 s until 2.85 s, and then it is 0 m/s to represent the immersed water level. The side walls are treated as adiabatic. Initial water pool temperature is 353.15K and initial temperature of cylinder head is 774.15K.

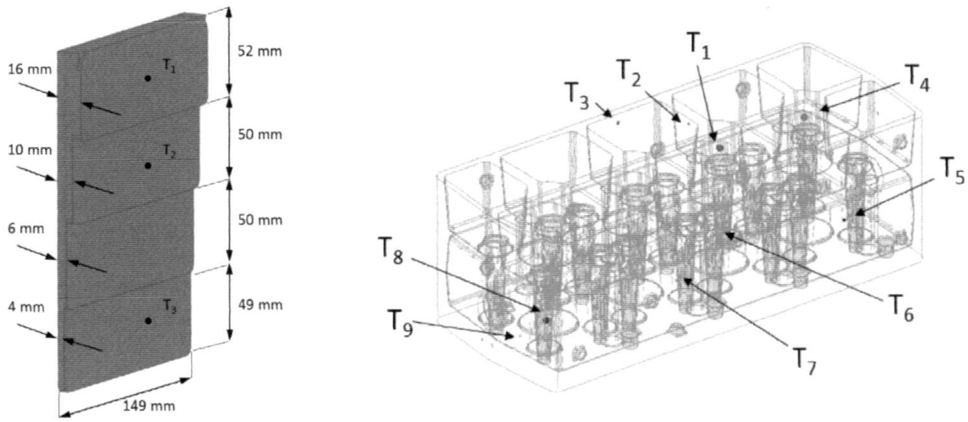

(a) Step plate test piece (b) cylinder head

Figure 2. Monitoring location and test piece details.

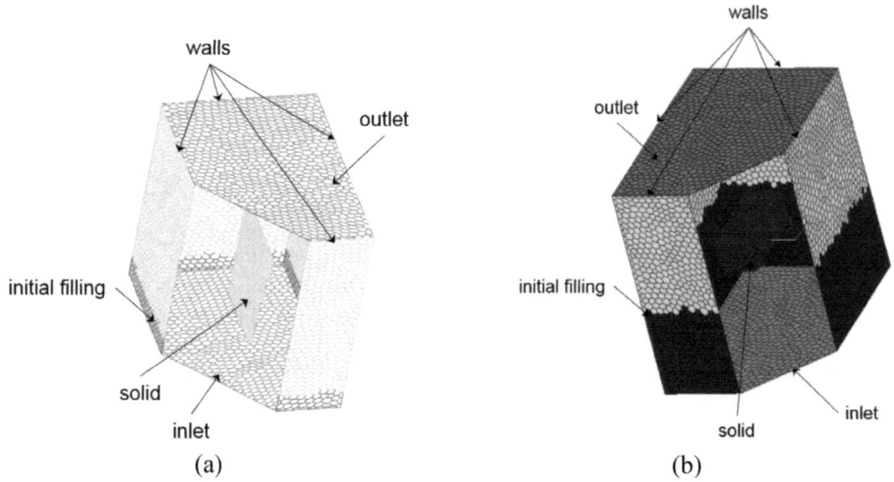

Figure 3. Computational details of simulation domains: (a) step plate case; (b) cylinder head.

4. Results and discussion

4.1. Step plate

The step plate test piece was immersed into a water pool with bulk temperatures of 353 K. The alloy's specific heat capacity, C_p varies from 884.8 J/(Kg.K) at 300 K to 1427.3 J/(Kg.K) at 1100 K. Leidenfrost temperature ranges from 560 K to 650 K, detailed variable Leidenfrost temperature model is referred to Kopun [10]. Different to the constant C_p used in the previous works [10, 13], the variable specific heat capacity as described Eq. (20) is used in this work.

Figure 4 illustrates continuous phase volume fraction distribution at different times. At the initial stage, i.e., from 1 to 3s, film boiling occurs as it is seen that water volume fraction surrounding the solid is still higher than 0.85. With time proceeding, more vapour bubbles are generated and travelling to the top due to buoyancy, which implies rapid transit boiling or nucleate boiling appears between 3 and 15 s. At 20 s, boiling happens only at the upper part of the work piece; whereas no bubbles are observed in the lower part of the plate, indicating that in this region, the solid surface temperature is already below the saturation temperature. It is noted in Figure 4 that, during quenching, in the lower part of the work piece, locally generated vapour bubbles escape the surface and rise up; in the upper part of the solid workpiece (anti-gravity direction), locally generated vapour bubbles and the upward travelling bubbles serve like a vapour film, and this blanket effect leads to the longer staying of film boiling regime in the upper part of the work piece. Therefore, during immersion quenching, Leidenfrost temperature increases in gravity direction.

Temperature snapshots in thickness and width directions are presented in Figure 5. It can be seen that intensive cooling is achieved, and across the entire plane, a highly non-uniform temperature distribution is established after only 3 seconds. The temperature drops substantially in the thinner part of the solid. After 20 seconds, the upper part, as it is thicker and heat transfer is still taking place. After 25 seconds, the metal is cooled down to 372K and natural convection begins, so, till 30 seconds, the temperature further drops to 367 K.

Figure 6 provides a quantitative comparison of the simulated and measured cooling curves at three monitoring points. In general, excellent agreement is achieved. It is obvious from the measured cooling curves in Figure 6 that, the Leidenfrost temperature at T3 is almost 740K and it is about 610 K at T1. As shown in Figure 2, monitoring point T3 locates in the lower tip of the plate and T1 is on the upper part of the plate, this again confirms that Leidenfrost temperature increases in the gravity direction. At monitoring point T3, the nucleate boiling regime is reached very early, almost immediately after immersion, and therefore it loses heat very fast; together with the fact that its thickness is thinner, it reaches the surrounding bulk water temperature very soon. The discrepancy between the measurements and simulated cooling curve at monitoring point T3 may come from the different Leidenfrost temperature. The simulated cooling curve first experiences a film boiling and then enters nucleate boiling so there is a lag between the measurements and simulated curve. The difference between the experiment and simulation at T2 could be partly contribute to measurement errors or the Leidenfrost temperature is not accurately predicted in the simulation. At monitoring point T1, simulated cooling curve perfectly fits with the measured ones and because it experiences longer film boiling and the workpiece thickness at this point is 16 mm, consequently, after about 25 seconds it reaches the surrounding bulk liquid temperature.

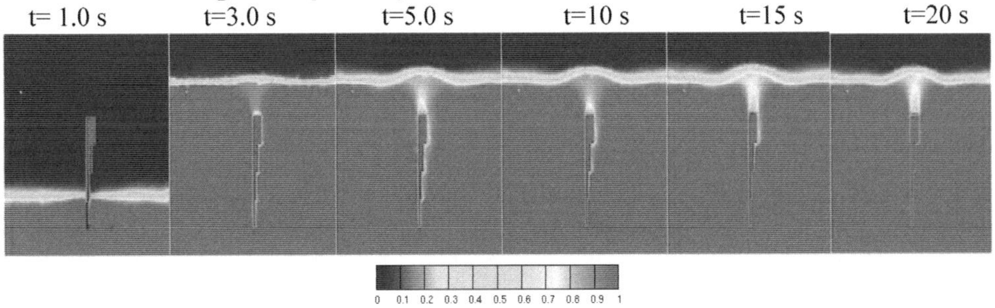

Figure 4. Continuous phase volume fraction distributions at different times.

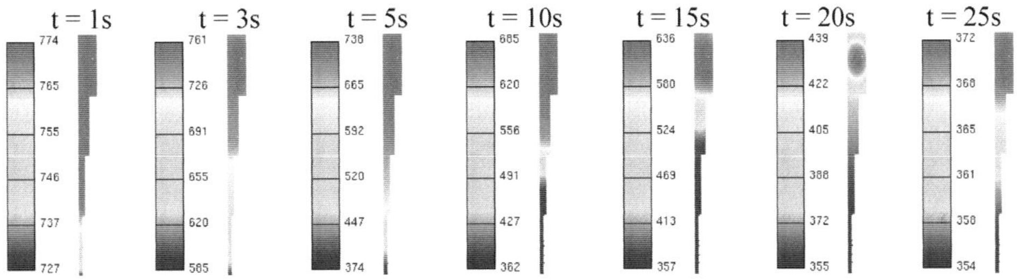

(a)

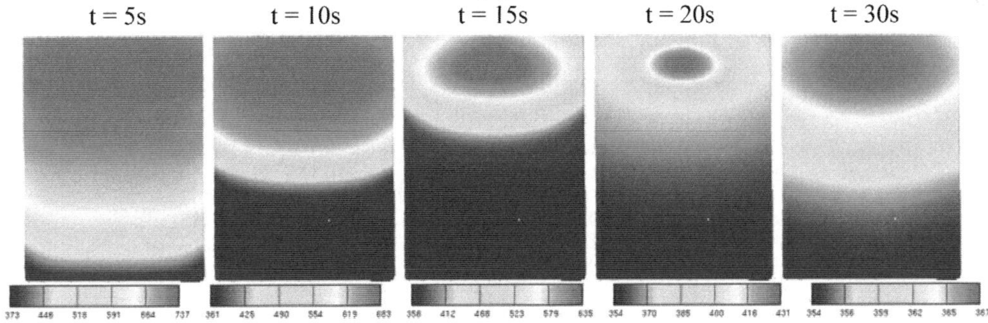

(b)

Figure 5. Snapshots of solid temperature distribution at different times: (a) y=0; (b) x=0.004 m.

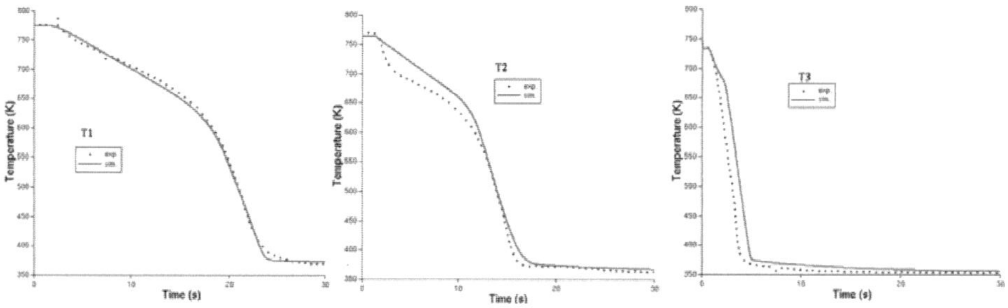

Figure 6. Cooling history on the three monitoring points.

4.2. Aluminium cylinder head

Immersion quenching of the current cylinder head was first numerically investigated by Kopun et al. [2]. The aluminium cylinder head is immersed into a water pool with a bulk temperatures of 353 K. The alloy's specific heat capacity, C_p varies from 880 J/(Kg.K) at 293.15 K to 1166 J/(Kg.K) at 775.15 K. Different to the work of Kopun et al. [2], where also temperature dependent specific heat capacity and thermal conductivity were used, in the present work the enhanced approach with the original definition of the static enthalpy, i.e. Eq. (16) is applied. Consequently, the specific heat capacity C_p is set to a constant value of 880 J/(Kg.K) when temperature is below 293 K and then varies from 880 J/(Kg.K) to about 1166 J/(Kg.K)).

Figure 7 shows the water (continuous phase) volume fraction in the mid-plane at different times. Immediately after the hot cylinder heat touches the sub-cooled water, boiling begins and a vapour film is formed surrounding the cylinder head. With time preceding, boiling becomes stronger and stronger.

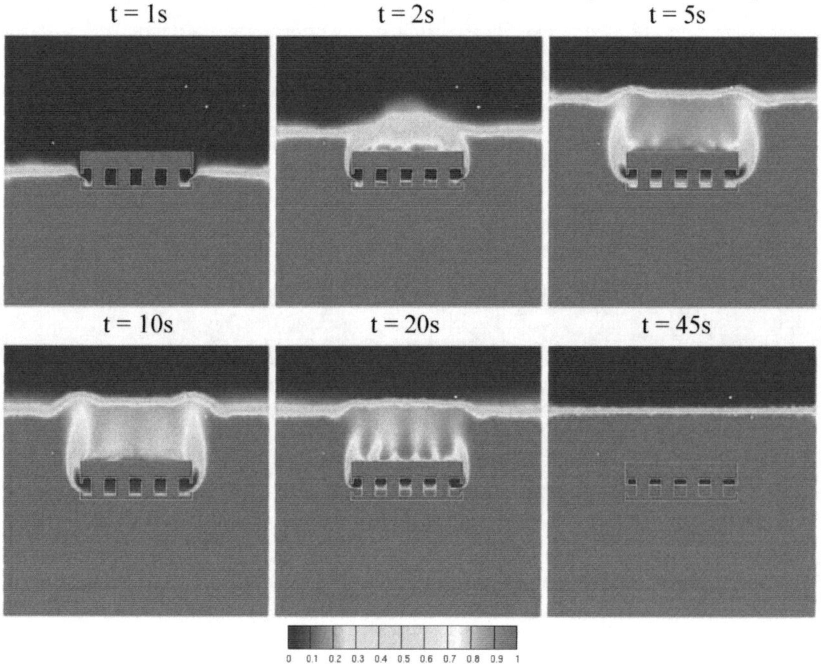

Figure 7. Snapshots of water volume distribution at different times.

Due to gravity, it is difficult for bubbles to escape from the cylinder head bottom, as it can be observed in Figure 7 that, a vapour film presents at the cylinder head bottom until approximately 45 s. This vapour blanket separates the hot solid from the sub-cooled water, only film boiling happens there, and therefore the heat transfer is weak during this period, and this behaviour should be avoided. Furthermore, it can be seen in Figure 7 that gas is captured in the hollow rooms. Gas occupies the upper part of the hollow room and cannot escape from these hollow rooms. After rapid boiling comes (at time = 5s), water is consumed and more vapour bubbles are generated and these vapour bubbles and air almost fill these hollow rooms, this significantly reduces the heat transfer. After 45 seconds, hardly no bubbles are generated.

Figure 8 shows the cylinder head temperature distribution in the mid-plane at different times. It is obvious that the temperature at the open corners drops faster since boiling happens there at all sides. As the vapour bubbles accumulate along the bottom of the cylinder head, and this vapour film prevents rapid boiling, the energy removal in the bottom region is weaker and so temperature drops slower. Again, due to vapour occupation in the hollow rooms, boiling only happens in the lower part of the hollow rooms and it removes heat there, this leads to much higher temperature in the upper part of the hollow room and greater temperature gradient. This is not desired during quenching as it produces higher thermal stress.

The quantitative comparison of the simulated and measured cooling curves at the twelve monitoring points is presented in Figure 9. Except at monitoring points 1, 6 and 9, the simulated cooling curves agree very well with the measured ones. At MP1 and MP6, the difference between simulation and experiments is related to the changing Leidenfrost temperature. The predicted Leidenfrost points are lower than the measured ones, and therefore longer film boiling is experienced at these two points in the simulation which leads to lagging to the measurements about 4 seconds. Furthermore, we had assumed a uniform initial temperature of 495 °C for the whole domain, but the measured one at MP9 is 510 °C. Consequently difference appears for this measurement point. It is further found in Figure 9 that, rapid boiling happens between 10 and 25 s and after 40 seconds, the cylinder head temperature is cooled down to the water saturation temperature, which is also observed in Figure 8.

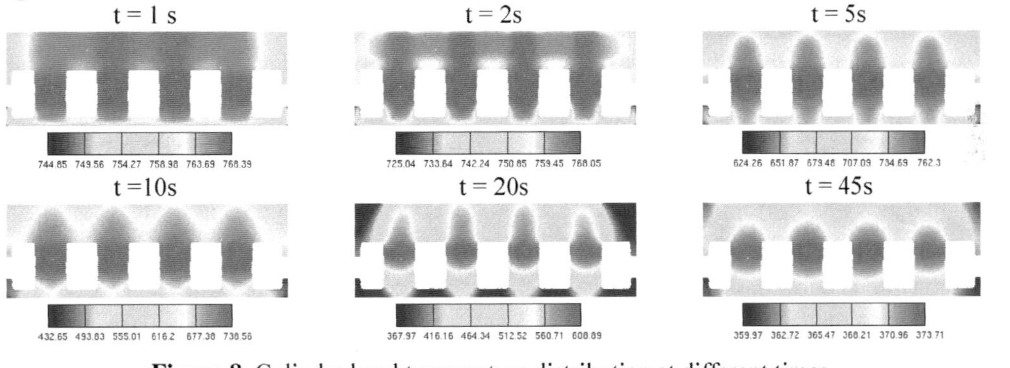

Figure 8. Cylinder head temperature distribution at different times.

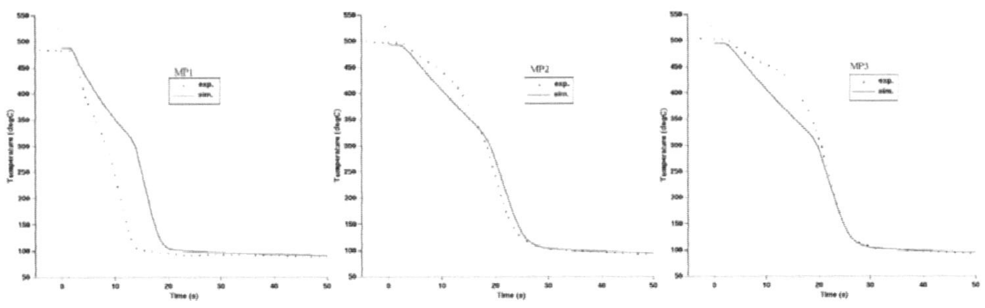

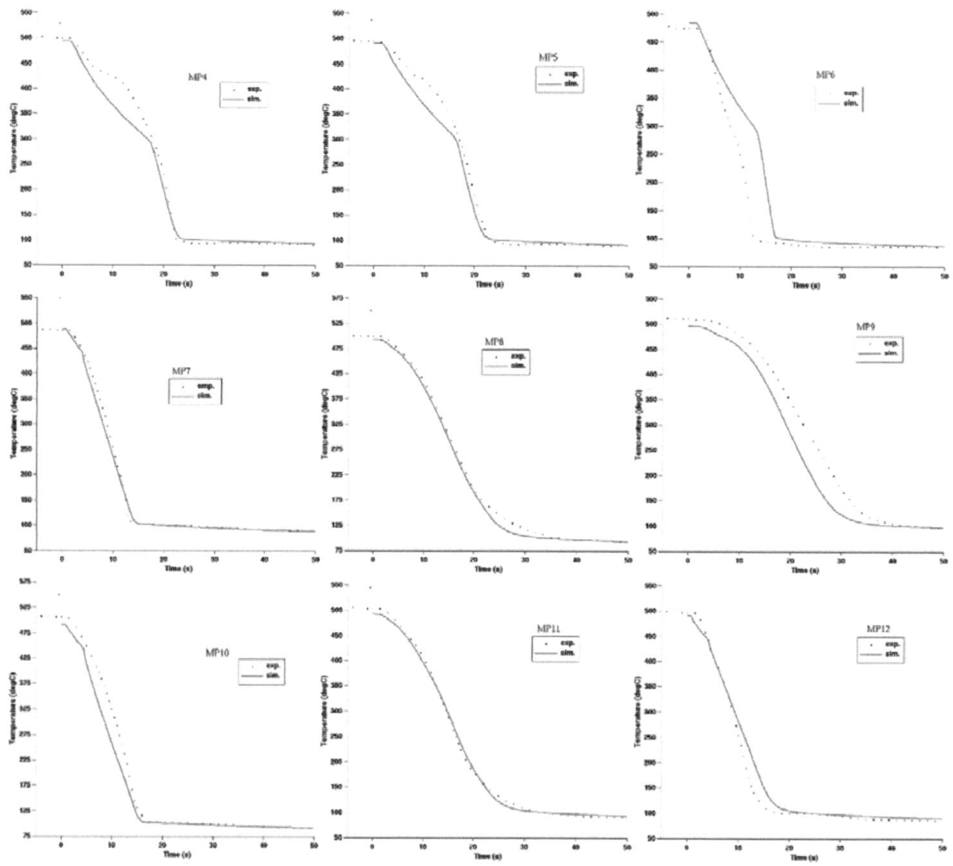

Figure 9. Comparison of simulated (straight red) and measures (dotted black) for the twelfe measurment points MP1 to MP12 of the cylinder head.

5. Conclusions

A numerical investigation of the heat and mass transfer during the immersion quenching process of two geometries is performed in the frame of Multi-fluid model using the CFD code AVL FIRE®. The treatment of the temperature dependent specific heat capacity is described in detail. Important physical phenomenon, e.g., different boiling regimes, vapour blankets and vapour accumulation in relatively closed region have been correctly predicted. For the immersion quenching of step plate, accumulation of vapour film in the upper region (anti-gravity direction) not only hinders heat transfer but also decreases the Leidenfrost point, thicker part downwards orientation quenching would be recommended to achieve a faster cooling. For the quenching of cylinder head, with current orientation, trapping of the vapour phase in the hollow room aggravates boiling and leads to highly non-uniformity of temperature gradient in hollow room's walls, immersion quenching should be performed in a way that vapour phase could easily escape from the hollow rooms. For almost all measurement points a very good agreement between numerical and experimental data have been achieved for both cases. It could be concluded that the robust improved CFD methodology provides a solid basis for detailed understanding and exploring into the essence of the quenching process in the workshop by utilizing CFD analysis. This methodology can be used to analyze fluid flow and heat transfer in immersion quenching processes with the objective of optimizing quenching performance for distortion control and reduction of cracking problems.

IOP Conf. Series: Materials Science and Engineering **164** (2017) 012004 doi:10.1088/1757-899X/164/1/012004

6. Nomenclature

d	bubble diameter [m]
g	gravity acceleration [m/s^2]
h	static enthalpy [J/kg]
h	heat transfer coefficient [W/(m^2.K)]
n	normal vector [-]
p	pressue [N/m^2]
q	heat flux [W/m^2]
t	time [s]
C	coefficient [-]
C	specific heat capacity [J/(kg.K)]
F	inter-phase force [N]
H	latent heat [J/kg]
K	model parameter [-]
M	inter-phase momentum exchange [N/m^3]
N	phase number [-]
S	source term
T	temperature [K]
V	velocity [m/s]
Γ	mass transfer rate [kg/(m^3.s)]
α	volume fraction [-]
ϕ	weighting factor [-]
λ	thermal conductivity [W/(m.K)]
ρ	density [kg/m^3]
σ	surface tension coefficient [N/m]
τ	stress tensor [N/m^2]

Index

0	lower limit
b	bubble
boil	boiling
fg	vaporization latent heat
h	enthalpy
i	interface
k	phase index
l	liquid
m	closure coefficient
ref	reference
sat	saturation
u	upper limit
v	vapor
w	wall
CHF	critical heat flux
D	drag force
H	enthalpy
L	lift force
MHF	minimum heat flux
T	temperature
TB	transition boiling
WL	wall lubrication force

IOP Conf. Series: Materials Science and Engineering **164** (2017) 012004 doi:10.1088/1757-899X/164/1/012004

References

[1] Abdulhay B, Bourouga B and Dessain C 2011 Experimental and theoretical study of thermal aspects of the hot stamping process *Appl. Therm. Eng.* **31** 674-685

[2] Kopun R, Kovačič Z, Škerget L and Hriberšek M 2014 *3rd Int. Conf. on Computational Methods for Thermal Problems (Lake Bled)* (Napoli: Giannini Editore)

[3] Krause F, Schüttenberg S and Fritsching U 2010 Modelling and simulation of flow boiling heat transfer *Int. J. of Num. Methods for Heat & Fluid Flow* **20** 312-331

[4] Babu K and Kumar T S P 2009 Mathematical modelling of surface heat flux during quenching *The Minerals, Metals & Material Society and ASM International* **41B** 214-224

[5] Oldshue J Y 1983 *Fluid Mixing Technology* (New York: McGraw-Hill) p 162

[6] Segerberg S O and Bodin J L 1988 Controlling the quench process for more consistent hardening *Heat Treat.* **20**(5) 26-28

[7] Garwood D R, Lucas J D, Wallis R A and Ward J 1992 Modeling of flow distribution in an oil quench tank *J. Mat. Eng. and Perf.* **1**(6) 781-786

[8] Kernazhitskiy S and Recktenwald G 2004 *ASME Heat Transfer/Fluids Engineering Summer Conference* (New York: American Society of Mechanical Engineers) p 129

[9] Kopun R, Zhang D S, Edelbauer W, Stauder B, Basara B and Greif D 2014 Immersion quenching simulation of realistic cylinder head geometry *SAE Int. J. Mater. Manf.* **7**(3) 520-529

[10] Kopun R, Škerget L, Hriberšek M, Zhang D S, Stauder B and Greif D 2014 Numerical simulation of immersion quenching processes for cast aluminium part at different pool temperatures *Appl. Therm. Eng.* **65** 74-84

[11] Greif D, Kovačič Z, Srinivasan V, Wang D M and Suffa M 2009 Coupled numerical analysis of quenching process of internal combustion engine cylinder head *BHM Berg-und Hüttenmännische Monatshefte* **154** 509-517

[12] Srinivasan V, Moon K, Greif D, Wang D M and Kim M 2010 Numerical simulation of immersion quench cooling process using an Eulerian multi-fluid approach *Appl. Therm. Eng.* **30** 499-509

[13] Schmalhorst C, Uhlig F, Smolnikar M and Kopun R 2015 *Proc. der 8. VDI-Tagung Gießtechnik im Motorenbau* (Düsseldorf: Verein Deutscher Ingenieure) p 59

[14] Drew D A and Passman S L 1998 *Theory of Multi-component Fluids* (New York: Springer) p 99

[15] Tomiyama A 1998 *ICMF'98, 3rd Int. Conf. Multiphase Flow (Lyon)* (New Jersey: Information Today, Inc.) p 1

[16] Frank T, Shi J M and Burns A D 2004 *3rd International Symposium on Two-Phase Flow Modelling and Experimentation* (Pisa: Edizioni ETS) p 2291

[17] Bromley L A 1950 Heat transfer in stable film boiling *Chem. Eng. Progr.* **46** 221-227

[18] AVL LIST GmbH 2013 *FIRE CFD Solver* (Graz: AVL LIST GmbH) pp 3-10

Influence of carbonation and freeze-thaw on macro-properties of concrete

M J Rao[1, 3, 4], Y Dong[1, 2], H Q Yang[1, 2], M X Li[1, 2] and Z Yu[1]

[1] Changjiang River Scientific Research Institute, Wuhan, China
[2] Collaborative Innovation Center for Geo-hazards and Eco-environment in the Three Gorges Area, Hubei Province, China
[3] School of Civil Engineering, Wuhan University, Wuhan, China

Email: raomeijuanding@163.com

Abstract. Concrete can be damaged under the effects of carbonation and the freeze-thaw cycle, which is a common cause of premature failure for concrete structures. In this study, range analysis was performed for compressive strength results after carbonation and freeze-thaw cycles for 90 days. A carbonation test, freeze-thaw test, and the alternating cycle of carbonation and freeze-thaws were performed to measure the relative dynamic elastic modulus and flexural strength of concrete samples. The results showed that specimens subjected to thawing first, then carbonation suffered the greatest impairment to strength due to expansion and osmotic pressures caused by water motion after thawing; this caused some of the original closed pores to link together, which loosened the internal organization and increased total porosity. The dynamic elastic modulus loss and weight loss of carbonized specimens were smaller than those of uncarbonized specimens within a reasonable range of carbonation. The flexural strength carbonation declined as the water-binder ratio or fly ash content increased.

Keywords: carbonation, freeze-thaw, macro-properties, fly ash.

1. Introduction

Concrete is the most common building material over the world [1-3]. Hydraulic concrete durability is one of the most important factors in the field of concrete studies because the performance and lifespan of many projects depend on concrete durability. Modern dams adopt a series of durability protection measures. Due to diversity of the application conditions and complexity of the concrete system itself, the durability problem of hydraulic concrete still exists; therefore, further discussion regarding hydraulic concrete, damaged caused by carbonation /freeze-thaw, and prevention measures are necessary [4-5]. In the 1960s, some developed countries began to focus on concrete durability, and a large number of experimental studies and theoretical analyses were conducted on concrete carbonation. Researchers have come to a relatively unified understanding of concrete carbonation mechanisms [6]. The influencing factors of concrete carbonation, artificially accelerated carbonation, and carbonation depth detection have been discussed. Concrete carbonation research has entered the molecular level, and the model for carbonation depth and life prediction under a single carbonation is relatively well developed. In recent years, multiple degradation factors on concrete durability are becoming more and more important among researchers [7]. As for hydraulic concrete, carbonation and freeze-thaw cycles are the most prominent factors. Currently, the study of single factor freeze-thaw or carbonation cycles are mature, but theoretical research of concrete durability under alternative freeze-thaw and carbonation cycles is not mature [8].

Concrete undergoes changes in macrostructure when it suffers carbonation and freeze-thaw cycles. In this study, carbonation, relative dynamic elastic modulus, mass loss rate and flexural strength were measured. Therefore, information that can lead to improving concrete performance under harsh carbonation and freeze-thaw cycles can be very beneficial.

2. Experimental

2.1. Raw materials

Chemical and physical properties of materials are listed in Table 1.

Table 1. Chemical, physical properties of materials.

	Chemical composition/%										Physical promerites	Compressive strength of cement/MPa	
	SiO_2	Al_2O_3	Fe_2O_3	CaO	MgO	K_2O	Na_2O	R_2O	SO_3	loss	Specific density /(kg/m^3)	7 d	28d
Cement	22.5	3.8	4.8	63.0	2.1	0.50	0.1	0.4	2.1	0.7	3120	25.4	45.1
Fly ash	52.4	24.0	9.4	3.2	1.2	—	—	—	0.7	2.3	2380	—	—
Slag	33.6	14.6	1.7	34.7	9.9	0.7	0.3	—	2.0	2.4	2860	—	—

2.2. Orthogonal experiments

The freeze-thaw, carbonation and content of fly ash were viewed as the main factors of experiment to study the composite action of freezing-thaw and carbonation on compressive strength of cement mortar. Due to the water-cement ratio was an important factor determining the concrete performance, which had a significant impact for concrete strength, durability, and workability. Thus the water-cement ratio (C), content of fly ash (F), carbonation time (B), and freeze-thaw times (T) were four relatively important elements as factors of the orthogonal experimental design. Table 1 showed the factors and levels, each factors in three levels. The cement mortar was cured for 90 days and measured its compressive strength after alternating action of carbonation and freeze-thaw.

Table 2. Factors and levels.

Level	Factors			
	C(W/C)	F(Content of fly ash)/%	B(Carbonation time)/d	T(Freeze-thaw times)/times
1	0.4	0	3	20
2	0.45	20	7	50
3	0. 5	50	14	100

According to the four factors with three levels as shown in Table 2, select L9 (34) orthogonal design table for test. The test was mainly on alternating action of carbonation and freeze-thaw impact on concrete durability. The test was divided into two parts, one was carbonation after freeze-thaw, and the other was freeze-thaw after carbonation.

2.3. Freeze-thaw test and carbonation test

Two graded aggregate concrete with six mix proportions, twelve group tests are chosen for the test. The specimen is conducted carbonation test followed by freeze-thaw test or freeze-thaw test followed by carbonation test. The test scheme is listed in Table 3 and Table 4. The air content of concrete is controlled between 4.5% and 5.0%, the collapsed slump is controlled between 50 mm to 70 mm and the mix proportion is shown in Table 4. Due to a larger number of test pieces, a standard curing age is 28d. Mix proportions are shown in Table 5.

Table 3. Freeze-thaw after carbonation.

Specimen	W/C	FA(%)	Curing Age (d)	Carbonation (d)	Freeze-thaw times
B1	0.35	0	28	14	100
B2		20			
B3	0.45	20	28	14	100
B4		40			
B5	0.55	20	28	14	100
B6		40			

Table 4. Carbonation after freeze-thaw.

Specimen	W/C	FlA (%)	Curing Age (d)	Freeze-thaw times	Carbonation (d)
T1	0.35	0	28	100	14
T2		20			
T3	0.45	20	28	100	14
T4		40			
T5	0.55	20	28	100	14
T6		40			

The molding number in each group is six. The specimen size is 100mm×100mm×400mm. After curing in standard curing room for a scheduledtime, three piece were transplanted into a closed carbonation box, then into a freezing and thawing box after accelerated carbonation for 28d. Measure the quality and relative dynamic elastic modulus, both before and after being transplanted into the freeze-thaw box for 100 times. At last measure the depth of carbonation. The other three pieces did freeze-thaw test first, and suffered from accelerated carbonation for 28d. Do the carbonation test according to Test Code for Hydro-concrete(DL/T 5150—2001). The specimens should be taken out from the standard curing room before two days of the standard curing age, then measure the depth of carbonation after drying at 60 ℃± 2 ℃ for 48h. Specimen B1 to B6 are conducted carbonation test after freeze-thaw test. Specimen T1 to T6 are conducted freeze-thaw test after carbonation test.

Table 5. Mix proportion of specimen.

Specimen	Gradation	W/C	FA (%)	Aggregate ratio (%)	Admixture PCA (%)	Admixture GYQ (%)	materials(kg/m^3) water	materials(kg/m^3) Middle stone	materials(kg/m^3) Small stone	Slump (mm)	Air content (%)
B1/T1	two	0.35	0	30	0.48	0.012	117	813	665	55	4.5
B2/T2	two	0.35	20	30	0.45	0.011	117	806	659	62	4.7
B3/T3	two	0.45	20	30	0.43	0.011	117	833	682	65	5.2
B4/T4	two	0.45	40	30	0.43	0.011	116	829	678	65	4.9
B5/T5	two	0.55	20	30	0.42	0.010	117	850	696	69	4.8
B6/T6	two	0.55	40	350	0.42	0.010	116	847	693	70	4.6

2.4. Dynamic elastic modulus

According to former research, the relative dynamic elastic modulus is usually calculated as follows

$$P_n = \frac{f_n^2}{f_o^2} \times 100 \quad (1)$$

P_n: relative dynamic elastic modulus after n-th time of freeze-thaw cycle %
f_0: the natural frequency before freeze-thaw (Hz)
f_n: the natural frequency after n-th time of freeze-thaw cycle(Hz.)

3. Results and analysis

3.1. Strength range analysis after carbonation and freeze-thaw

Range analysis was done for compressive strength results after carbonation and freeze-thaw for 90 days. Table 6 and Figure 1 showed the results of range analysis

Table 6. Range analysis (freeze-thaw after carbonation).

Specimen	C	F	B	T	Compressive strength in 90d(MPa)
S1	1	1	1	1	37.1
S2	1	2	2	2	49.4
S3	1	3	3	3	41.5
S4	2	1	2	3	46.3
S5	2	2	3	1	38.1
S6	2	3	1	2	36.3
S7	3	1	3	2	34.8
S8	3	2	1	3	50.6
S9	3	3	2	1	48.9
(K1)	127.3	120.9	121.5	123.6	
(K2)	119.2	136.6	144.1	123.9	
(K3)	139.4	126.1	117.2	138.6	
[$K1_a$]	42.1	40.9	41.2	40.3	
[$K2_a$]	39.0	45.2	48.7	41.6	
[$K3_a$]	46.1	41.4	40.2	46.7	

It can be seen that sample 8 had the highest 90-day compressive strength for 51.0 MPa under the C3F2B1T3 test condition; the best combination was B2C3T3F2 from the range analysis graphs.The range test values in each column were used to measure the effects of the respective factors in the test. It can be seen through the above table that in the 90-day concrete compressive strength test, the factors that affected strength from strongest to weakest were carbonation time (B), water-cement ratio (C), freeze-thaw times (T), and fly ash content (F). Of these, the best combination was B2C3T3F2. Compared with the intuitive view angle (B1C3T3F2), the main difference was the carbonation time length (B). According to range analysis graphs, the carbonation time had the greatest influence on mortar strength. Select carbonation for 3 days as experimental factor, then the best combination B1C3T3F2 could be drawn, namely water-cement ratio of 0.50, content of fly ash of 20%, carbonation time of 3 days, and freeze-thaw times of 100 times.

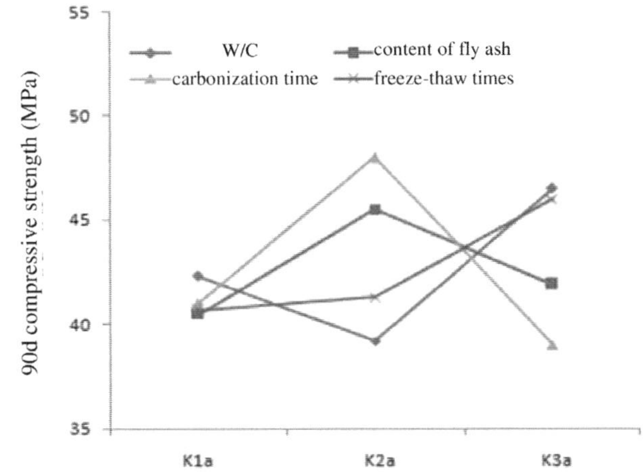

Figure 1. Range analysis curves (freeze-thaw after carbonation).

The reason why carbonation had a greater impact on mortar strength might be that during carbonation, solid material were made and became blocked inside of concrete pores, thus reducing the concrete's internal porosity; this weakened carbon dioxide diffusion that improved concrete density and strength. However, concrete alkalinity decreased after carbonation, and when the carbonation over the protective layer of concrete, it would make concrete lose protection for steel bar under conditions of water and air and the steel began to rust. The carbonation shrinkage cracks promoted rust expansion, which damage the protective layer of concrete; this created a self-perpetuating cycle that continuously degraded concrete strength.

As seen from the test results, the freeze-thaw cycle had the lowest impact on mortar intensity. One reason might be fewer freeze-thaw times which withstood freeze-thaw cycles of specimens, and the other was the generated calcium carbonate deposited in the pores of specimens during carbonation. After freeze-thaw cycle, the reduction in capillary pore volume meant a reduction in total expansion pressure for specimens. In addition, moisture migration became relatively difficult due to the compact structure, and the increase in seepage pressure correspondingly lagged behind, so concrete after carbonation could withstand more freeze-thaw cycles. However, it was not the case that longer carbonation times lead to better freezing resistance. The carbonation process was a precipitation process of calcium carbonate crystals, which lead to gradual contraction and crack formation. Therefore, carbonation improved the concrete's frost resistance to a certain extent. Under a certain time range, concrete freezing resistance would become stronger when carbonation time is longer.

Range analysis was also conducted for compressive strength results after freeze-thaw cycles and carbonation for 90 days. Figure 2 shows the water-cement ratio, fly ash content, carbonation times, freeze-thaw times, and impact on hydraulic concrete strength. Range analysis are shown is Table 7.

Table 7. Range analysis 2.

Specimen	C	F	T	B	Compressive strength in 90d (MPa)
S1	1	1	1	1	45.9
S2	1	2	2	2	39.7
S3	1	3	3	3	0
S4	2	1	2	3	53.9

S5	2	2	3	1	0
S6	2	3	1	2	38.2
S7	3	1	3	2	0
S8	3	2	1	3	41.4
S9	3	3	2	1	0
K1	85.6	99.8	125.5	45.9	
K2	92.1	81.1	93.6	77.9	
K3	41.4	38.2	0	95.3	
$K1_a$	28.5	33.3	41.8	15.3	
$K2_a$	30.7	27.0	31.2	26.0	
$K3_a$	13.8	12.7	0	31.8	

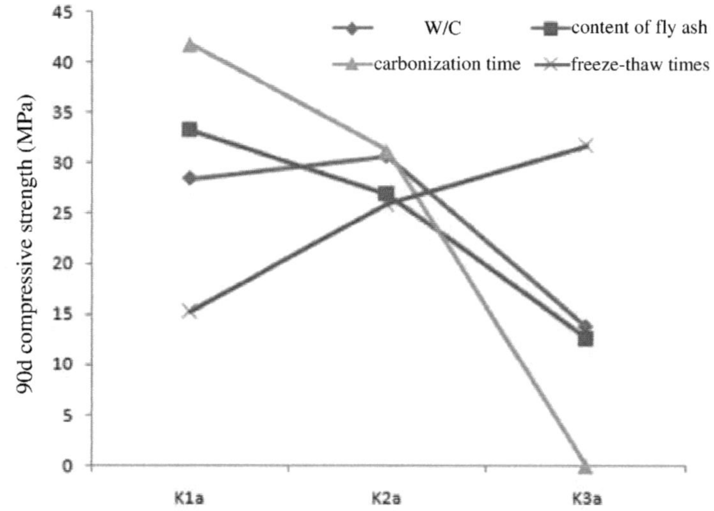

Figure 2. Range analysis curves (carbonation after freeze-thaw).

Sample S4 showed the highest 90-day compressive strength for 53.9 MPa under the C2F1B3T2 test condition; the best combination was B1F1T3C2 from the trend graphs. Range calculation showed that the factors affecting the 90-day compressive strength of concrete in order from greatest to weakest impact were carbonation time (B), fly ash content (F), freeze-thaw cycle times (T) and water-cement ratio (C). Compared with the intuitive view angle (B3F1T2C2), the main difference was the carbonation time length (T) and the freeze-thaw cycle time length. According to range analysis graphs, carbonation time had a greater influence on cement mortar compressive strength, followed by freeze-thaw time. Considering the test objectives and test results, choose carbonation for 14 days and freeze-thaw times of 100 times as experimental factors, then the best combination was with water-cement ratio of 0.45, none content of fly ash, carbonation time of 14 days, and freeze-thaw times of 50 times.

For specimens that suffered freeze-thaw cycles first, the carbonation time had the greatest impact on strength due to the expansion and osmotic pressures caused by water motion after thawing. This linked some of the original closed pores together, which loosened the specimen's internal organization and increased its total porosity. The longer the freeze-thaw time, the greater damage the specimen suffered, the greater the porosity was because carbon dioxide could quickly spread into the specimen. In addition, there were small cracks on the specimen's surface after the freeze-thaw cycle. Carbon dioxide entered the specimen through cracks, which made carbonation depth after freeze-thaw cycles much greater than those of specimens that had not endured freeze-thaw cycles. According to the mechanism of carbide, the carbonation generated calcium carbide; this and other solid materials clogged mortar pores, resulting in decreased porosity for the internal specimen, which ultimately improved mortar strength.

3.2. Dynamic elastic modulus

From the results in Table 8, for the specimen that endured carbonation after freeze-thaw cycle, the elastic modulus loss and loss rate gradually increased as the water-binder ratio increased when the fly ash content was 20%. When the specimen was carbonation only, the elastic modulus loss increased more than other specimen. Otherwise, the dynamic elastic modulus increased when the content of fly ash increased. When mixing with fly ash, the frost resistance of specimen became worse, but the dynamic elastic modulus increased.

Table 8. Relative dynamic elastic modulus of samples.

Specimen	W/C	FA(%)	Air content (%)	Relative dynamic elastic modulus	
				P_{50}	P_{100}
B1	0.35	0	4.5	97.41	96.04
B2	0.35	20	4.7	94.94	92.11
B3	0.45	20	5.0	92.08	89.26
B4	0.45	40	4.9	90.33	88.08
B5	0.55	20	4.8	89.96	86.43
B6	0.55	40	4.6	88.61	84.05
T1	0.35	0	4.5	96.06	93.99
T2	0.35	20	4.7	92.02	89.38
T3	0.45	20	5.0	90.09	85.87
T4	0.45	40	4.9	84.62	78.17
T5	0.55	20	4.8	83.43	80.30
T6	0.55	40	4.6	80.32	71.67

Under a certain condition, concrete frost resistance can improve after carbonation; solid materials were made during the carbonation process and deposited in specimen pores, which reduces porosity, and compact the structure. In the freeze-thaw process, the total swelling pressure is reduced, moisture migration is hindered, and the osmotic pressure increase correspondingly lags which lead to a improvement of frost resistance.

3.3. Flexural strength

(1) Table 9 shows the flexural strength of samples under the alternating action of freeze-thaw and carbonation. It can be seen that the flexural strength drop rate increases gradually, and, specimens conducted carbonation test followed by freeze-thaw test have larger flexural strength. When fly ash content was 20%, the specimen weight loss of the two groups increased as the water-binder ratio increased, and the loss rate also gradually increased.

(2) For specimens that endured carbonation test followed by a freeze-thaw test, the internal structure becomes dense, and the frost resistance was increased due to the filling effect of the calcium carbonate from carbonation. During the freeze-thaw process, the osmotic and swelling pressures of water can cause calcium carbonate to gradually decompose; in addition, some original closed pores begin indirect within one another, the internal tissues fluid becomes loose, and the compressive strength slightly declines than before. According to the previous measured dynamic elastic modulus, the frost resistance of the carbonized specimens was stronger than that of uncarbonized specimens

within a reasonable range of carbonation; as a result, the flexural strength was higher than that of specimens without carbonation.

(3) For specimens that endured a freeze-thaw test followed by a carbonation test, tiny cracks appeared on the specimen surface due to the osmotic and swelling pressures caused by water movement. This loosened the internal structure and increased total porosity.

Table 9. Flexural strength of samples under the alternating action of freeze-thaw and carbonation.

Samples	Gradation	W/C	Fly ash (%)	Air content (%)	Flexural strength(MPa)	
B1	Two	0.35	0	4.5	9.69	freeze-thaw followed by carbonation
B2	Two	0.35	20	4.7	8.19	
B3	Two	0.45	20	5.0	8.06	
B4	Two	0.45	40	4.9	6.56	
B5	Two	0.55	20	4.8	5.94	
B6	Two	0.55	40	4.6	5.31	
T1	Two	0.35	0	4.5	12.81	carbonation followed by freeze-thaw
T2	Two	0.35	20	4.7	11.25	
T3	Two	0.45	20	5.0	10.67	
T4	Two	0.45	40	4.9	9.69	
T5	Two	0.55	20	4.8	9.02	
T6	Two	0.55	40	4.6	5.94	

4. Conclusions

(1) For specimens that endured a freeze-thaw cycle first, the carbonation time had the greatest impact on strength due to the expansion and osmotic pressures caused by water motion. After thawing, some of the closed pores linked together, which loosened the internal organization and increased total porosity.

(2) When the fly ash content was constant, the dynamic elastic modulus loss, loss rate, and weight loss increased as the water-binder ratio gradually increased. The dynamic elastic modulus loss and weight loss of the carbonized specimens were smaller than those of uncarbonized specimens within a reasonable carbonation range. Specimens without fly ash content had the highest frost resistance, the lowest dynamic elastic modulus, and the lowest weight loss.

(3) For specimens that endured the carbonation test followed by a freeze-thaw test, and for specimens that endured a freeze-thaw test followed by carbonation test, the flexural strength declined as the water-binder ratio or fly ash content increased.

Acknowledgments

This material is based upon work supported by the central non-profit scientific research fund for institutes (No. CKSF2016003/CL) Projects (No: 51279017 and 51139001), supported by the National Natural Science Foundation of China. China Postdoctoral Science Foundation (No. 2015M582213).

References

[1] Zhao X L 2003 The Comparison of Evaluation Indexes for Durability Degradation of Concrete in Freezing Thawing Cycles *Journal of Huazhong University of Science and Technology*

31(2) 103-105

[2] Wang W Y 2009 Effect of Fly Ash on Frost-resistance and Chloride Ions Diffusion Properties of Marine Concrete *China Ocean Engineering* **23**(2) 367-377

[3] Khunthongkeaw J 2006 A Study on Carbonation Depth Prediction for Fly Ash Concrete *Construction and Building Materials* **29**(9) 744-753

[4] Liu H K 2010 Three-demensional Mesoscopic Numerical Simulation of Concrete Based on CT Scan *Journal of Architecture and Civil Engineering* **27**(1) 54-59

[5] Chen X 2011 Study on Deformation Characteristics of Concrete Mixed with MgO *Post Doctor Report of Wuhan University*

[6] Ge X R 2000 Primary Study of Real-time CT Testing of Unloading Damage Evolution Law of Rock *Chinese Journal of Rock Mechanics and Engineering* **19**(6) 697-70

[7] Wittmann F H 2006 Influence of Combined Environmental Loads on Durability of Reinforced Concrete Structures *International Journal for Restoration of Buildings and Monuments* **12**(4) 349-362

[8] Niu D T 2013 Deterioration Mechanism of Sulfate Attack on Concrete under Freeze-thaw Cycles *Journal of Wuhan University of Technology-Mater. Sci.* **28**(6) 1172-1176

IOP Conf. Series: Materials Science and Engineering **164** (2017) 012006 doi:10.1088/1757-899X/164/1/012006

A study of the conversion of ultrasonic energy and their transducers

Q L Li[1, 2, 3] and J Wu[1]

[1]Institute of Applied Acoustics, School of Physics and Information Technology, Shaanxi Normal University, Chang'an District, Xi'an, Shaanxi 710119, China
[2]Shaanxi Key Laboratory for Ultrasonics, Chang'an District, Xi'an, Shaanxi 710119, China

Email: Lql0314@snnu.edu.cn

Abstract. This paper essay offers a survey of research on interconvertibility of the conversion of ultrasonic energy with other energy, and their transducers, the conversion materials. Furthermore, it mainly gives piezoelectric ultrasonic transducer and piezoelectrical ultrasonic transducer material. Finally, some problems of ultrasonic energy and transducers, especially microelectronic materials and its acoustoelectric devices used for the ultrahigh-frequency (UHF) ultrasonic which need further research are pointed out.

1. Introduction

Ultrasonic, study and applications of the energy of sound waves vibrating at frequency greater than 20kHz, i.e. beyond the range of human hearing. The oscillator that can converts high-frequency electric current into vibrating ultrasonic waves. Ultrasonic has found wide industrial use. In the ultrasonic, the power ultrasonic can be used as a means of mechanical vibration energy to produce emulsification and dispersion effects, etc., in related solutions. Generally, power ultrasonics is also a process in which the energy of ultrasonic vibration is made use of to change the structure, states, characteristics, and functions, etc., of substantial organizations or can be used to speed up these changes. To have an active applied technology of dynamic applications is an outstanding characteristic of power ultrasonics, which has larger acoustic power and higher acoustic intensity. The first successful and ripe application of power ultrasonics was the use of ultrasonic cleaning machines (techniques) as early as the 1940s to 1950s. In broad applications of ultrasonic treatment, generally electroacoustical ultrasonic equipment or instruments are used. The schematic diagram of the working principle of the ultrasonic biomedical engineering with electroacoustical ultrasonic equipment is illustrated in Figure 1.

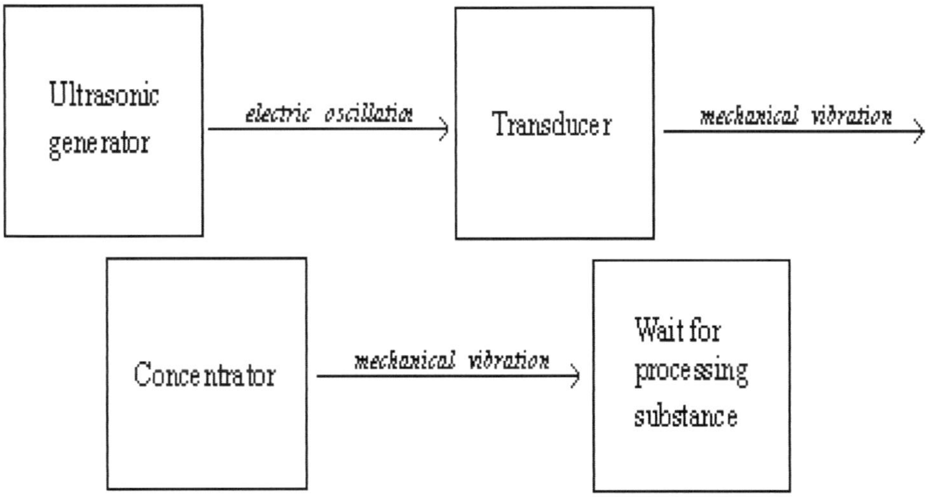

Figure 1. Schematic diagram of working principle of electroacoustical ultrasonic equipment.

2. The interconvertibility of ultrasonic energy with other energy

Generally, in the role of state phenomena in electric, electronics and microelectronic engineering technologies, the energy conversion, storage and finding new energy sources [1], etc. are the core effects. Hereinafter some examples of solid-state phenomena all are so. Electrical energy only, (1) Conduction processes: normal metals, superconductors, semiconductors (thermistors, Junction devices, and Hall Effect Gyrators, etc...), photoconductors; (2) Dielectric phenomena: normal dielectrics, frère-electrics (electrets, modulators, and storage devices, etc.); (3) magnetic behaviors: conventional magnetic metals, Ferrites (storage devices, gyrators, etc.), other spin resonances, masers, etc.. In energy conversion, (1) Mechanical~electrical (electronic, microelectronic): piezo-electric, electro-striction, magnetic-striction, strain effects; (2) Thermal~electrical(electronic, microelectronic): Ohmic heating, thermoelectric effects (generation of electricity, heat pumping and cooling), (3) Radiation~electrical (electronic, microelectronic): Photo-valtaic (photo-electric transfer device [2]), electroluminescence, recombination processes, (4) particle bombardment (X-rays, cathodoluminescence, etc.). In present paper, mainly research on ultrasonic energy conversion with other energy, their related transducer materials and transducers.

In broad applications of ultrasonic treatment, generally electroacoustical ultrasonic equipment or instruments are used. In this work, the electroacoustical ultrasonic equipment, where there is a key device used as an electroacoustic transducer (for receiving waves from an electrical system and delivering waves to an acoustic system, or vice versa.). These include magnetostrictive ultrasonic transducers (see Figure 2. schematic diagram of structural principle of the magnetostrictive ultrasonic transducer) and piezoelectric ultrasonic transducers (see Figure 3. schematic diagram of working principle of the piezoelectric ultrasonic transducer), used in electric energy, electromagnetic energy conversion ultrasonic energy to achieve the purposes and effects of ultrasonic treatment processing, such as, ultrasonic biochemical effect, ultrasonic accelerate chemical reaction, etc. belong to the applications of ultrasonic energy conversion [3].

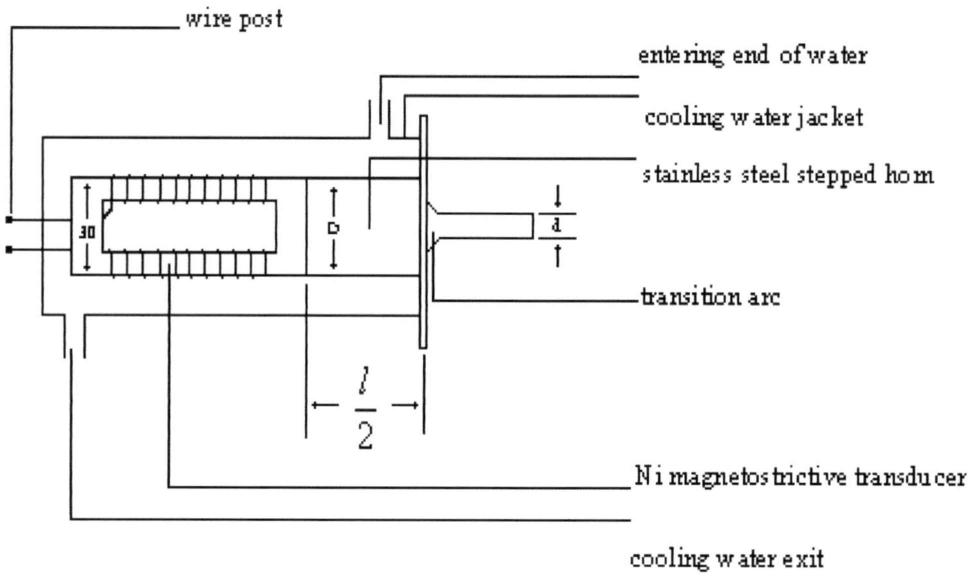

Figure 2. Schematic diagram of structural principle of magnetostrictive ultrasonic transducer.

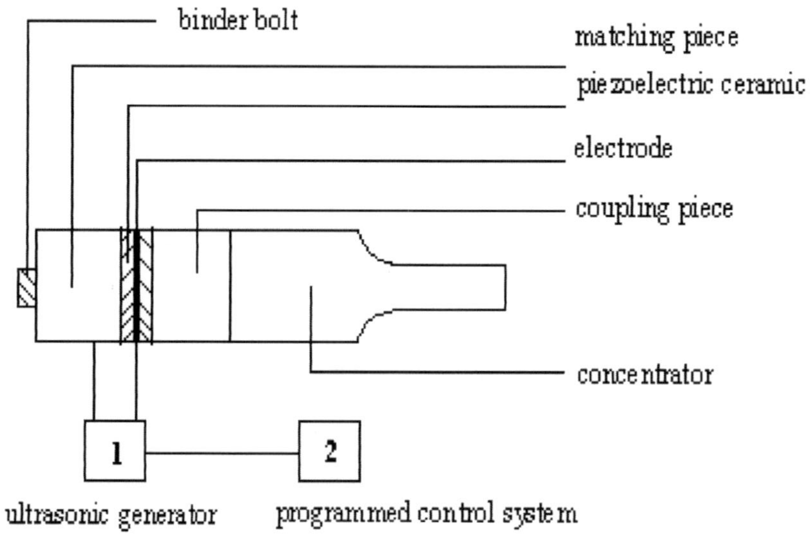

Figure 3. Schematic diagram of working principle of piezoelectric ultrasonic transducer.

3. The piezoelectrical ultrasonic transducer materials and transducers

All piezoelectrical ultrasonic transducer are on the basis of the linear theory of piezoelectricity of piezoelectrical materials (including quartz, piezoelectric ceramics, piezoelectric polymers, piezoelectric single crystals, piezoelectric semiconductors, piezoelectric films and piezoelectric biofilms, etc.) to designs, process and complete. Varied piezoelectrical materials and the transducer are mainly application in vest fields, such as: Non-liner ultrasound, Medical, biological and chemical ultrasound, Underwater acoustics (underwater communication; sonar and hydrophones), High power acoustics (industrial processes utilizing high power sonics such as welding, wire drawing, filtering, drilling, cutting, cleaning, emulsification, etc.), Ultrasonic NDT (ultrasonic non-destructive Test), etc. In present work, the piezoelectric ultrasonic diagnostic transducer is the component which, when connected to the ultrasonic equipment, transmits the ultrasonic wave and receives its reflections. A transducer can also be used solely as a receiver or a transmitter, depending on how it is connected. In addition, the piezoelectric ultrasonic transducers have multiple series, such as, in power ultrasonic: ultrasonic processing transducer, laminated transducer, single-window transducer, immersion transducer, sandwich transducer, etc.; in medical ultrasonic have ultrasonic therapeutic transducer, etc.; also, ultrasonic testing series transducers etc.

4. Microelectronic materials and its acoustoelectric devices used for the ultrahigh-frequency (UHF) ultrasonic

The microelectronics study the solid materials (main semiconductors) constitute the microminiaturization electro-circuit, the circuits and its subsystems are a branch of the electronics. It explores the laws of the movement and the applications of the electronics (or ions) which display the function of signal processing. In microelectronics, the spatial size of the devices, the components and the electrocicuits are between $\mu m(1\mu m = 10^{-6} m)$ and $nm(1nm = 10^{-9} m)$ as a unit generally. The Physical basis of microelectronic techniques depends on the structure of a semiconductor device and its electrical behavior [4]. This paper studied the microelectronic materials and its processing, their high-frequency ultrasonic transducer and its applications etc., especially, microelectronic materials and devices which are used for the ultrahigh-frequency (UHF) ultrasonic including microwave ultrasonic (with frequency $10^{10} - 10^{12} Hz$) and light-wave ultrasonic (with frequency $10^{12} - 10^{14} Hz$), etc..

The microelectronics is broadly applied in the ultrahigh-frequency (UHF) ultrasonic treatment and processing. Generally, in the ultrahigh-frequency (UHF) ultrasonic equipment or the instruments, the microelectronic techniques must be used. The schematic diagram of the working principle of the ultrahigh-frequency (UHF) ultrasonic equipment use in the microelectronic techniques is illustrated in Figure 1. In fact, the ultrasonic transducer frequency is particularly high when it uses surface wave techniques. For example, on the face of the metal sprayed CdS film to obtain piezoelectric transducer. And, on the end of $\alpha - SiO_2$ glass rod sprayed Ni film which made-up the magnetostriction type ultrasonic transducer. In these transducers, their two faces of the transducers must be smooth and mutual parallel, and its dimensions must be less than the wavelength of the acoustic waves. The limit of the frequency of these transducers are approximately $100GHz(10^{11} Hz)$, the wavelength of the acoustic waves is less than $0.1\mu m$ .The transducers which frequency exceeds $100GHz(10^{11} Hz)$ are used for the researches and the developments of the phonon and the quantum fields of the superconductors [5]. When the frequency of the film transducers is higher, the transducers are used the micro-sonic ($10^{10} - 10^{12} Hz$) and the light-wave ultrasonic ($10^{12} - 10^{14} Hz$), etc.. The direct amplifications of the ultrahigh-frequency (UHF) ultrasonic mean that in propagation process of the ultrahigh-frequency (UHF) ultrasonic waves, the waves could absorb the energy from the microstructure of the acoustic transmission medium continually and increase self-amplitudes. Studies on the direct amplifications of the ultrahigh-frequency (UHF) ultrasonic are as follows: the ultrahigh-frequency (UHF) ultrasonic wave propagating in piezoelectric semiconductors, because their carriers (i.e. electronics and holes) suffer from the driving of the applied DC (direct current) electric field, additionally, the carriers and the acoustic wave in same direction motion, when the movement speed of the carriers is greater than the velocity of the ultrahigh-frequency (UHF) ultrasonic wave, in this way, the ultrahigh-frequency (UHF) ultrasonic wave will absorb the energy from the carriers (i.e. electronics and holes) of the piezoelectric semiconductors. In that way, it will come out the amplifications of the velocity of the ultrahigh-frequency (UHF) ultrasonic, which have two amplifying ways: the body wave and the surface wave.

At the present, in nanotechnology propelling the research and development of microelectronic materials is unimaginable. The authors put forward the following points that need further research:

(1) The relationship of low dimension microelectronic material structures and nanotechnology applications are extremely important [6];

(2) The applications and developments of acoustic wave amplifiers (which used the principle of acoustoelectric effect) are completely dependent on microelectronic material;

(3) In the preparation of the microelectronic material films we need to control theirs the thickness accurately, to manufacture the ultrahigh-frequency (UHF) ultrasonic transducers. On the other hand, by control the frequency and the light signal to control the thickness of the microelectronic material films accurately.

References

[1] Li Q L 1996 New energy sources of 21st century: hydrogen energy and controlled nuclear fusion *New Energy Sources* **18**(2) 32-39
[2] Li Q L 2000 Integrated photoelectric device made of a piezoelectric ceramic exhibiting pyroelectricity and an internal photoeffect *J. Acoustical Soc. Amer.* **108**(2) 608-611
[3] Li Q L, Zhang Y H and Wu J 2013 Researches and applications of the ultrasonic emulsifications and dispersions *Journal of the Acoustical Society of America* **134**(5) 3546-3550
[4] Ghandhi S K 1968 *The Theory and Practices of Microelectronics* (New York: John Wiley) p 1
[5] Ma D Y and Shen H 2004 *Handbook of Acoustics* (Beijing: Science Press) p 522
[6] Peng Y C, Zhao X W and Fu G S 2011 *Low Dimensions Semiconductor Physics* (Beijing: National Defense Industry Press) p 95

Mechanical properties of gutta-percha sulfide modified asphalt

X Y Zou[1], X Y Gu[2] and X W Wang[2, 3]

[1] Jinhua Highway Administration Bureau, Jinhua 321000, China
[2] School of Transportation, Southeast University, Nanjing 210096, China

Email: wswxw2011@163.com

Abstract. Gutta-percha is the isomer of caoutchouc and can be used to enhance the performance of asphalt. In this paper, the produce proceedings of gutta-percha sulfide and gutta-percha sulfide modified asphalt are introduced. The performance indices of gutta-percha sulfide modified asphalt samples with different proportions are examined based on laboratory tests and the optimum ratio of gutta-percha and sulfur is decided.The micromechanism, temperature sensitivity, high and low temperature properties and viscoelasticity of the polymer modified asphalt are analyzed to discuss the modified mechanism and to decide the optimal polymer content. Low temperature bending tests are carried out to verify the low temperature performance of gutta-percha sulfide modified asphalt mixture. Research results showed that gutta-percha sulfide modified asphalt has good low temperature performance and a promising application prospect in the cold regions.

Keywords: gutta-percha sulfide, polymer modifier asphalt, vulcanization, micromechanism, viscoelasticity, ductility.

1. Introduction

Nowadays modified asphalt has been widely applied in surface course and middle course of pavement with the rapid development of China's social economy as well as heavy traffic. Because of its superior low-temperature performance and elastic recovery capability, styrene-butadiene-styrene (SBS) has been widely used in asphalt modification [1-4]. However, the cost of SBS modified asphalt remains high and SBS does not have a good compatibility with asphalt [5-6]. Taking saving social resources and mitigating the conflict between supply and demand into account, it is of great importance to find new asphalt modifiers.

Gutta-percha tree is a kind of propolis plant which grows only in the semitropical and warm temperature regions of China. Gutta-percha is the isomer of caoutchouc:gutta-percha is trans-polyisoprene, while caoutchouc is cis-polyisoprene [7] (Figure 1).

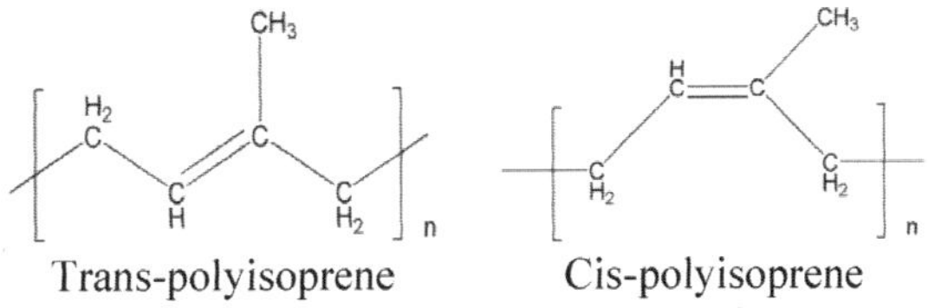

Figure 1. Molecular structure of gutta-percha and caoutchouc.

The objective of this paper is to explore the mechanism and application of gutta-percha in asphalt modification. This was accomplished by conducting an indoor tests to evaluate the performance of modified asphalt.

2. Vulcanization mechanism of gutta-percha

The glass transition temperature of gutta-percha is -70℃, which is close to that of caoutchouc. Gutta-percha does not take on elasticity under normal temperature because of its crystallizability. So the elasticity of its molecular chain can not be released until its crystallization is restrained.

Similar to caoutchouc, there are plenty of double bonds in gutta-percha molecule chains. So gutta-percha has the potential to be vulcanized. During the vulcanization process, the planar structure of double bonds are broken up and then connected by a chemical bond and form a new cross linked structure, therefore vulcanization is a kind of cross link reaction. Two broken bonds from two different chains can link the two chains by forming a junction. Along with the vulcanization process, more and more junctions turn separated chains into a network structure. Consequently crystallization in gutta-percha is restrained and elasticity is gained.

According to the different degrees of cross linking, vulcanization of gutta-percha is commonly divided into three phases [8] and three different materials:

Phase A: zero cross linking degree, gutta-percha sulfide in this phase is thermoplastic material.

Phase B: low cross linking degree, gutta-percha sulfide in this phase is thermoelastic material.

Phase C: high cross linking degree, gutta-percha sulfide in this phase is rubbery material.

When all crystals are broken up, the sulfide becomes typical rubber. Under this circumstance, cross link junctions will become closer with further vulcanization. This structure will prevent the movement of gutta-percha molecular chains; therefore the rigidity of the chains is enhanced while the flexibility is lost. As a result, to choose a reasonable cross linking degree in vulcanization imposes a great effect on the low temperature performance of modified asphalt.

3. Materials preparation and methods

High speed sheering, a common laboratory modified asphalt preparation method, is employed in the present study to prepare the modified asphalt test samples.

3.1. Instruments and materials

Matrix asphalt made by Ningbo Baoying Asphalt Storage Co. Ltd. is used in this study. Its performance indices are outlined in Table 1:

Table 1. Main performance indices and technical requirements of the matrix asphalt.

Test Items		Performance Index	Technical Requirements
Penetration(25℃)/0.1mm		65.3	60～80
Penetration Index		-1.92463	-1.5～+1.0
Ductility(10℃)/cm		57.5(>150)	≮20
Ductility(15℃)/cm		71.5(>100)	≮100
Softening Point TR&B /℃		47.8	≮46
RTFOT (163℃,5h)	Weight Loss/%	-0.0543	±0.8
	Penetration Ratio/%	61.26	≮61
	Ductility(10℃)/cm	>100	≮15
	Ductility(15℃)/cm	12.75	≮6
Dynamic Viscosity(60℃)/Pa.s		197.13	≮180
PG Grade		64-22	—

Parts of the modifiers and instruments used in this study are outlined in Table 2:

Table 2. Trademark and manufacturer of materials and instruments.

Name	Trademark/Type	Manufacturer
Styrene-butadiene Block Copolymer	SBS/YH-791	Baling petrochemical Corp.
Gutta-percha	—	—
Mixer	—	Brabender
High Speed Sheering Machine	L4RT	Silverson machine Inc.

3.2. Vulcanization process of gutta-percha

The Brabender mixer is used to blend sulfur and gutta-percha in this study (Figure 2). The compatibility between gutta-percha and the matrix asphalt is not so good, so direct blending of sulfur and gutta-percha does not enhance the performance of asphalt significantly [9]. When a few of asphalt is mixed, the compatibility turns much better. In this way, the asphalt added plays the role of compatibilizer here. This method is similar to the matrix method in production of modified asphalt.

Vulcanization process is as follows: turn on the mixer and warm it up till the temperature of feed tank reaches 80℃. Put small pieces of gutta-percha into the feed tank, heat and soften these pieces and then blend them slowly. Then load sulfur and asphalt, close the feed tank and blend at a speed of 60r/min and keep the temperature at 80℃. After 5 minutes' blending, peel off the gutta-percha sulfide from the mixer.

Figure 2. The Brabender mixer.

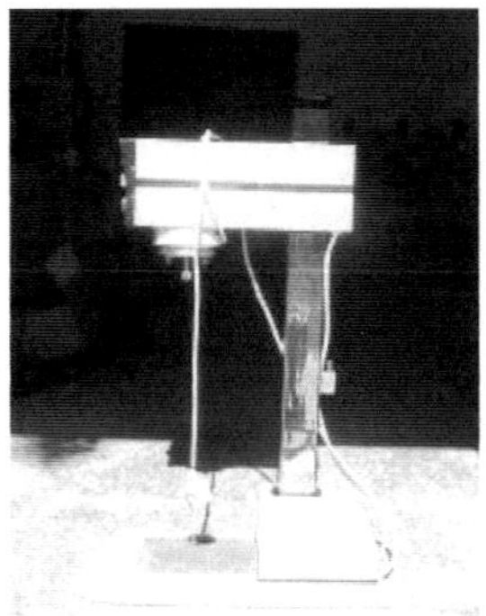

Figure 3. High speed shearing machine.

3.3. Preparation process of modified asphalt

The L4RT high speed shearing machine (Figure 3) is used to prepare the modified asphalt samples in this study with an electric furnace and an infrared thermometer which can measure the liquid surface temperature are needed as well.

Put the pieces of gutta-percha sulfide into an oven with a set temperature of 60℃ for 15 to 20 minutes until they are soft, and then cut them into small pieces. Put the iron barrel filled with matrix asphalt into the oven of 120℃ for one hour to make it free-pouring.

Place the barrel on the electric furnace, put modifier slices inside and then hand mix for about 20 minutes to make the modifier swollen in the matrix asphalt. Make sure to keep the temperature of the asphalt below 170℃ by infrared thermometer. After hand mixing, place the barrel and the furnace below the shearing machine for two hours with the shearing speed slowly reaching 6000r/min and the temperature around 170℃.

3.4. Decision of optimal ratio

When SBS is the modifier and its content is low, to increase the content of SBS can apparently promote the performance of modified asphalt; however with a high content, the change of SBS content has little influence on the performance of modified asphalt. Exsisting studies show that the most reasonable content of SBS exits [10, 11]. It refers to a critical state when SBS in asphalt begins to turn from a disperse phase to a continuous phase. Test results indicate that the most reasonable content weights approximate 5% of asphalt.

By the same token, it is hypothesized that there may be an optimal ratio for gutta-percha sulphide in asphalt. According to the most reasonable content of SBS, set the content of gutta-percha to 5 wt% of the matrix asphalt, and prepare samples containing 0.1 wt%, 0.4 wt% and 0.6 wt% sulfur. The Softening Point and ductility (5℃ and 10℃) are tested which represent the high and low temperature performance of the asphalt. The results of the performance indices are outlined in Table 3.

Table 3. Main performance indices of gutta-percha sulphide.

Matrix Asphalt	Gutta-percha /wt %	Sulfur /wt %	Softening Point TR&B /℃	Ductility(5℃) /cm	Ductility(10℃) /cm
100	0	0	47.8	0	57.5
100	5	0	52.4	6.05	11.25
100	5	0.1	50.1	19	53
100	5	0.4	50.1	35.2	66.8
100	5	0.6	49.2	10	38.5

Figure 4 and Figure 5 are illustrated to present the results. As the sulfur content increased, ductility of both 5℃ and 10℃ went up and then down. When the sulfur content was 0.4%, ductility got its maximum value. However, to change the sulfur content had little effect on softening point. Hence, the optimal ratio of gutta-percha and sulfur is 5:0.4.

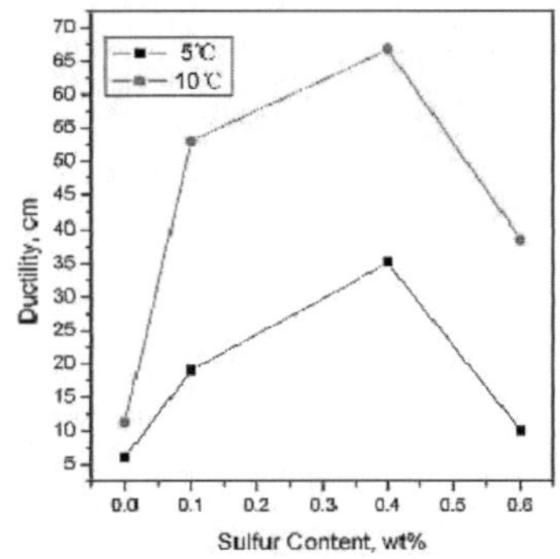

Figure 4. Ductility versus sulfur content.

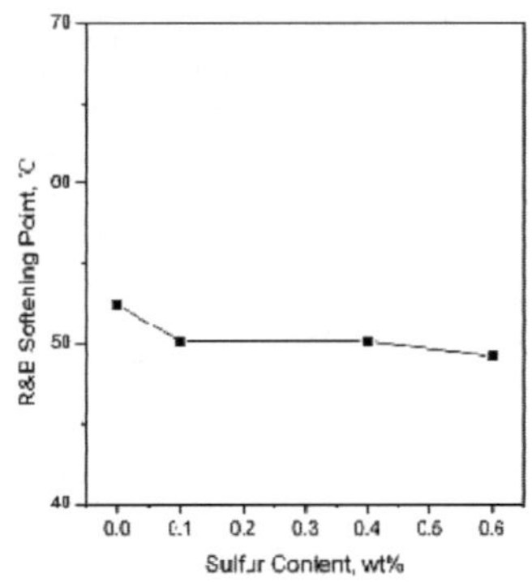

Figure 5. Softening point versus sulfur content.

4. Performance of gutta-percha sulphide modified asphalt

To explore the optimal content of gutta-percha sulfide, the performances of three modified asphalt samples designated as A, B and C were tested in every aspect. The three samples had respective gutta-percha sulfide contents, namely 3%, 5% and 7% with the same ratio of gutta-percha and sulfur (12.5:1) as previously concluded.

4.1. Micromechanism

Microscope was applied to study the micromechanism of the underlying polymer modifier asphalt(PMA). Figure 6 (a) ~ (c) shows the morphology of three samples with different polymer contents. In these pictures, the dark phase represents asphalt and the light phase polymer. The legends in these pictures represent 200 um. The particles of asphalt were big and that of the polymer phase seemed difficult to disperse in the asphalt as shown in Figure 6(a). As the sulfur content went up, the asphalt phase became increasingly small, and the polymer phase dispersed more uniformly. The decrease in asphalt particles indicated that the stability of PMA was promoted due to vulcanization.

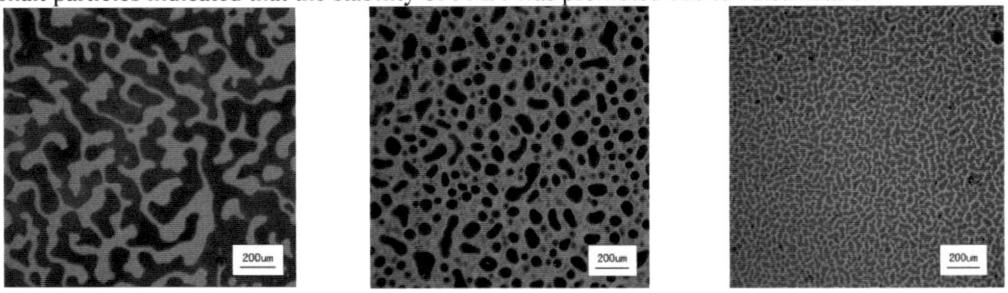

(a) 3%Gutta-percha Sulfide (b) 5%Gutta-percha Sulfide (c) 7% Gutta-percha sulfide

Figure 6. Micrographs of gutta-percha sulfide modified asphalt.

4.2. Temperature sensitivity

Figure 7 shows the Penetration index of sample A, B and C. Penetration index (PI) reflects the temperature sensitivity and colloidal structure of asphalt. High PI value reflects temperature stability, while low PI value indicates instability as temperature changes. According to the asphalt colloid theory, asphalt containing high asphaltene and low oil is less sensitive to temperature. As shown in Picture 7 PI value went up as more sulfides were added. Gutta-percha is of weak polar, so it can be swollen by nonpolar molecules as saturated fragrance, therefore a colloidal like structure is formed and the asphaltene content is relatively increased. As a result, high Gutta-percha Sulfide content leads to higher temperature stability.

4.3. High temperature property

Figure 8 and Figure 9 show the penetration and R&B softening point of the samples. The two indices represent the high temperature property in Chinese standard. As shown in the figures, gutta-percha slightly promoted high temperature property of asphalt, but as the content increased, the asphalt appeared to be softer. As previously mentioned, sulfuration is a cross link reaction, which makes the vulcanized material softer due to the restraint to crystallization, so it explains why asphalt became softer as gutta-percha sulfide content increased. The melting point of gutta-percha is only 65℃, while that of SBS is nearly 180℃, therefore particulates of gutta-percha are easy to be softened and lose binding force to asphalt particulates because of the increase of temperature. So gutta-percha sulfide can hardly improve the high temperature property.

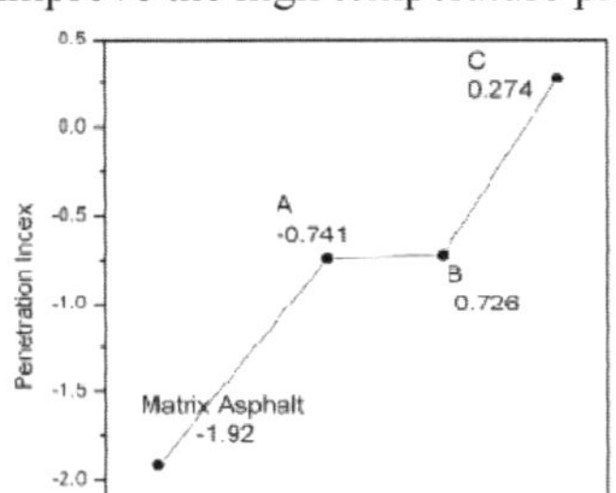

Figure 7. PI value of gutta-percha sulfide modified asphalt.

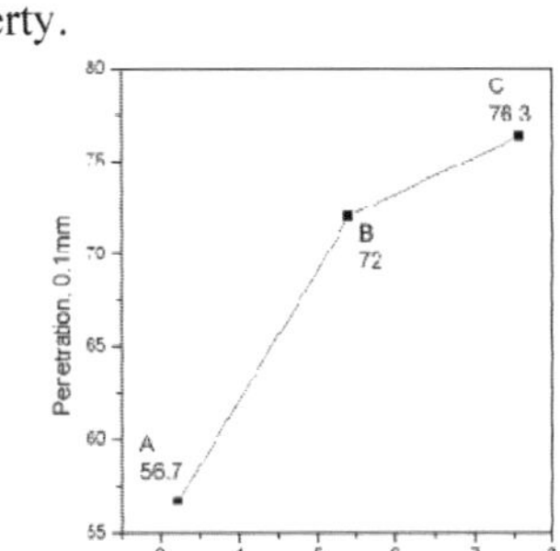

Figure 8. Penetration of gutta-percha sulfide modified asphalt.

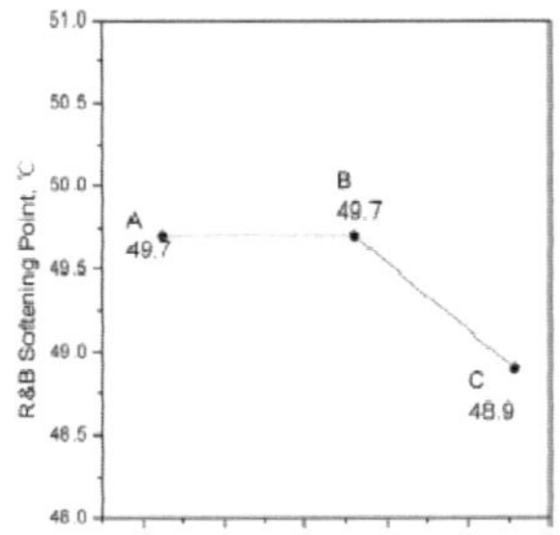

Figure 9. R&B softening point of gutta-percha sulfide modified asphalt.

4.4. Low temperature property

Figure 10 indicates that gutta-percha sulfide can improve low temperature property of asphalt. As the polymer content rose, 10℃ ductility kept elevating but the increase rate decreased when the polymer content reached 5.4 wt% of matrix asphalt. 5℃ ductility rose first and fell later at the turning point of 5.4 wt%. The glass state temperature of gutta-percha is -65℃, so when the temperature gets low, particles of gutta-percha can maintain at a certain elasticity to disperse the stress. Moreover, after the vulcanization, reticular structure is formed, which enables the gutta-percha an additional resilience capability. As is mentioned above, vulcanization can promote the compatibility of gutta-percha and asphalt; all these factors produce a significant promotion to the low temperature property of asphalt. If gutta-percha has been over cross-linked, the elasticity and ductility of the sulfide will drop, this can explain why the increasing rate of ductility slowed down when the polymer content increased.

4.5. Viscoelasticity property

As shown in Figures 11 and 12, the viscoelasticity of asphalt was slightly improved. As the gutta-percha sulfide content increased, the viscoelasticity of PMA kept rising. It proves that vulcanization can enhance the resilience and stretch property of gutta-percha, however under normal and high temperature, the promotion is not so marked at low temperature.

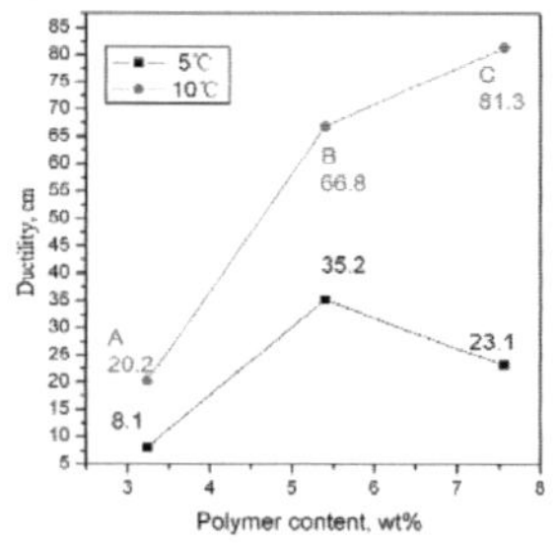

Figure 10. Ductility of gutta-percha sulfide modified asphalt.

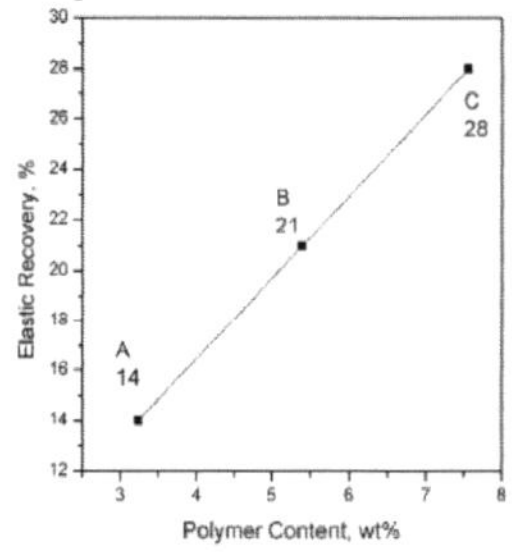

Figure 11. Elastic recovery of gutta-percha sulfide modified asphalt.

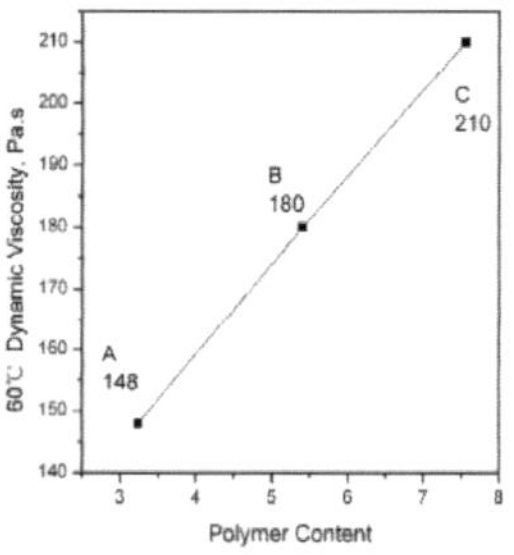

Figure 12. 60 ℃ Dynamic Viscosity of gutta-percha sulfide modified asphalt.

5. Low temperature bending test of gutta-percha sulphide modified asphalt mixture

According to the testing results, gutta-percha sulfide can obviously improve the low temperature property of asphalt. Low temperature properties of the mixtures were also investigated to evaluate the modified effects of gutta-percha sulfide.

In this study, the low temperature bending beam test at -10 ℃ with a loading rate of 50 mm/min was used. The beam specimen size was 250mm×30mm×35mm. Tensile strength and maximum tensile strain of failure could be collected and calculated by the experimental data as well as bending stiffness modulus and strain energy density.

The aggregate gradation of the mixture is a NMAS of 13 mm labeled as AC-13 with the optimum asphalt content 5.1% and the performance of coarse aggregate and fine aggregate all meet the specification requirements. Test results are shown below as Table 4.

Table 4. Low temperature bending test results.

Asphlat mixture Type	Tensile strength /Mpa	Maximum tensile strain /με	Bending stiffness modulus /MPa	Strain energy density /KJ/m3
General Asphalt mixture	6.346	1774	5395	1.983
Gutta-percha sulphide modified asphalt mixture	8.763	3621	2420	2.602

In China, the specification requirement for maximum tensile strain of low temperature bending test is no less than 3000με. Gutta-percha sulphide modified asphalt mixture have reached the requirement with higher tensile strength, higher strain energy density and much lower bending stiffness modulus, which means gutta-percha sulphide modified asphalt can greatly enhance the low temperature performance and anti-cracking abilities of the asphalt mixture.

6. Conclusions

According to the analysis, the main conclusions have been summarized as follows:

(1) Gutta-percha sulfide can obviously improve the low temperature property of asphalt, but only has a slight influence on the high temperature and viscosity properties, thus it possesses great application prospect in low temperature regions.

(2) Vulcanization enables gutta-percha a better performance on resilience and stretch. However, the breaking up of cross link due to vulcanization will reduce the rigidity of gutta-percha, and over vulcanization leads to the loss of elasticity.

(3) Taking consideration of all the factors, the optimal content is 5 wt% gutta-percha and 0.4 wt% sulfur of the matrix asphalt.

References

[1] Fu H, Xie L and Dou D Y 2007 Storage Stability and Compatibility of Asphalt Binder Modified by SBS Graftcopolymer *Construction and Building Materials* **21**(7) 1528-1533

[2] Cong Y, Liao K and Huang W 2006 Study on Techoical Parameters of SBS Modified Asphalt *Petroleum Science and Technology* **24**(5) 507-512

[3] Airey G D 2004 Fundamental Binder and Practical Mixture Evaluation of Polymer Modified Bituminous Materials *The International Journal of Pavement Engineering* **5**(3) 137-151

[4] Fawcett A H and McNally T 2003 Polystyrene and Asphaltene Micelles within Blends with a Bitumen of an SBS Block Copolymer and Styrene and Butadiene Homopolymers *Colloid and Polymer Science* **281**(3) 203-213

[5] Yildirim Y 2007 Polymer Modified Asphalt Binders *Construction and Building Materials* **21**(1) 66-72

[6] Liu Z, Xuan M, Zhao Z, Cong Y and Liao K 2003 A Study of the Compatibility Between Asphalt and SBS *Petroleum Science and Technology* **21**(7) 1317-1325

[7] Yan R 1995 Prospects and Research Progress on Eucommia Ulmozdes Gum *Progress in Chemistry* **7**(1) 65-71

[8] Xue Z, Xie L and Yan R 1992 Study on NR/Gutta-Percha and PB/Gutta-Percha Bends *Journal of China Elastomerics* **2**(1) 12-17

[9] Chen J S and Huang C C 2006 Fundamental Characterization of SBS Modified Asphalt Mixed with Sulfur *Journal of Transportation Research Board* **103**(5) 2817-2825

[10] Zhao K and Yuan J 2001 The Study on the Mechanism of Polymer Modified Asphalt (Part 1) ——Absorbing-action of Modifier to Light Constituent in Asphalt and the State of Aggregation *Journal of Xi'an Highway University* **21**(2) 34-37

[11] Zhao K and Yuan J 2001 The Study on the Mechanism of Polymer Modified Asphalt (part2) --Summarizing on the Submicroscopic Shape of Modifier Particles *Journal of Xi'an Highway University* **21**(3) 21-24

Precision machining of advanced materials with waterjets

H T Liu[1]

OMAX Corporation, 21409 72 Avenue S., Kent, WA 9832, USA

Email: peter.liu@omax.com

Abstract. Recent advances in abrasive waterjet technology have elevated to the state that it often competes on equal footing with lasers and EDM for precision machining. Under the support of a National Science Foundation SBIR Phase II grant, OMAX has developed and commercialized micro abrasive water technology that is incorporated into a MicroMAX® JetMachining® Center. Waterjet technology, combined both abrasive waterjet and micro abrasive waterjet technology, is capable of machining most materials from macro to micro scales for a wide range of part size and thickness. Waterjet technology has technological and manufacturing merits that cannot be matched by most existing tools. As a cold cutting tool that creates no heat-affected zone, for example, waterjet cuts much faster than wire EDM and laser when measures to minimize a heat-affected zone are taken into account. In addition, waterjet is material independent; it cuts materials that cannot be cut or are difficult to cut otherwise. The versatility of waterjet has also demonstrated machining simulated nanomaterials with large gradients of material properties from metal, nonmetal, to anything in between. This paper presents waterjet-machined samples made of a wide range of advanced materials from macro to micro scales.

1. Introduction

Waterjet technology has made significant advancements in the past decade. For many applications of precision machining, it is now often competing on equal footing with existing machine tools such as lasers and electro-discharge machining (EDM) [1].[2] In particular, waterjet has several technological and manufacturing merits that cannot be matched by most existing tools. A list of such merits is given elsewhere [2, 3]. For a number of applications, waterjet is advantageous over other tools in terms of fast cutting speed and turnaround, setup simplicity, material independence, and preservation of integrity of parent materials. For example, as waterjet is a cold cutting tool that creates no heat-affected zone (HAZ), waterjet does not induce heat damage to parts as lasers, EDM, and plasma cutting do. Figures 1a and 1b illustrate a pair of titanium tweezers and a miniature stainless steel butterfly cut with a CO_2 laser and waterjet. The CO_2 laser induced considerable heat damage to the parts as indicated by the presence of discoloring and slag, leading to part warpage and surface hardening [1, 3, 4]. Waterjet, on the other hand, induced no such damage on the parts and therefore preserved the structural and chemical integrity of the parent materials (Figures 1a and 1b). For the miniature butterfly with many narrow slots and thin webs, the heat generated by the laser simply melted the thin webs, which were then vaporized (Figure 1c). Again, waterjet did not induce any heat damage to the parts. Besides, waterjet

[2] Waterjet is often referred to either the waterjet-only-jet (WJ) or abrasive waterjet (AWJ) depending on the properties of the materials to be cut.

exerted a minimal side force to the part during cutting. As a result, it is capable of cutting very thin webs without distorting them.

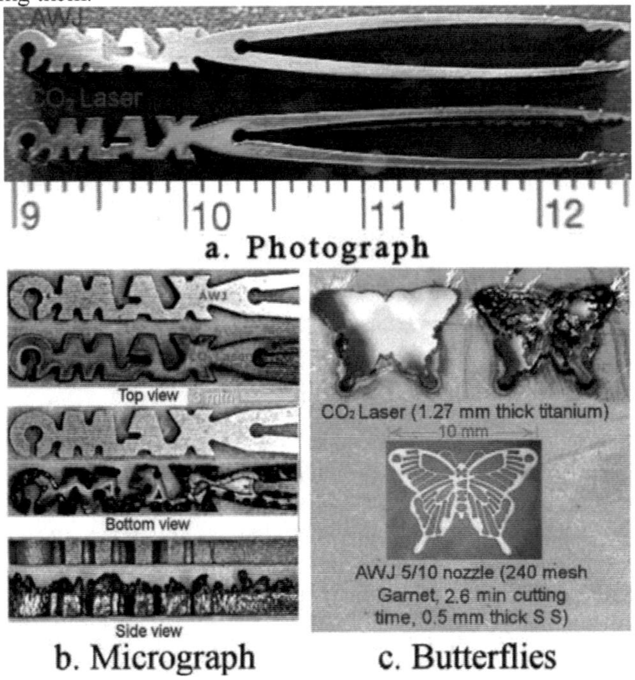

Figure 1. Typical heat damage by CO_2 laser [3].

[Courtesy of Center for Bits and Atoms (CBA)]

In order to minimize the HAZ, solid state lasers pulsed at high frequencies and wire EDM cut with multiple passes must be used. Such measures significantly reduce their cutting speeds. As a result, waterjet cuts much faster than lasers and EDM [3]. Also, waterjet is material independent; it cuts materials that cannot be cut with lasers (e.g., reflective materials) and EDM (nonconductive materials). The versatility of μAWJ has also demonstrated for machining simulated nanomaterials with large gradients of material properties from metal, nonmetal, to anything in between [3]. Even for extreme precision parts, the versatile waterjet has been proven to be a preferred near-net shaping tool. Particularly for materials that are difficult to cut with other tools, waterjet net shaping would remove most of the bulk materials. The net-shaped part is readily to be finished by light trimming with a qualified precision tool. As such, the operating life of that tool is significantly extended [5].

AWJ is capable of 3D machining but must be carried out with discretion. One of the properties of the spent abrasives of AWJ, if not "tamed" or captured, still possess considerable residual cutting power that could induce damage to other parts of the workpiece and pose a potential hazard to the operators. In other words, AWJ is not inherently suitable for 3D machining by simply mounting the nozzle on a multi-axis manipulator. Because the simplest and most effective means to dissipate the residual energy of spent abrasives is to let the spent abrasives shoot generally downward into a column of still water, most AWJ systems are built on top of a water tank that also serves to support the traversing mechanism. Such AWJ systems that are operating within the limitations of safety are mainly designed for 2D machining [6]. Novel approaches have been developed to facilitate AWJ 3D machining while ensuring operation safety, by either manipulating the workpiece or incorporating accessories to the 2D

AWJ platform (Liu and Olsen, 2013). 3D sample parts cut with these processes are presented to demonstrate the extended applications of 2D AWJ platforms for 3D machining.

Waterjet is amenable for micromachining as the nozzle can be downsized to generate jets with micron-size diameter [2, 7]. Supported by an NSF SBIR Phase II grant, OMAX Corporation developed and commercialized micro abrasive waterjet (μAWJ) technology for meso-micro machining. The new product was the MicroMAX JetMachining Center in which the μAWJ technology was incorporated.[3] With the addition of the MicroMAX, OMAX has established a full capability of precision machining of most materials from macro to micro scales for a wide range of part sizes and thickness.

In this paper, the waterjet technology with emphasis on the μAWJ technology will be described briefly. Detailed description of the technology is available elsewhere [8, 9].[4] Examples of 2D and 3D waterjet-machined parts that are made from advanced and delicate materials and with a wide range of part size and thickness are presented to demonstrate the versatility of the technology. The potential of waterjet for precision machining of various industrial and engineering components will be discussed and demonstrated.

2. Objective

The main objective of this paper is to demonstrate the versatility of waterjet technology for machining most materials from macro to micro scales for a wide range of part size and thickness. Demonstration will be made by presenting a wide range of waterjet-cut examples. Performance comparison of waterjet and other machine tools will be made when appropriate.

3. Technical approach and methods

Abrasive waterjet technology (AWJ) was commercialized in the late 1980s. Refer to a time line of the technological development presented briefly by Liu [3]. At that time, in the absence of precision controlling software and hardware, it was used merely as a rough cutting tool to take advantage of its material independent property. Material removal is accomplished by eroding away the materials by either the high-speed water of a water-only-nozzle (for relatively soft materials) or the abrasives (for hard materials) that were accelerated by the high-speed waterjet (WJ) through the mixing tube. Such a mode of material removal differs from other machine tools. It was perceived that there was considerable potential for developing the technology into a precision machine tool provided the above merits could be fully exploited. With such a vision, major efforts were devoted to advance the AWJ technology in the past two decades. Advancement included but was not limited to development of a cutting model for common engineering materials, position and cutting accuracies, control software and hardware, operational simplicity, cost effectiveness and fast turnaround, and material independence [2, 3]. The cutting model was developed based on extensive cutting of many materials. For each material, unique indices of machineability and pierceability are uniquely defined [2]. These indices are used to the cutting and piercing of various materials. The cutting model is one of the key parameters of the PC-based CAD/CAM for automation, greatly enhanced the user friendliness of the technology for precision machining. There was no steep learning curve, unlike CNC tools, for new users to master the operations of waterjet machining. New users could be trained in a week to learn how to design/cut parts and perform system maintenance proficiently.

Under the support of an NSF SBIR Phase II grant, OMAX developed and commercialized μAWJ technology for meso-micro machining [1]. The commercialized product was the MicroMAX JetMachining Center, which was released for production in September 2013. With the addition of the MicroMAX, OMAX has two product lines of JetMachining Centers capable of machining most materials from macro to micro scales (the 5M advantage) with a wide range of part size and thickness.

As a premier manufacturer of waterjet equipment, OMAX invested heavily on R&D for product improvement and demonstration for broadening applications of waterjet precision machining. In the

[3] The MicroMAX (https://www.omax.com/omax-machine/micromax) has made the Finalist of the 2016 R&D 100 Awards.
[4] www.waterjets.org

R&D and Demonstration laboratories, new processes are routinely being investigated for test cuttings of new materials and part geometries. For extremely precision parts, waterjet could serve beneficially as a near-net shaping tool. The near-net shaped parts could then be finished by light trimming them with appropriate precision tools. As such, the operating life of these tools are greatly extended.

A large portion of the test cutting was from requests by existing and prospective customers who were interested in how waterjet would benefit their operations. Other examples presented in this paper represent a collection of results of the test cutting on advanced materials using several JetMachining Centers equipped with various nozzles and accessories described in Section 4. Others are the results of collaboration between OMAX and its collaborating research institutes.

4. Facilities and equipment

At OMAX's Headquarters, there are a R&D Laboratory and a Demonstration Laboratory equipped with several JetMachining Centers JMCs). Figure 2 illustrates two JMCs – a MicroMAX and a Model 160X. In this section, waterjet nozzles together with the most common accessories are described for precision machining of 2D and 3D parts made from a wide range of materials from macro and micro scales.

a. MicroMAX b. Model 160X JMC

Figure 2. Two JetMachining Centers with different work envelops and accuracy.

4.1. Nozzles

One of the unique advantages of waterjet is that it only needs one tool to qualify for multiple modes of machining - parting, drilling, turning, milling, facing, beveling. The single tool is the waterjet nozzle that has difference sizes to cover a wide range of part thickness. Typical nozzles include the Water-Only-Nozzle, MAXJET 5 nozzle, the MINIJET nozzle, and the 7/15 nozzle[5]. Each nozzle can accommodate orifices and mixing tubes with different sizes of IDs. Figure 3 illustrates a schematic of the AWJ nozzle together with photographs of the four nozzles described above.

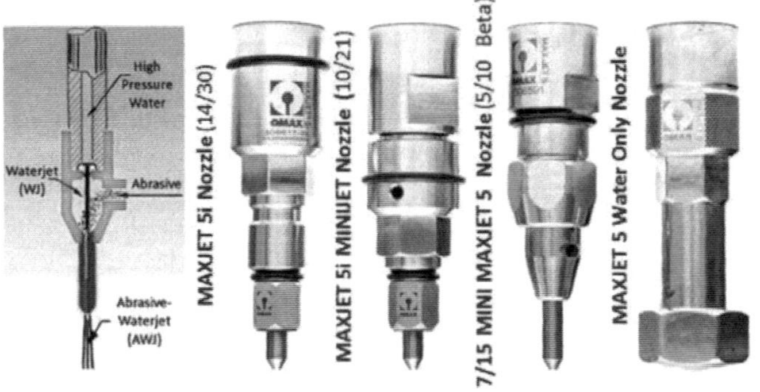

Figure 3. AWJ nozzles.

[5] 7/15 nozzle refer to the AWJ nozzle with a 0.007" (0.18 mm) ID orifice and a 0.015" (0.38 mm) ID mixing tube.

4.2. Accessories

There are three popular accessories: (a) Tilt-A-Jet for taper compensation, (b) Rotary Axis for cutting axisymmetric features, and (c) A-Jet for beveling and countersinking. 3D parts with complex 3D geometry can be cut by combining the use of both the Rotary Axis and A-Jet. Figure 4 illustrates photographs of the Tilt-A-Jet, Rotary Axis, and A-Jet. 3D parts with complex geometry can be readily machined by combining the operations of the Rotary Axis and the A-Jet.

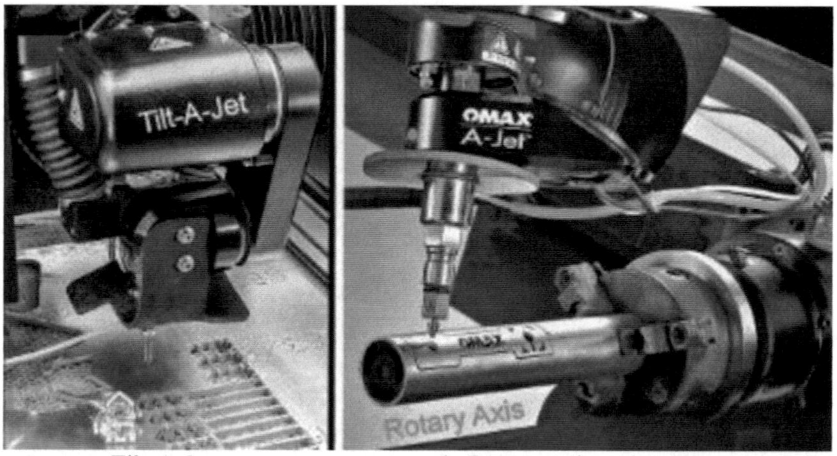

a. Tilt-A-Jet b. Rotary Axis and A-Jet

Figure 4. Accessories.

Another useful accessory is the vacuum assist kit which is used during low-pressure piercing and cutting where the Venturi vacuum or the jet pump effect is too week to entrain fully all the abrasive fed from the hopper. The kit was particularly designed for low-pressure piercing of delicate materials such as composites, laminates, and brittle materials such as glass particularly when miniature nozzles are used. Piercing damage often occurs when the tensile strength of the target materials is lower than the stagnating pressure developed inside the blind hole [2, 10, 11].

4.3. PC-based CAD/CAM

Waterjet machining is controlled by a PC-based Intelli-MAX Software Suite that is powerful while remaining intuitive and easy to use with no steep learning curve. From creating a drawing to importing part files with the PC-based CAD (LAYOUT), Intelli-MAX easily takes a design and prepares it with the PC-based CAM (MAKE) for machining. Cutting enhancements are achieved through the incorporation of a cutting model consisting of a lookup table by assigning machineability and pierceability indices to each common engineering material [2]. MAKE controls the traverse and several accessories to enable precision machining of 2D and 3D parts cost effectively with fast turnaround.

5. Results

5.1. Material independence

One of the technological merits of waterjet is its material independence as compared with other machine tools that are often material selective. Figure 5 illustrates photographs of AWJ-cut parts made from a wide range of materials that include metals, nonmetals, composites, laminates, and brittle materials. Such a versatility is superior of most other machine tools. For example, CO_2 lasers cannot cut reflective materials (e.g., copper) and EDM cannot cut nonconductive materials (e.g., PEEK and Lexan).

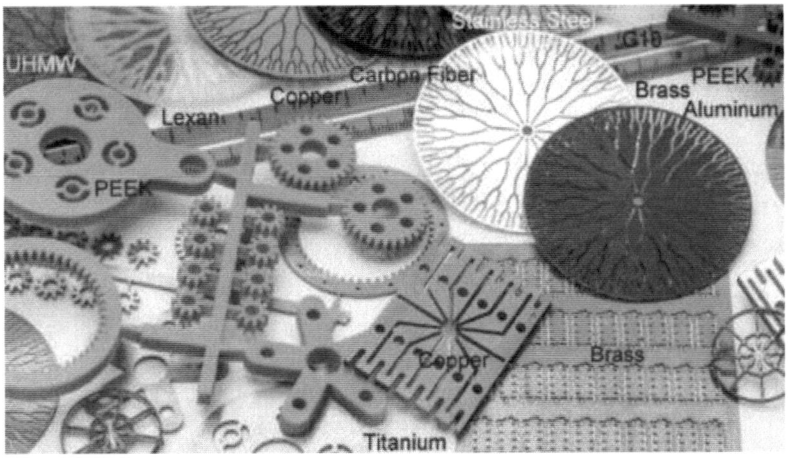

Figure 5. Material independence of AWJ [12].

5.2. Performance comparison

The performances of several machine tools and the waterjet were discussed by Liu [3]. In this subsection, examples are given to compare the performances of waterjet, wire EDM, and solid state lasers, respectively. As illustrated in Figure 1, thermal damage to parts cut with CO_2 lasers and EDM is caused by the presence of the HAZ. Remedies to minimize the thermal damage is to slow down the cutting speed of wire EDM and to employ solid state lasers pulsed at high frequencies. As a result, the cutting speed of waterjet is considerably fast than those of solid state lasers and wire EDM when thermal damage by the presence of the HAZ must be minimized to preserve the structural and chemical integrity of parent materials. Figure 6 illustrates photographs of two tweezers cut with the AWJ equipped with a 7/15 nozzle and a wire EDM with a 0.15-mm-diameter wire. It took 38 min and 32 sec for the EDM and AWJ to cut the parts; the ratio of the cutting time is 70. Since the wire diameter is nearly 3 times finer than the AWJ diameter, the cutting accuracy of the EDM is better than that of the AWJ as can be seen from Figure 6. The tweezers were also cut with a solid state laser pulsed at 5 kHz to minimize the thermal damage to the part. Figure 7 shows photographs of the AWJ- and laser-cut tweezers. There was still a faint trace of discoloring was observed on the laser-cut tweezers. The cutting accuracy of the laser is considerably better than that of the AWJ as the laser has a spot size of 50 μm that is 5 times smaller than the 250 μm diameter of the AWJ. However, the AWJ cut over 200 times faster than the laser as the times to cut the part were 45 s and nearly 3 h, respectively.

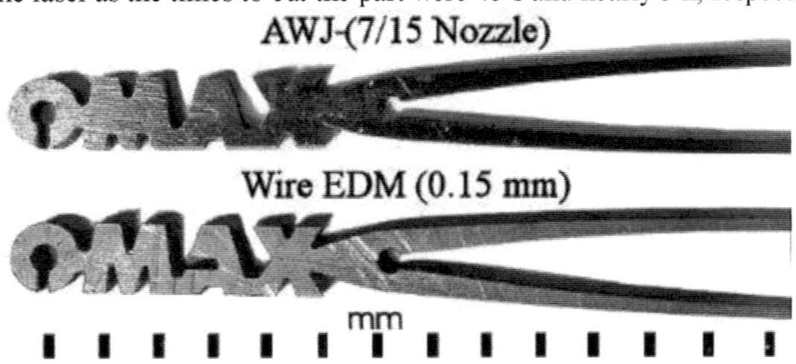

Figure 6. AWJ- and wire EDM-cut titanium tweezers [3] (courtesy of MIT CBA).

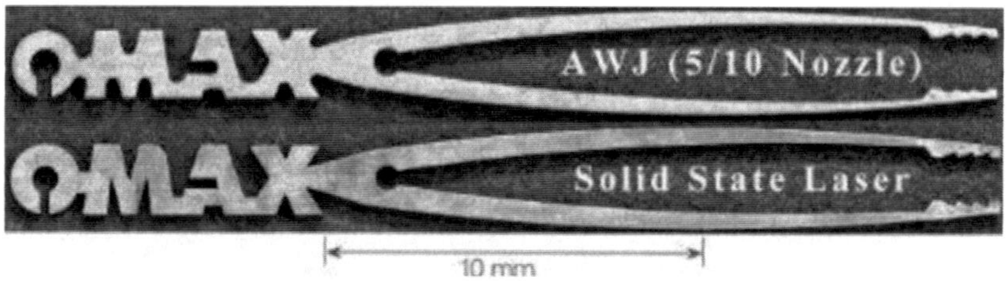

Figure 7. AWJ- and wire solid-state-laser-cut titanium tweezers [3] (courtesy of MIT CBA).

The solid state laser was also used to cut the miniature butterfly shown in Figure 1, as illustrated in Figure 8. The thickness of the part reduced from 0.5 mm to 0.25 mm. The cutting times for the AWJ and the laser were 2.2 min and about 60 min, respectively, a ratio of cutting time of 30. Again, the cutting accuracy of the laser is considerably better than the AWJ due to the fine spot size of the laser. No attempt was made to cut the butterfly with the wire EDM; its geometry is so complex with many piercings and internal features that it would be very difficult and time consuming to cut with the wire EDM.

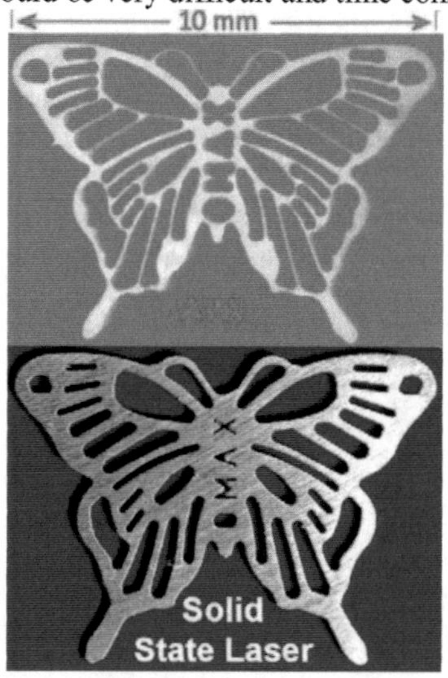

Figure 8. μAWJ versus solid state laser.

5.3. Wide range of part size and thickness

The addition of the MicroMAX to the two product lines of OMAX's JetMachining Centers has established a full capability of machining most materials from macro to micro scales. In practice, waterjet has been successfully machining aluminum blocks 200 mm and thicker. With taper compensation, Figure 9 illustrates several AWJ-cut aluminum blocks with nearly taper free edges. On the other hand, the μWJ technology has made considerable progress toward micromachining. Figure 10 illustrates a set of stainless steel and titanium tweezers to demonstrate the recent trend of the above progress in recent years.

Figure 9. AWJ-machined aluminum blocks.

Figure 10. Progress in AWJ meso-micro machining.

5.4. Parametric shape generator

One of the built-in parametric shape generators available in the Intelli-MAX software suite is a "Gear, Rack, and Sprocket Shape Generator" for designing tool paths of those components. For sample, the tool path of a gear can be readily generated by inputting the following five parameters: number of teeth, pitch, root fillet radius, pressure angle, and tip fillet radius. Waterjet was used to repair a damage segment of bisalloy wind turbine gear segment (100 mm thick and around 1.5 m in diameter), as illustrated in Figure 11. On the other hand, Figure 12 illustrates a set of miniature titanium planetary gears machined with the 5/10 micro nozzle. The components were nested on a titanium sheet (right). Tiny tabs were used to hold on the components that could be retrieved by breaking the tabs. The as-cut components were assembled into an operating planetary gear that was driven by a micro motor.

Figure 11. Repair of 100 mm thick bisalloy gear (OD > 1.5 m).

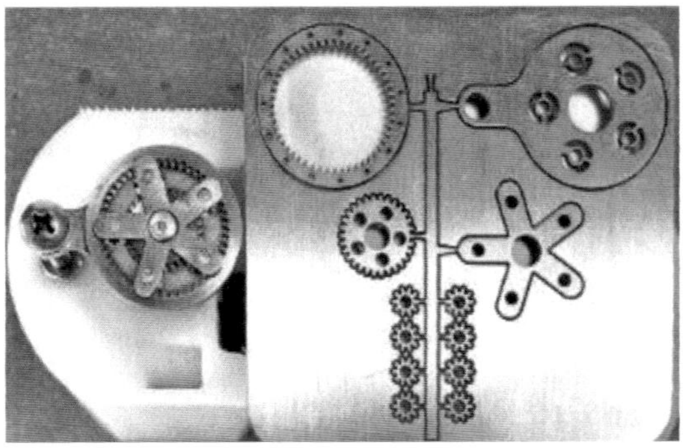

Figure 12. Miniature planetary gear set – assembled (left) and nested components (right).

5.5. Machining delicate materials

For cutting internal features, the first step is to drill a starting hole. AWJ if untamed would build up large stagnating pressure inside the blind hole before breakthrough takes place. Delicate materials such as campsites, laminates, and brittle materials with weak tensile strength, the buildup of stagnating pressure inside the hole would cause piercing damage provided the stagnating presses exceeds the tensile strength of the material [11, 13, 14]. Figure 13 illustrates typical AWJ piercing damage in terms of cracking/chipping (left) and delamination (right) for laminated glass and GX1 composite.

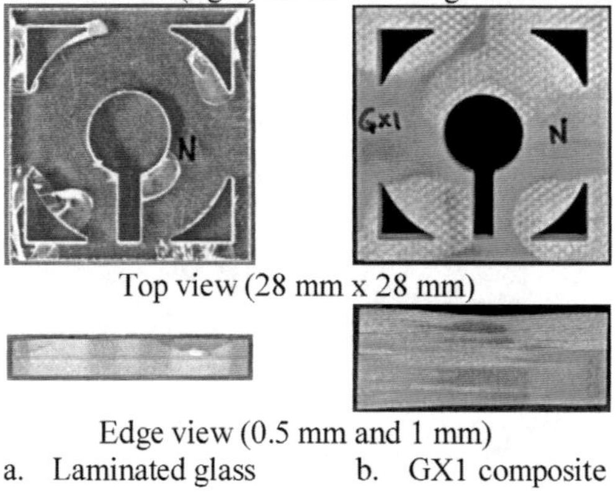

Figure 13. Typical AWJ piercing damage.

Several remedies were developed by the author and his colleagues to minimize and/or mitigate AWJ piercing damage. Abrasive cryogenic jet (ACJ) using liquefied nitrogen as the working fluid was first developed [15]. Subsequently, a super-heated abrasive waterjet (FAWJ) was developed [10]. The working fluids of both the ACJ and the super-heated FAWJ evaporated as the jets exiting the mixing tube. Only a small percentage of the working fluids enters the blind hole, resulting in considerable reduction in the stagnating pressure. Both the jets succeeded in minimizing/mitigating the piercing damage. Figure 14 illustrates photographs of FAWJ-cut samples that are free of piercing damage.

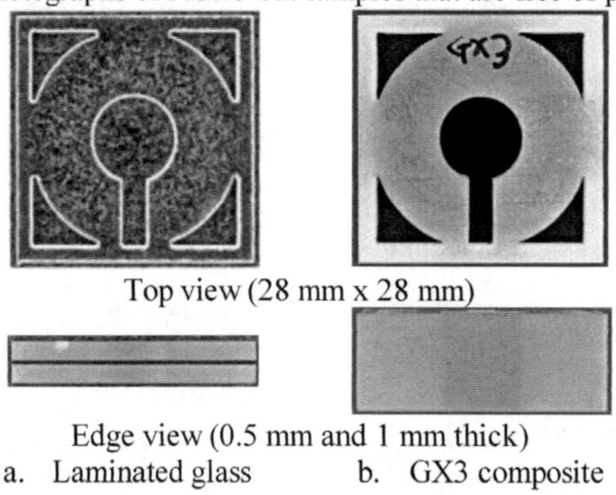

Figure 14. Mitigation of piercing damage with FAWJ.

The ACJ and FAWJ, however, were merely research tools that were too complex and not reliable for industrial deployment. Based on the concept of these tools, reliable Turbo Piercer and Mini Piercer were developed for drilling large and small holes, respectively. Figure 15 illustrates two samples cut with the Turbo and Min Piercers on an aerospace aluminum laminate (BAC1534-63F), respectively [2]. The laminate consisted of 19 layers of 0.076-mm-thick aluminum shims. The bonding strength of the adhesive between shims was relatively weak. It is imperative that stagnating pressure induced inside the blind hole must be lower than the adhesive strength to avoid delamination.

IOP Conf. Series: Materials Science and Engineering **164** (2017) 012008 doi:10.1088/1757-899X/164/1/012008

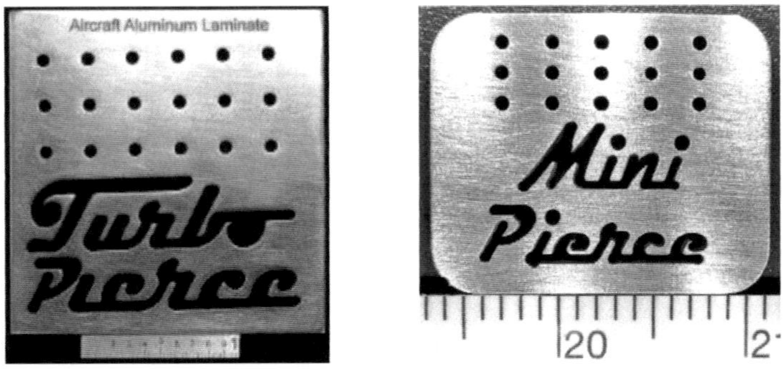

Figure 15. Samples of internal features cut with the turbo and mini piercers on aluminum laminate.

Carbon fiber is another material that was difficult to cut with conventional AWJ. By using the vacuum assist kit, low-pressure cutting at pressures as low as 35 MPa was used successfully to pierce this material without damage. Figure 16 illustrates photographs of a carbon fiber sheet and tube with internal features machined with the AWJ. The Rotary Axis was used to cut the internal features on the tube wall. A sacrificial carbide rod was placed inside the tube to protect the opposite wall of the tube from damaging by the spent abrasives.

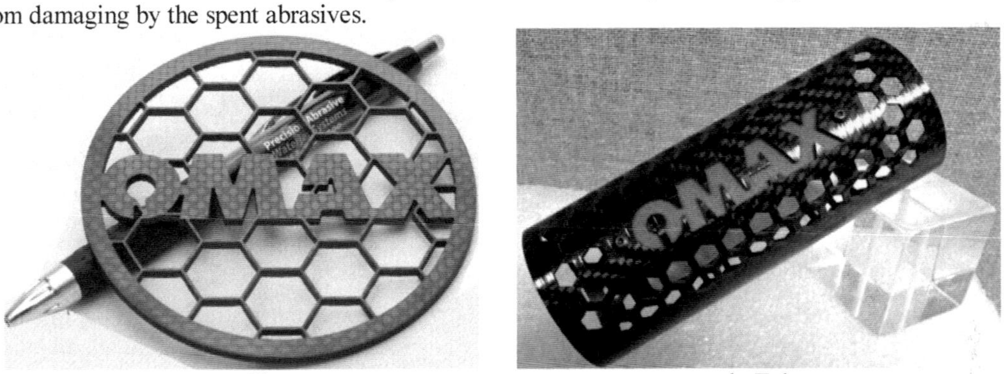

a. Sheet b. Tube

Figure 16. Internal features cut on carbon fiber sheet/tube.

5.6. Machining difficult materials

The machineabilities of stainless steel, Inconel, and titanium are 80.8, 83.6, and 108, respectively [2]. In other words, waterjet cuts Inconel and titanium 3% and 34% faster than cutting stainless steel. Such a trend is opposite to machining these "difficult" materials with conventional tools such as CNC hard tools and EDM. As a result, waterjet are often used to machine these difficult materials. Even for extremely precise part, waterjet can be advantageously used as the near-net shaping tool. The net-shaped part can then be precisely trimmed with qualified tools. The operational life of these tools is greatly extended since the bulk of the material is removed by the waterjet.

Heat treatment are common practice to improve the performance of steel parts. For example, tool steel is often hardened by heat treatment to increase the Rockwell hardness from RC20 to as high as RC65. The hardened tool steel, however, is very difficult to machine with CNC tools. Steel parts are usually machined in its annealed state (RC20) and then are heat treated to the required Rockwell hardness. Unfortunately, the heat treatment brought above distortion of the finished parts that would degrade its tolerance. For waterjet, the machineabilities of annealed steel with RC20 and hardened steel with RC60 are 70.2 and 54.1, respectively. In other words, waterjet is capable of machining the hardened steel simply by slowing down the cutting speed by 77%. There is considerable benefit by machining the steel in its hardened state as the degradation of the part tolerance due to post heat treatment is mitigated.

For industrial applications, OMAX recently collaborated with Competitive Engineering to cut narrow slots on a 2.1 mm thick 440C stainless steel plate hardened to RC 58, as illustrated in Figure 17. The pockets and patterns were precut on the blank before waterjet machining. The current method to machine the part is by wire EDM. The EDM process requires three passes to cut each slot in order to minimize the HAZ. As a result, it takes over 6 hours to cut the part. By comparison, the MicroMAX using the 5/10 nozzle cuts the same features, to almost the same quality without the HAZ in only 23 minutes. In other words, the cutting speed of waterjet is better than 15 times faster than that of the wire EDM.

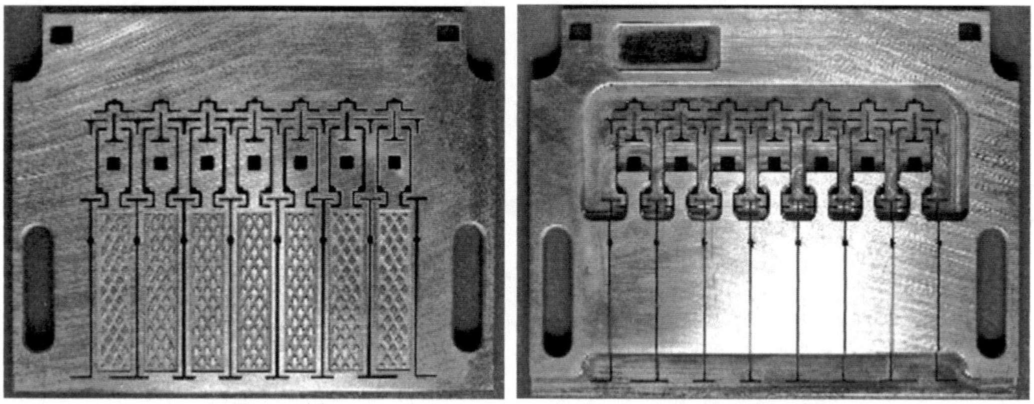

a. Entry side b. Exit side

Figure 17. μAWJ-machined slot patterns on a bonding extender for lapping thin-film ceramic substrates – courtesy of competitive engineering.

For biomedical applications, titanium has been used extensively for fabricating orthopedic implants and prosthetic components. With the μAWJ technology commercialized, waterjet has been potentially applied to fabricate many orthopedic and prosthetic components [16]. For example, the titanium mesh cage shown in Figure 18c is machined with the μAWJ 5/10 nozzle on the Rotary Axis.

5.7. 3D parts

In order to take advantage of the 3D capability of AWJ while operating the machine within the limitation of safety [6], accessories were developed enable 3D machining using a 2D AWJ platform. The Rotary Axis as illustrated in Figure 4b was designed to machine many axisymmetric and 3D features on tubes, pipes, and other cylindrical parts. The A-Jet provided two tilt angles – one perpendicular to the X axis and the other perpendicular to the Y axis. The two tilt axes were controlled to a specific limited distance so that the abrasive waterjet stream is always shooting in a generally downward direction toward a machine's water tank.

The A-Jet articulating head, for instance, tilts at any angle within its "120-degree cone of safety." By combining the Rotary Axis and the A-Jet, 3D parts with complex geometry can be readily fabricated [17]. Figure 18 illustrates several 3D parts cut with the Rotary Axis and/or the A-Jet. Figure 18a illustrates the setup for cutting a slotted copper tube that serves as the stator/rotor of a novel low-resistance slot-less armatures for a high-efficiency motor/generator [18]. The Rotary Axis was rotating while cutting slots that are not parallel to the axis of the tube enabling the curvature of the slots conformed to that of the tube. Figure 18b is a photograph of the pair of stator and rotor. The width of the slots was around 0.300 mm. The μAWJ is the only tool that can be used cost effectively to machine the slotted copper tubes since lasers cannot cut copper because of its high index of refraction and the nonsymmetrical patterns of the slots present considerable challenge to EDM. Figure 18c presents additional photographs of parts cutting from various materials. The Rotary Axis was used to cut most of them except the "fish mouth" weld joint where its nonaxisymmetric geometry required the pairing the Rotary Axis and the A-Jet.

a. Setup of Rotary Axis and A-Jet

b. AWJ-cut slotted tubes

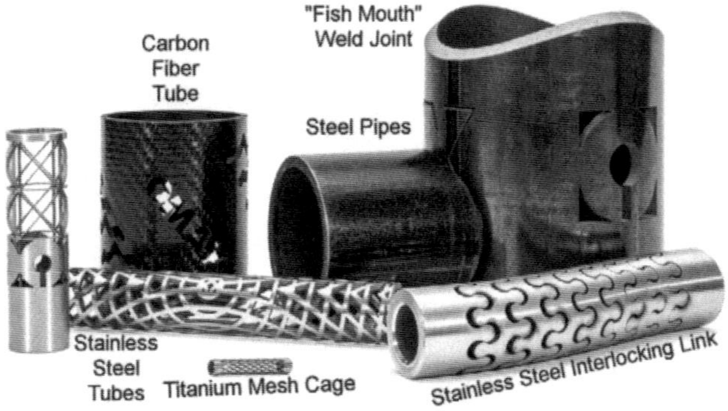

c. Assortment of 3D parts

Figure 18. 3D parts with complex geometry.

5.8. Taper free parts

For certain precision parts, such as mechanical flexures, it is imperative that the edges be taper free for optimum performance. Figure 19 illustrates a monolithic aluminum flexure with six narrow flexural hinges 0.25 mm wide in plane (MIT patent pending). This part, which cannot be fabricated with CNC machining, laser, or EDM, serves as a part of a novel laparoscopic trocar and blade retraction mechanism to mitigate damage to internal organs caused by over puncturing [19]. Another application is for fabricating nonlinear load cells which were designed to be highly sensitive (1% changes in the force) with a large force range (five orders of magnitude). Figure 20 illustrates two such load cells, one with large-aspect-ratio flexures of constant width of 1 mm and the other with tapered flexures. The large-aspect-ration flexures were over 100 mm long. The waterjet exerts a minimum side force onto the workpiece and is most suitable for machining narrow webs with large aspect ratios [1]. Laboratory tests were conducted to demonstrate successfully that the performance of the load cell agreed well with the theory [20].

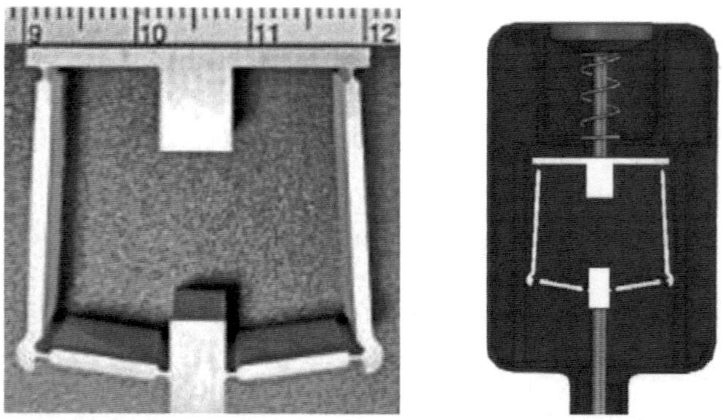

Figure 19. AWJ-cut flexures (courtesy of MIT Mechanical Engineering Department).

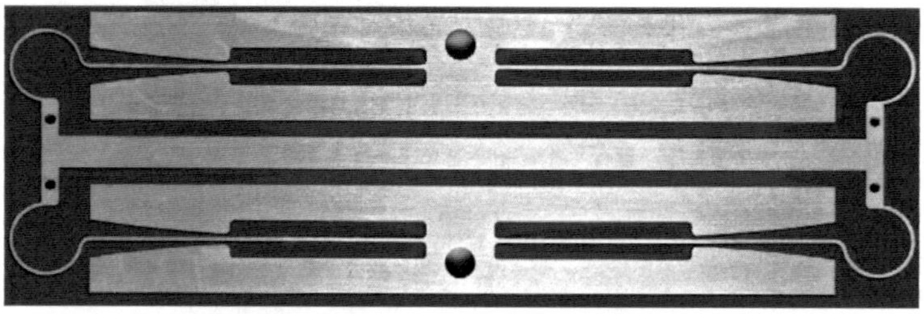

a. Large-aspect-ratio flexures

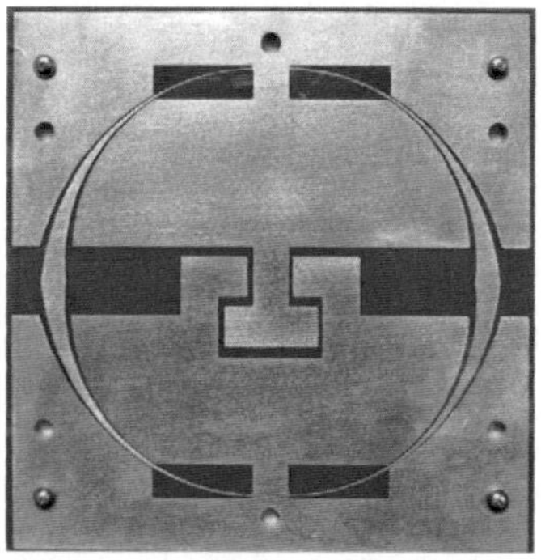

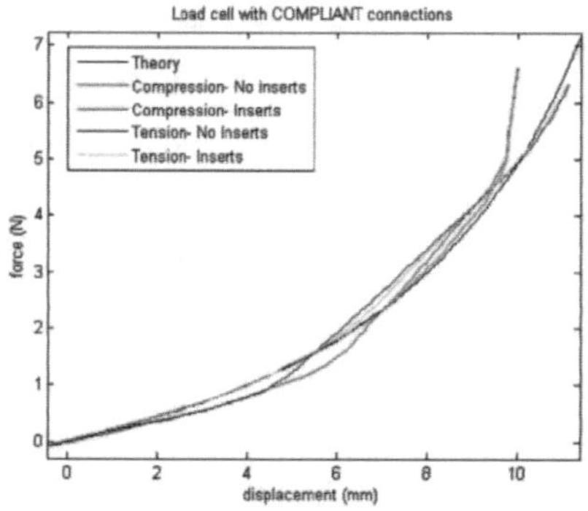

b. Tapered flexures c. Performance comparison

Figure 20. Nonlinear Load Cells (courtesy of MIT Mechanical Engineering Department).

5.9. R&D at educational institutes

The user friendliness and the 5M advantage of waterjet technology have attracted the attention of educational institutes. For example, MIT has installed more than 15 JetMachining Centers including a MicroMAX on its campus.[6] The Hobby Shop at MIT offers training classes for students, faculty members and alumnae. Trained users have used the machine for R&D, hobby projects, and prototyping. As a result, the JetMachining Centers have been the most used machine tool on campus. In particular, graduate students used the MicroMAX to build models and prototypes for their extracurricular activities and academic work. For example, the MicroMAX was used to machine 1-mm-thick electrically isolated alumina (Al_2O_3) clamping rings for a deep reactive ion etcher of silicon wafer. The μAWJ-cut ring and the setup of a deep reactive ion etcher are illustrated in Figures 21a and 21b, respectively.

[6] OMAX offers considerable discount of its products to educational institutes for education purpose. This serves as an effective means to raise the awareness of waterjet technology.

a. Alumina ring b. Ion etcher of silicon

Figure 21. μAWJ-cut electrically isolated alumina ring (courtesy of MIT Hobby Shop).

Meanwhile OMAX has collaborated with MIT extensively to apply waterjet for various R&D projects. Selected projects are presented in Figures 1, 6-8, and 18-20, respectively.

5.10. Nanomaterials

One of the advantages of nanotechnology is its capability of manufacturing nanomaterials with large gradients of material properties from metal, nonmetal, and anything in between. Such nanomaterials would present considerable challenge to conventional tools because of their material selective limitation. At present, there are no commercial nanomaterials available. Simulated nanomaterials consisting of stacks of several thin sheets with different material properties were assembled. These stacks were then machined with the μAWJ to demonstrate the versatility of waterjet technology, as illustrated in Figure 22. The model consisted of a stack of eight materials: titanium, float glass, G10 composite, aluminum, Polycarbonate, stainless steel, carbon fiber, to copper. Few tools, if any, would be capable of cutting such a stack effectively.

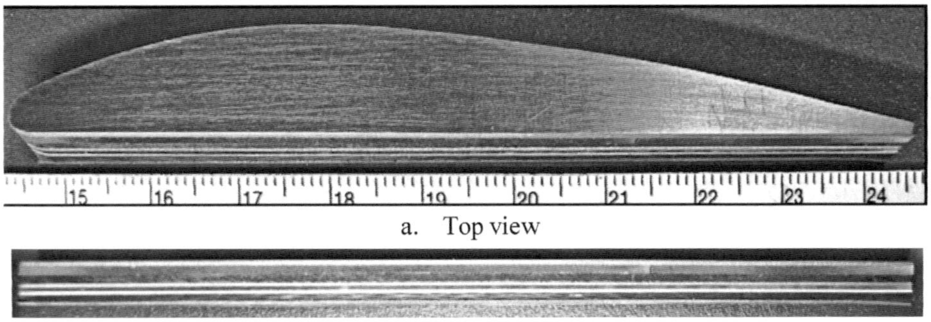

a. Top view

b. Edge view

Figure 22. AWJ-cut simulated nanomaterial with large gradient of material properties.

6. Conclusions

The versatile waterjet technology has inherent technological and manufacturing merits that are unmatched by most existing machine tools. With the commercialization of μAWJ technology supported under the support of an NSF SBIR Phase II/IIB grant, waterjet has achieved the full capacity of machining most materials from macro to micro scales for a wide range of part size and thickness. Among the merits, waterjet requires only a single tool to qualify it for multiple machining modes from parting, drilling, turning, milling, facing, to beveling. Waterjet removes materials through erosion by either the high-speed waterjet or abrasives depending on the property of the workpiece. Such a material removal mechanism is the reason that waterjet is a material independent and cold cutting process. It also facilitates cutting advanced materials that are difficult or even impossible to cut otherwise. For each material, indices of machineability and pierceability are defined in the cutting model; these indices are one of the key parameters for automating the waterjet machining process.

In this paper, a collection of 2D and 3D samples cut with waterjet from various advanced materials for a wide range of part size and thickness is presented. The merits of cold cutting and low side force exertion to the workpiece enable waterjet to cut certain part geometries, such as large-aspect-ratio slots separated by thin webs that are too difficult to cut otherwise. This capability is most suitable for machining micro fluid channels used in heat exchangers, reactors, and reformers for energy production and storage [21]. As a cold cutting machine tool that does not induce the HAZ, waterjet cuts much faster than solid state lasers and wire EDM (up to 2 orders) as lasers must be pulsed at high frequencies and the wire EDM must cut multiple times to minimize the HAZ.

The versatility of waterjet was further demonstrated by machining delicate (composites, laminated, and brittle materials such as glass and silicon wafer) and difficult materials (alloys and hardened steel). Novel processes were developed to mitigate piercing damage to delicate materials based on the results of considerable R&D effort in understanding mechanism of AWJ piercing. Machining hardened steel helps mitigate part tolerance degradation resulted from the post heat treatment process. One of the highlights of the paper is to demonstrate cutting with the μAWJ nozzle a simulated nanomaterial with large gradients of material properties from metal, nonmetal, to anything in between. The simulated nanomaterials was assembled by stacking eight sheets of different materials - from titanium, float glass, G10 composite, aluminum, Polycarbonate, stainless steel, carbon fiber, to copper. As such, waterjet would serve as a preferred machine tool for cost-effective machining of nanomaterials when they are ready for commercialization.

Acknowledgment
This work is supported by an OMAX IR&D fund and an NSF SBIR Phase II grant #1058278. Any opinions, findings, and conclusions or recommendations expressed in this material are those of the authors and do not necessarily reflect the views of the NSF.

References

[1] Liu H T 2015a Abrasive-waterjet machining of most materials from macro to micro scales *Proc 2015 TechConnect World Innovation Conference and Expo* (Washington DC: Taylor & Francist) pp 39-42
[2] Liu H T and Schubert E 2012 Micro Abrasive-Waterjet Technology *Micromachining Techniques for Fabrication of Micro and Nano Structures* ed Mojtaba Kahrizi (Rijeka: INTECH Open Access Publisher) chapter 10 pp 205-234
[3] Liu H T 2016 Versatility of micro abrasive waterjet technology for machining nanomaterials To appear in *Encyclopedia of Nanoscience and Nanotechnology* 3rd Ed. (Boca Raton: CRC Press)
[4] Liu H T, Gnäupel-Herold T, Hovanski Y and Dahl M E 2009 Fatigue Performance Enhancement of AWJ-Machined Aircraft Aluminum with Dry-Grit Blasting *Proc. 2009 Amer. WJTA Conference* (St. Louis: American WJTA) Paper 3-A
[5] Liu H T and McNiel D 2010 Versatility of waterjet technology: from macro and micro machining for most materials *Proc. 20th Int. Conf. on Water Jetting* (Cranfield: BHR Group) pp 419-433
[6] Olsen J H 2012 Abrasive waterjets move into 3-D shapes, including pipe intersections (http://www.thefabricator.com/article/waterjetcutting/abrasive-waterjets-move-into-3-d-shapes-including-pipe-intersections)
[7] Miller D S 2005 New abrasive waterjet systems to complete with lasers *Proceedings of the 2005 WJTA Conference and Exposition* (St. Louis: American WJTA) Paper 1A-1 pp 1-11
[8] Momber A W and Kovacevic R 1998 *Principles of Abrasive Water Jet Machining* (Berlin: Springer-Verlag) p 394
[9] Summers D A 1995 *Waterjet Technology* (New York: Taylor & Francis) p 900

[10] Liu H T, Schubert E, McNiel D and Soo K 2010 Applications of abrasive-waterjets for precision machining of composites *Proceedings of the SAMPE 2010 Conference and Exhibition* (Lancaster: DEStech Publications, Inc.)

[11] Liu H T 2006a Collateral damage by stagnation pressure buildup during abrasive-fluidjet piercing *Proceedings of the 18th International Conference On Water Jetting* (Cranfield: BHR Group) pp 47-61

[12] Liu H T 2015b Novel Processes for Improving Precision of Abrasive-Waterjet Machining *Proc 2015 WJTA-ICMA Conference and Expo* (St. Louis: American WJTA) pp 1-14

[13] Liu H T 2006b Hole Drilling with Abrasive Fluidjets *Int. J. of Adv. Manuf. Tech.* **32** 942-957

[14] Liu H T and Schubert E 2009 Piercing in Delicate Materials with Abrasive-Waterjets *Int. J. of Adv. Manuf. Tech* **42**(3-4) 263-279

[15] Liu H T, Fang S and Hibbard C 1999 Enhancement of Ultrahigh-Pressure Technology with LN_2 Cryogenic Jets *Proc. 10th Amer. Waterjet Conference* (St. Louis: American WJTA) Paper 21 pp 1-17

[16] Liu H T 2012 Abrasive-Waterjet Technology for Biomedical Applications *Proc 21st Int Conf on Water Jetting* (Cranfield: BHR Group) pp 99-112

[17] Liu H T and Olsen J H 2013 Application of AWJ for 3D Machining *Proc 2013 WJTA-ICMA Conference and Expo* (St. Louis: American WJTA) Paper A2 pp 1-15

[18] Trimble A Z 2011 Vibration Energy Harvesting of Wide-Band Stochastic Inputs with Application to an Electro-Magnetic Rotational Energy Harvester Ph.D. Dissertation Massachusetts Institute of Technology p 183

[19] Begg N 2011 Blind Transmembrane Puncture Access: Design and Development of a Novel Laparoscopic Trocar and Blade Retraction Mechanism Master Thesis Massachusetts Institute of Technology p 103

[20] Kluger J M, Sapsis T P and Slocum A H A 2016 High-resolution and large force-range load cell by means of nonlinear cantilever beams *Precision Eng.* **43** 241-256

[21] Liu H T, Hovanski Y, Caldwell D D and Williford R E 2008 Low-Cost Manufacturing of Flow Channels with Multi-Nozzle Abrasive-Waterjets: A Feasibility Investigation *Proc. 19th Int. Conf. on Water Jetting* (Cranfield: BHR Group) pp 337-351

A numerical analysis on forming limits during spiral and concentric single point incremental forming

M L Gipiela[1], V Amauri[1], C Nikhare[2, 4] and P V P Marcondes[3]
[1]Faculty of Technology – SENAI – CIC, National Service for Industrial Apprenticeship in Paraná, 298 Senador Accioly Filho Street, Curitiba, 81310-000, Brazil
[2]Mechanical Engineering, The Pennsylvania State University, Erie, PA-16563, USA
[3]Graduate Program in Mechanical Engineering - PPGEM, Federal University of Paraná, 210 Coronel Francisco H. dos Santos Avenue, Curitiba, 81531-990, Brazil

Email: cpn10@psu.edu

Abstract. Sheet metal forming is one of the major manufacturing industries, which are building numerous parts for aerospace, automotive and medical industry. Due to the high demand in vehicle industry and environmental regulations on less fuel consumption on other hand, researchers are innovating new methods to build these parts with energy efficient sheet metal forming process instead of conventionally used punch and die to form the parts to achieve the lightweight parts. One of the most recognized manufacturing process in this category is Single Point Incremental Forming (SPIF). SPIF is the die-less sheet metal forming process in which the single point tool incrementally forces any single point of sheet metal at any process time to plastic deformation zone. In the present work, finite element method (FEM) is applied to analyze the forming limits of high strength low alloy steel formed by single point incremental forming (SPIF) by spiral and concentric tool path. SPIF numerical simulations were model with 24 and 29 mm cup depth, and the results were compare with Nakajima results obtained by experiments and FEM. It was found that the cup formed with Nakajima tool failed at 24 mm while cups formed by SPIF surpassed the limit for both depths with both profiles. It was also notice that the strain achieved in concentric profile are lower than that in spiral profile.

1. Introduction

The sheet metal forming industry often use different forming methods which are based on die and punch geometry to manufacture automotive parts. These methods are usually for mass production but when small series or prototypes are required, conventional forming processes can be replaced by new manufacture methods in order to decrease the final costs [1]. In this context, the incremental sheet forming (ISF) processes can replace some of classical sheet metal forming [2].

ISF is a low cost forming process and its use is growing for rapid prototyping, biomechanics and small batch production of sheet metal parts. Considering the following comprehensive market for the ISF technology: small batches of automotive parts, prostheses of biocompatible materials and complex geometries formed from small thickness sheets, it is clear that the ISF is a manufacturing process that needs to be studied [3].

The ISF consists on deform metal blanks by the move of a ball nose punch (spherical) over a blank properly steady in a blank holder [4]. The single point incremental forming (SPIF) is one technique of ISF. In the SPIF, the sheet is held in position with the help of the blank holder which is fixed in a particular position. A simple geometry tool, fixed to the CNC tool is moved along the surface and metal is formed [5]. The principle of the process is shown in Figure 1.

The SPIF depends on a properly programmed path using Computerized Numerical Control (CNC) technology what allows the tool to form a blank by small deformations during the punch tool move [3].

The range of equipment available to perform the ISF techniques is still very limited [6] and because this, some researchers generally are studying ISF using CNC machines. Furthermore, computer aided manufacturing softwares (CAM) are helpful to set the tool path in three axis of the CNC machine and it also permits to test suitable forming parameters in the experiments with reasonable accuracy from CAD files.

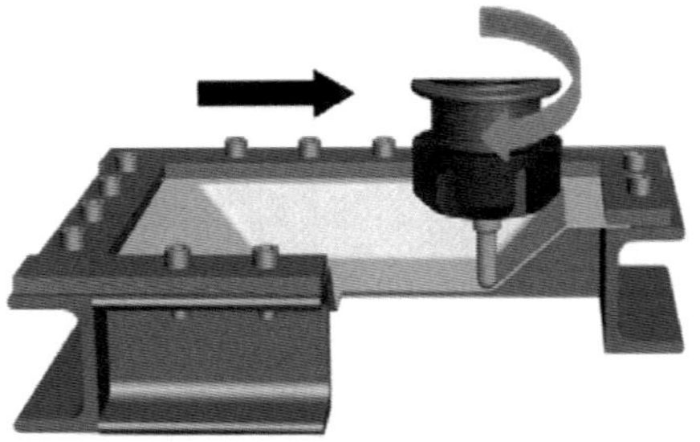

Figure 1. Single Point Incremental Forming principles [4].

Currently, the implementation and understanding of the ISF process focuses on two forms of analysis: numerical and experimental [7]. In the experimental test most of the equipment used for ISF result of the adaptation of CNC machines and running the process with a range of applications in terms of materials and geometries [8]. The use of robots has also been observed in research and it has a great advantage in terms of working area and flexibility. Nevertheless, in addition of error due to the robots joints, it aggravates the IST in terms of final accuracy of the product geometry [9].

Efforts in numerical analysis has been made to develop a better understanding of the fundamental mechanisms involved to predict the materials behavior during the ISF. The modeling of the process by Finite Element Method (FEM) remains a demanding task due a refined mesh element is necessary to achieve convergence and accuracy in the simulations [10].

In the last years, a large number of research related to ISF has been focused on low carbon steels [10-13] and non-ferrous alloys [14-16]. For this reason, there is a great lack of information about process parameters for ferrous materials such as stainless steel, HSS and AHSS steels [3].

Under the circumstances previous mentioned, the main goal of this study is to compare the experimental and numerical results obtained in classical Nakajima test and SPIF numerical results for high strength low alloy steel HSLA440.

2. Material and methodology

The material used for this study is HSLA 440. The experimental true stress-strain curve alongwith the power law is shown in Figure 2. The power law is given in equation 1. The strength coefficient "K" and strain hardening exponent "n" was found by fitting with the experimental curve from 3% to 10% strain and material properties are given in Table 1. For simulation purpose the mix curve was used.

The mix curve is the combination of actual material data upto 3% strain and power law for onwards strain. This curve was used with isotropic hardening law.

$$\sigma = K\varepsilon^n \tag{1}$$

where: σ – True stress and ε – True strain

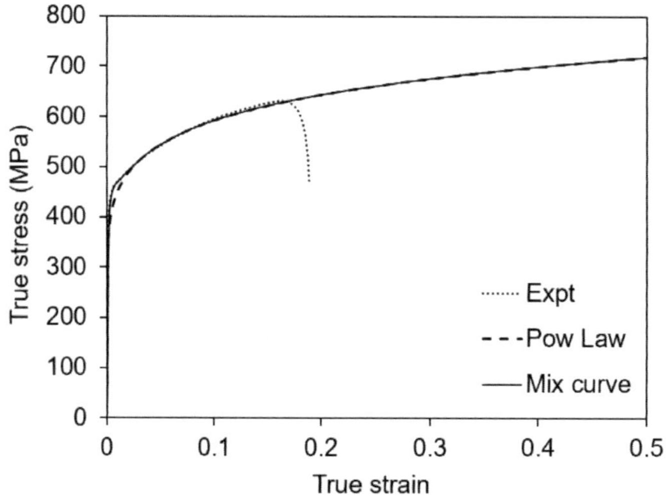

Figure 2. True stress strain curve along with fitted power law.

Table 1. Mechanical properties for HSLA 440.

Yield strength (MPa)	Tensile Strength (MPa)	Total Elongation (%)	K (MPa)	n
437	540	20.8	780	0.12

2.1. Nakajima test and model

The Nakajima experiment was performed to form the hemispherical cup till failure. The Nakajima experimental set up is shown in Figure 3 along with sample dimension. The thickness of the sheet was 1.5 mm. The tests were conducted using 1000 kN capacity hydraulic press. The specimens was oriented with their principal axis along the rolling direction.

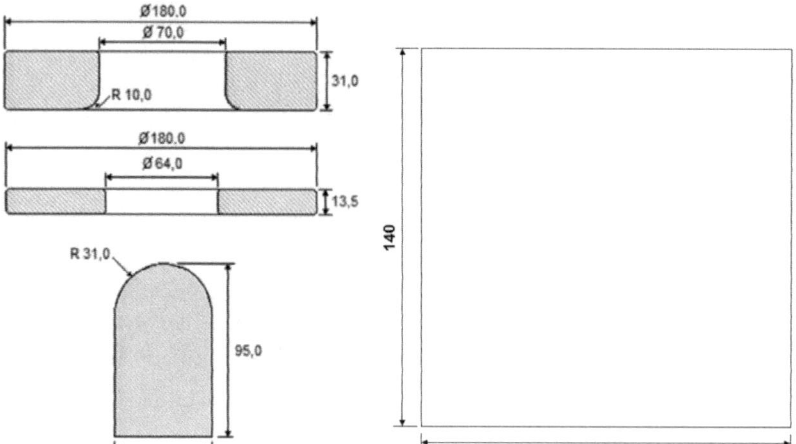

Figure 3. Nakajima test set-up section drawing with dimension in mm (left) and sheet sample for deformation with dimension in mm (right).

Circle grids of 4.2 mm diameter was etched on the specimen. The specimens were mounted with the grid side opposite to the punch and stretched with the punch velocity of 3.99 mm/s ISO VG68 mineral oil was used to reduce the friction between punch and sheet during test. After test, major and minor dimensions of the ellipses, with uniform, localized and fracture deformations, were measured with a Mitutoyo digital caliper.

The Nakajima test was then modeled using ABAQUS/Explicit 6.13-2. The test was investigated with a three-dimensional model approach (Figure 4). The tooling was assumed as a rigid surfaces, while S4R shell elements (4-node quadrilateral, reduced integration) were used to mesh the blank with mesh size of 1.5 mm same as thickness dimension (Figure 4). Full model was used to reduce any discrepancy in the test as single point incremental forming (SPIF) models were needed to done with 3D approach. The average sheet thickness measured experimentally for the HSLA steel was detailed in the model. The true stress-strain data, determined in the tensile tests, was applied to define the material properties using isotropic hardening as explained in material section.

All conditions and process parameters as present in the experimental tests were applied in the numerical model. The interaction between the blank and the rigid bodies were trialed to match the cup height at failure.

Figure 4. Nakajima test model.

2.2. *SPIF models*

To model the SPIF the punch was replaced by a 8 mm diameter hemispherical pin. The model is shown in Figure 5. In this paper two profiles were considered to model the SPIF a) spiral profile and b) concentric profile. The spiral x-y profile to create a hemispherical cup with 24 mm height is shown in Figure 6a and the z-depth profile is given in Figure 6b. The profile was created from CAM software Edgecam 2014 R2, which considers the z-increment.

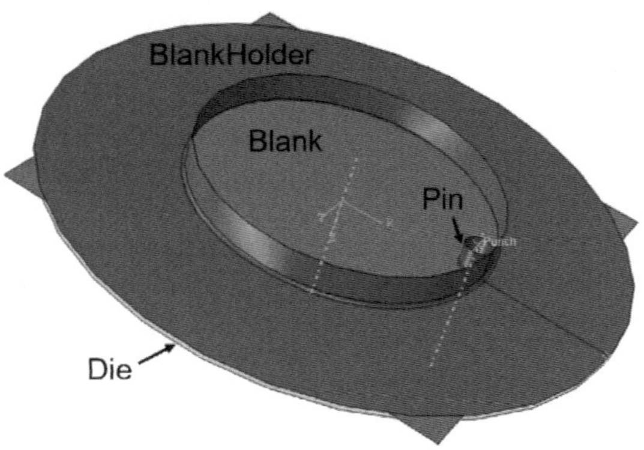

Figure 5. Single point incremental forming model.

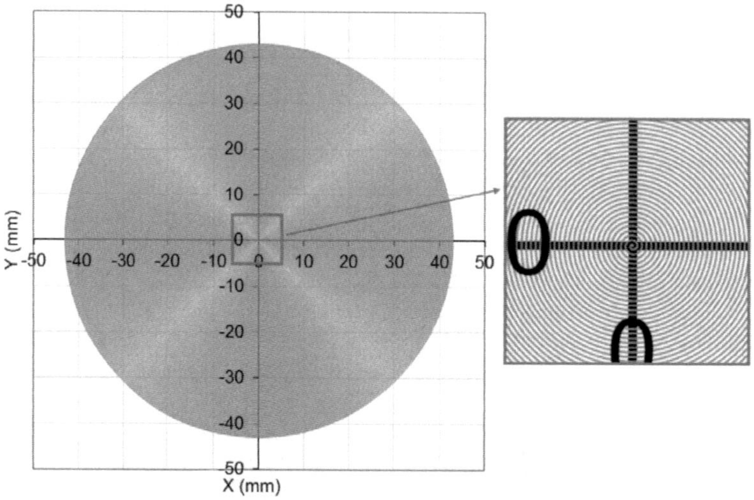

(a)

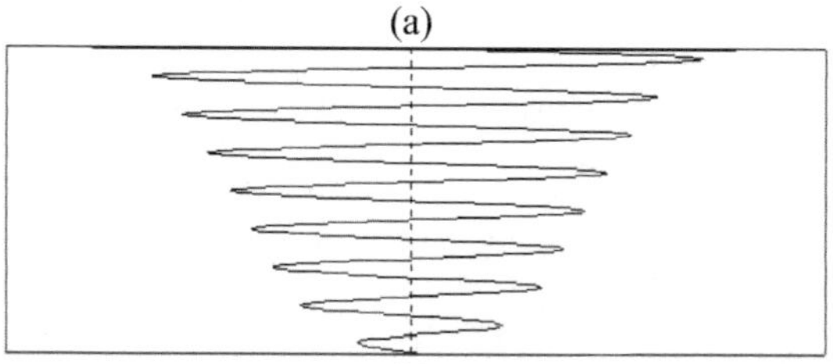

(b)

Figure 6. Single point incremental forming spiral path: (a) x-y profile view and (b) z-depth view.

Similarly the concentric x-y profile is shown in Figure 7a and the z-depth in Figure 7b. In both spiral and concentric SPIF models the tools were assumed as rigid surface and sheet metal as deformable with S4R shell elements. All details are similar to the Nakajima model.

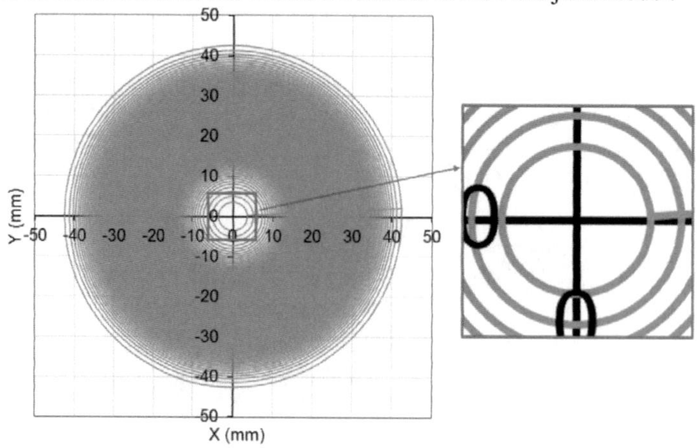

(a)

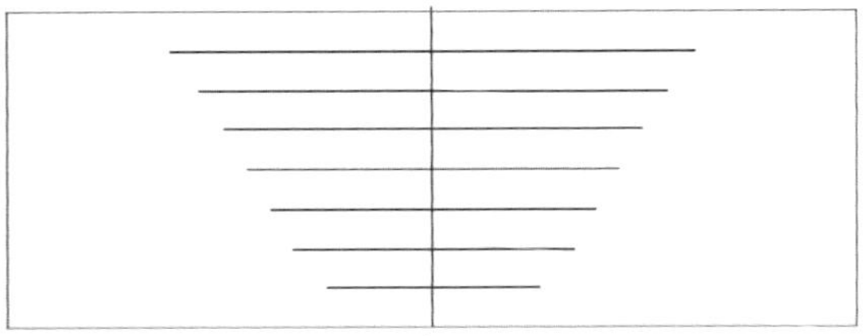

(b)

Figure 7. Single point incremental forming concentric path: (a) x-y profile view and (b) z-depth view.

3. Results and discussion

3.1. Nakajima test and model comparison

The Figure 8 shows the specimen after Nakajima test. The specimen got failure at 24 mm cup depth. A good agreement between experimental and numerical Nakajima results was observed. The results converged in terms of the dome depth and the instability region where the crack occurs. The tested specimen failed at the dome when the punch displaced 24 mm (failure point: engineering minor strain = 0.13 and engineering major strain = 0.37) while in the simulated cup (see Figure 9), the plastic strain distribution reached the instability at the dome when the punch displaced until 24.34 mm. To match the simulation with the experimental cup depth and instability region different coefficient of friction between the punch and blank was tried. The Figure 10 provides the simulated force displacement curve for different friction value. It can be noted that with coefficient of friction of 0.3 the force dropped at 24.34 mm creating the instability in the elements for failure.

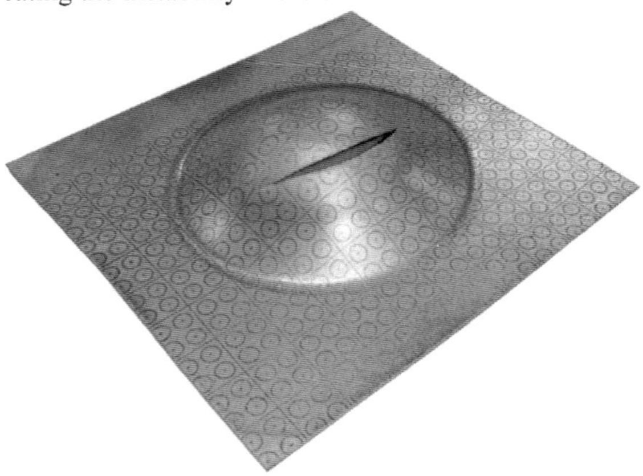

Figure 8. Deformed cup through Nakajima test.

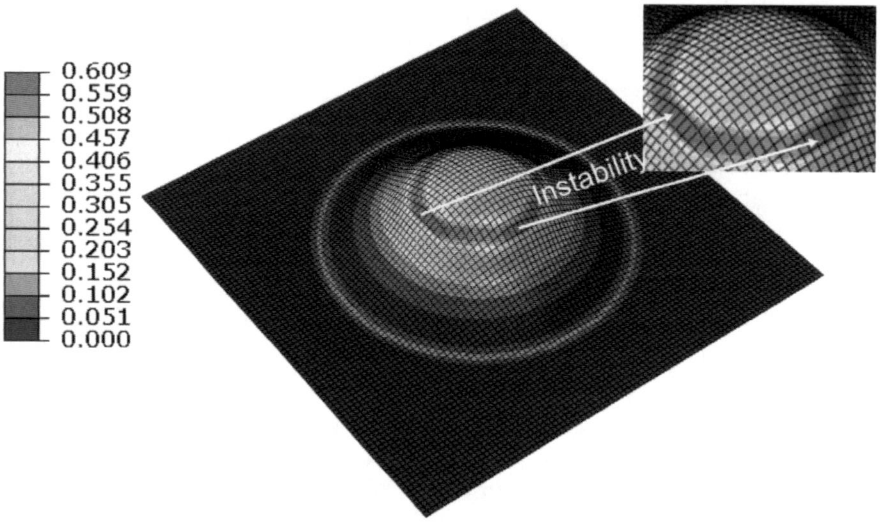

Figure 9. Equivalent plastic strain distribution on cup formed using Nakajima dome test.

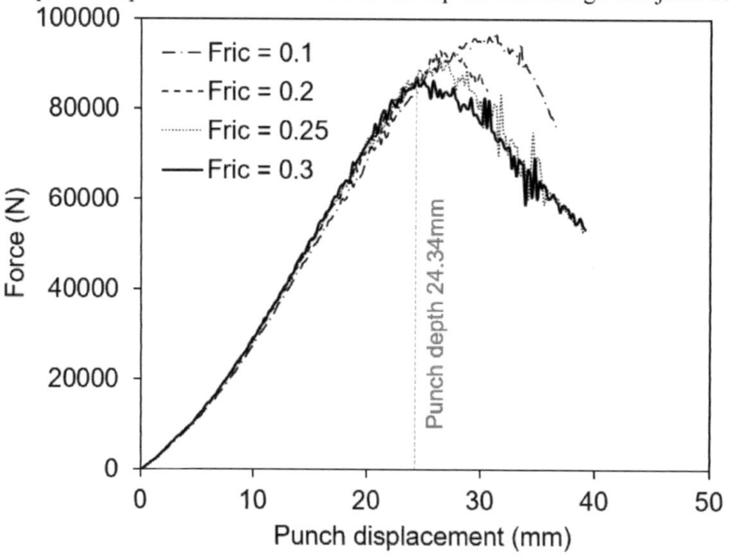

Figure 10. Simulated Force-displacement curve.

3.2. SPIF – Spiral path

Figure 11 and Figure 12 provides the simulated cups by SPIF using spiral path for 24 mm and 29 mm cup height. From equivalent plastic strain contour it can be noted that the cups realizes very high strain of 179.6% for 24 mm cup height and 241% for 29 mm cup height. With these high strain values it can be assumed that these cups are already realized the failure as conventional cups failed at 60.9% equivalent plastic strain (refer Figure 9). However due to unique deformation technique in SPIF that at any time only one point gets deformed and thus can sustain higher strain before failure.

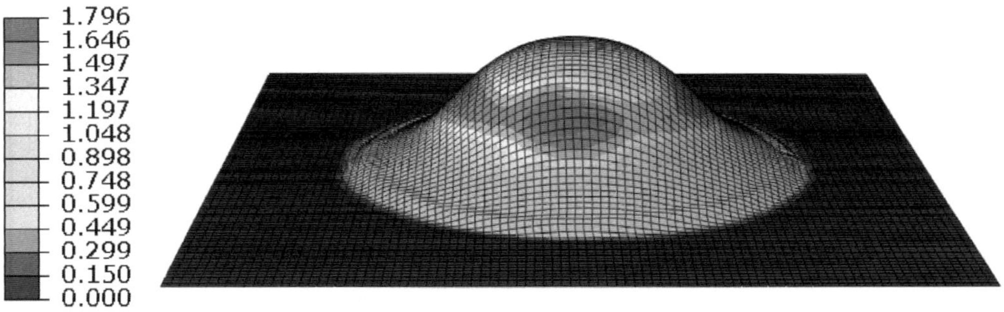

Figure 11. Equivalent plastic strain distribution on 24mm height cup formed using single point incremental forming with spiral path.

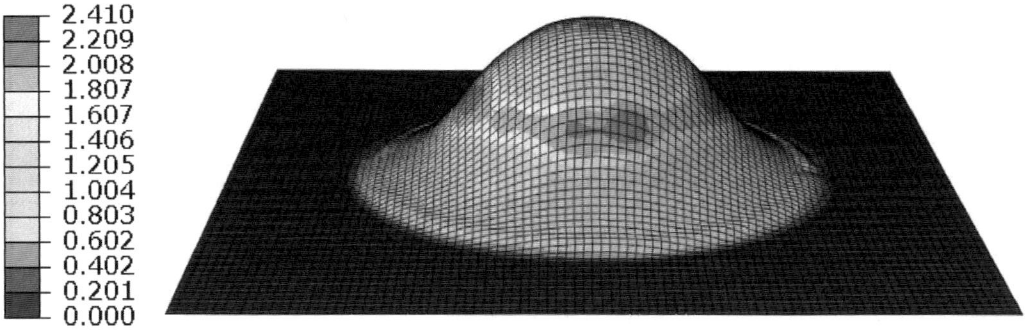

Figure 12. Equivalent plastic strain distribution on 29mm height cup formed using single point incremental forming with spiral path.

3.3. SPIF – Concentric path

Similarly, Figure 13 and Figure 14 provides the simulated cups by SPIF using concentric profiles. It can be observed that these cups also realizes the higher strain of 182.4% and 242% for 24 and 29 mm cup depth. Similar prediction can be made that these cups could have realized failure if they would have been deformed using conventional method.

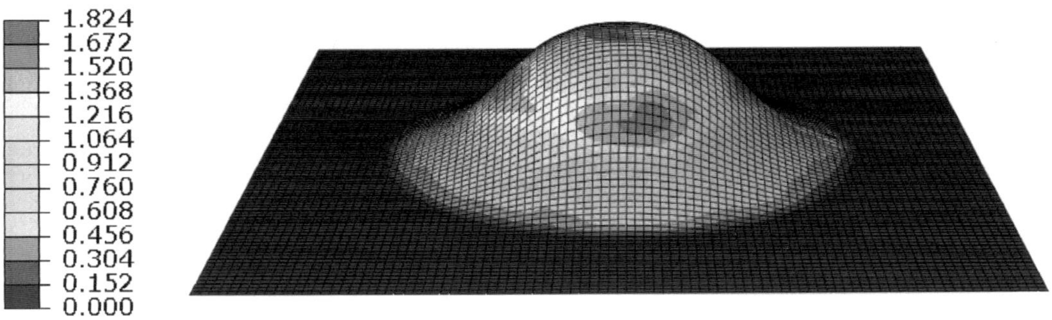

Figure 13. Equivalent plastic strain distribution on 24mm height cup formed using single point incremental forming with concentric path.

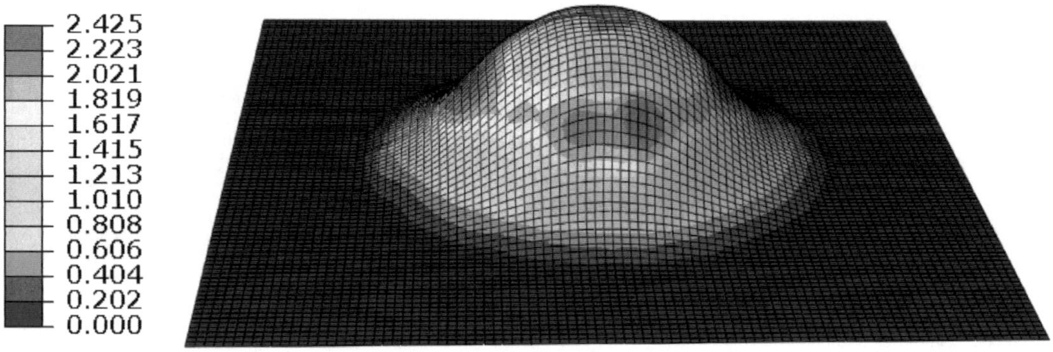

Figure 14. Equivalent plastic strain distribution on 29mm height cup formed using single point incremental forming with concentric path.

3.4. Process comparison

To understand the forces on pin during SPIF, the force with respect to the process time was plotted in each axis. It was found that the forces were similar with both spiral and concentric profile SPIF process in all x, y and z axis (Figure 15-17).

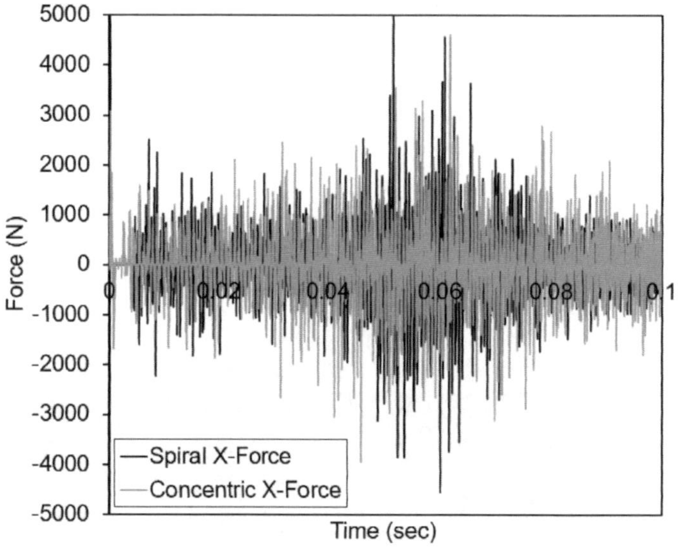

Figure 15. X-axis force with respect to time for 29 mm cup depth.

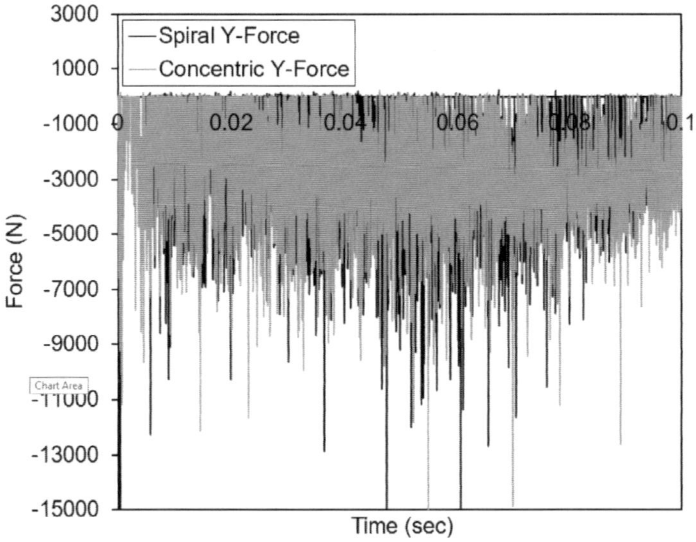

Figure 16. Y – axis force with respect to time for 29 mm cup depth.

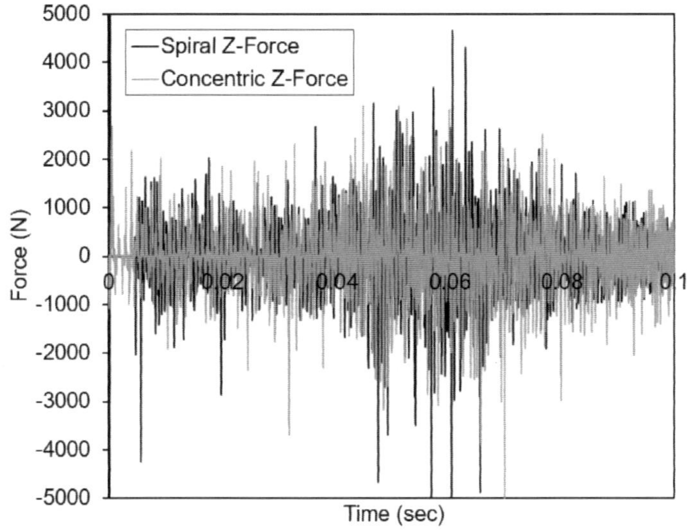

Figure 17. Z – axis force with respect to time for 29 mm cup depth.

Further the critical elements strain paths were plotted in all simulated cups, i.e., Nakajima model, SPIF – Spiral with 24 and 29 mm cup depth, and SPIF – Concentric with 24 and 29 mm cup depth alongwith the failure point from Nakajima test (0.13, 0.37). In addition the literature [17] forming and fracture limit curve for HSLA steel was plotted. It can be seen that the Nakajima test failure point and the simulated instability point matches with the literature forming limit curve.

However as it concluded in previous literature [18] that the failure in SPIF will not be able to predict through conventional forming limit curve and thus fracture limit curve is needed. As can be seen that the strain paths from the SPIF simulated cups with both spiral and concentric profiles for 24 and 29 mm depth could not realized the failure and that they can deform more for further higher depths. It was also observed that for similar depths the spiral profile provides higher major strain value than the concentric profile and would be more prone to failure than the concentric profile for higher

IOP Conf. Series: Materials Science and Engineering **164** (2017) 012009 doi:10.1088/1757-899X/164/1/012009

depths. This can also be observed in Figure 19 which provides the flow of engineering major strain with respect to the process time for the critical element. Thus concentric profile would be preferred to reach more formability.

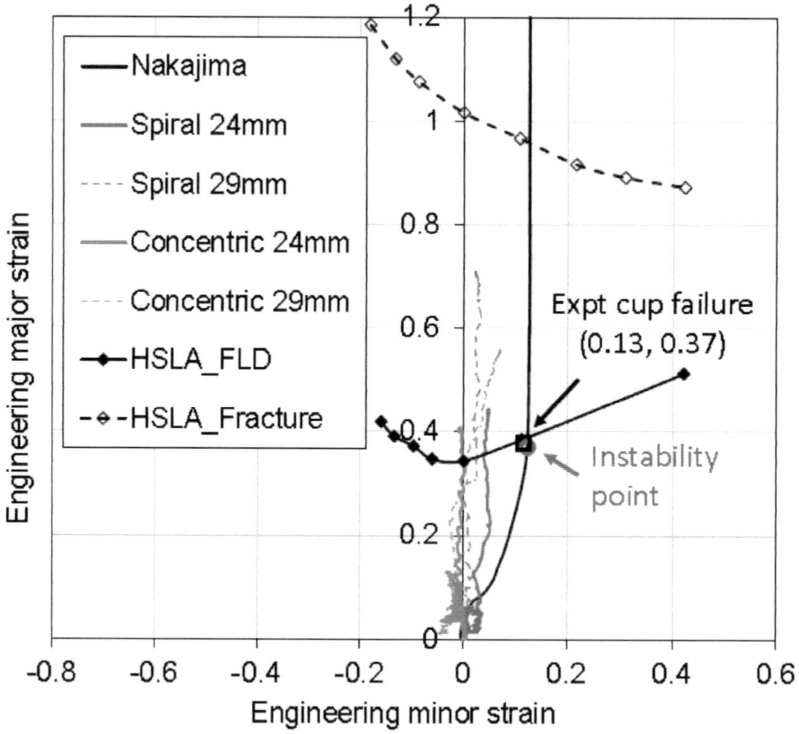

Figure 18. Strain path and forming limits.

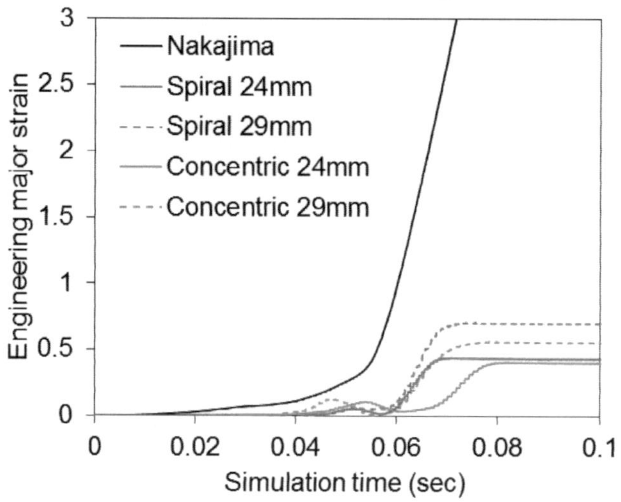

Figure 19. Engineering major strain with respect to simulation time.

4. Conclusion

In this paper the formability and the limits of a hemispherical cups formed by single point incremental forming was studied. For this, the Nakajima test was performed and further modeled the test in ABAQUS. It was observed that the prediction of formability and forming limits was in good agreement with the test. With strain path the deviation was right at the limit curve and thus validates the model. Further single point incremental forming models were prepared and simulated for spiral and concentric path for 24 and 29 mm cup depth. As discussed in previous literature the forming limit curve doesn't realize the failure and thus fracture limit is needed. After plotting the critical element strain path it was observed that both 29 mm cups with spiral and concentric profiles are much below the fracture limit curve and can go further for higher depth. After comparison between the spiral and concentric profile cups for same depth it was found that the concentric profile cup strain path is lower than the spiral profile cup and would realize the failure after spiral profile cup with higher depth. Thus it can be concluded that the formability can be achieved higher in single point incremental forming with concentric profile.

Acknowledgements

The authors would like to thank the ArcelorMittal S. A. by providing HSLA440 blank and SENAI-PR – Faculdade de Tecnologia – Curitiba – Brazil, UFPR – Universidade Federal do Paraná – Curitiba – Brazil and The Pennsylvania State University – Erie – USA for all support.

References

[1] Cerro I, Maidagan E, Arana J, Rivero A and Rodríguez P P 2006 Theoretical and experimental analysis of the dieless incremental sheet forming process *Journal of Materials Processing Technology* **177** 404-408

[2] Bambach M, Hirt G and Ames J 2004 Modeling of optimization strategies in the incremental CNC sheet metal forming process *AIP Conference Proceedings 712: Materials Processing and Design: Modeling, Simulation and Applications, NUMIFORM 2004 Proceedings of the 8th International Conference on Numerical Methods in Industrial Forming Processes (Columbus/Ohio, USA)* pp 1969-1974

[3] Cavaler, L C D C 2010 Parâmetros de conformação para a estampagem incremental de chapas de aço inoxidável AISI 304L Tese de Doutorado, Universidade Federal do Rio Grande do Sul *(Porto Alegre/Brasil)* p 152

[4] Ingarao G, Ambrogio G, Gagliardi F and Lorenzo R D 2012 Sustainability point of view on sheet metal forming operations: material wasting and energy consumption in incremental forming and stamping processes *Journal of Cleaner Production* **29-30** 255-268

[5] Khare U and Pandagale M 2014 A review of fundamentals and advancement in incremental sheet metal forming *IOSR Journal of Mechanical and Civil Engineering* 42-46

[6] Marabuto S R, Sena J I V, Afonso D, Martins M A B E, Coelho R M, Ferreira J A F, Valente R A F and Sousa R J A 2011 Numerical studies and equipment development for single point incremental forming *The 14th International ESAFORM Conference on Material Forming* 1701-1706

[7] Marabuto S R, Sena J I V, Afonso D, Martins M A B E, Coelho R M, Ferreira J A F, Valente R A F and Sousa R J A 2011 Estampagem incremental: estudo numérico e experimental *Congresso de Métodos Numéricos em Engenharia (Coimbra/Portugal)*

[8] Jeswiet J, Micari F, Hirt G, Bramley A, Duflou J and Allwood J 2005 Asymmetric single point incremental forming of sheet metal *CIRP Annals – Manufacturing Technology* **54** 88-114

[9] Meier H, Dewald O and Zhang J 2005 A new robot-based sheet metal forming process *Proc. Shemet* (Erlangen: Germany Trans Tech Publications Ltd) pp 465-470

[10] Fontanari V, Benedetti M, Bruschi S and Fuganti A 2012 A Numerical and experimental analysis of the single point sheets incremental forming process *15th International Conference on Experimental Mechanics (Porto/Portugal)* pp 1-12

[11] Guzmán C F, Bettaieb A B, Sena J I V, Sousa R A, Habraken A M and Duchêne L 2012 Evaluation of the enhanced assumed strain and assumed natural strain in the SSH3D and RESS3 solid shell elements for single point incremental forming simulation *Key Engineering Materials* **504-506** 913-918
[12] Petek A, Kuzman K and Kopac J 2009 Deformations and forces analysis of single point incremental sheet metal forming *Archieves of Materials Science and Engineering* **35** 107-116
[13] Lora F A and Schaeffer L 2014 Incremental forming process strategy variation analysis through applied strains *Brazilian Journal of Science and Technology* **1**(5) 1-8
[14] Taleb Araghi B, Manco G L, Bambach M and Hirt G 2009 Investigation into a new hybrid forming process: incremental sheet forming combined with stretch forming *CIRP Annals – Manufacturing Technology* **58** 225-228
[15] Kopac J and Kampus Z 2005 Incremental sheet metal forming on CNC milling machine-tool *Journal of Materials Processing Technology* **162-163** 622-628
[16] Pohlak M, Majak J and Küttner R 2007 Incremental sheet forming process modelling – limitation analysis *Journal of Achievements in Materials and Manufacturing Engineering* **22** 67-70
[17] Nikhare C, Hodgson P D and Weiss M 2011 Necking and fracture of advanced high strength steels *Materials Science and Engineering A* **528** 3010-3013
[18] Silva M B, Skjødt Martin, Atkins A G, and Martins P A F 2008 Single-point incremental forming and formability-failure diagrams *Journal of Strain Analysis for Engineering Design* **43** 15-35

Determination of an acoustic reflection coefficient at the inlet of a model gas turbine combustor for power generation

W J Song[1] and D J Cha[2, 3]

[1] Department of Mechanical Systems Engineering for Energy Convergence, Dongshin University, Naju, 58245, Republic of Korea
[2] Department of Building and Plant Engineering, Hanbat National University, Daejeon 34158, Republic of Korea

Email: djcha@hanbat.ac.kr

Abstract. A phenomenon that potentially influences the reliability of power generation systems is the presence of thermo-acoustic oscillations in the combustion chamber of a land-based gas turbine. To develop specific measures that prevent the instability, it is essential to predict and/or evaluate the underlying physics of the thermo-acoustics, which requires the acoustic boundary condition at the exit of the burner, that is, at the inlet of the combustor. Here we report a procedure for calculating acoustic reflection coefficients at the burner exit by utilizing two microphone method (TMM) for dynamic pressure signals. The procedure has been verified by comparing its results with reported ones and further successfully employed to determine the acoustic boundary condition of the burner of a partially-premixed model gas turbine combustor.

1. Introduction

Combustion instability (CI) in practical combustion systems including land-based gas turbines of the lean premixed combustion technique, which is employed for the reduction of the emission of pollutants such as nitrogen oxides, is one of the ongoing issues requiring proper solutions yet to be prepared for through further studies [1]. The phenomenon is generated from the excitation of certain resonance mode of the acoustic field in combustion chamber induced by the feedback originated from the perturbed heat release rate attributable to the fluctuation of equivalence ratio and/or of the velocity of the pre-mixture of oxidizer and fuel of natural gas in most cases.

Fuel flexibility has also become a very keen issue for the utility of the power generation [2]. Alternative natural gases such as coal-derived gas, synthetic natural gas (SNG), biomass and landfill gases, refinery off-gas, etc., whose chemical compositions are more likely to vary, may exacerbate CI, because most land-based gas turbine combustors are sensitive to change in operation conditions. When the systems experience the CI, thereby, the magnitude of dynamic pressure wave of the chamber may nonlinearly increase and end up with the limit cycle.

In general, the dominant frequency, the growth rate, and the amplitude of the limit cycle of the instability are of concern during from the design stage of gas turbine combustors to their commercial operations including tuning after regular overhauls. Some of the predictive approaches for this issue including the lumped network method [1, 3], finite difference method (FDM) [4], etc., deal with the Helmholtz equation, instead of the full Navier-Stokes equation [5]. The Helmholtz equation, time-independent version of the wave equation, mostly uses the flame transfer function which is determined either analytically or experimentally, as a flame response model [6, 7]. Besides, the acoustic boundary conditions of the calculation domain which are also determined either analytically [3] or experimentally [8, 9], in terms of either reflection coefficient or impedance, are needed for the prediction of CI. Kim et al. [8] measured spatially distributed flame transfer functions and reflection coefficient at the single dominant frequency and then predicted combustion dynamics of lean-premixed combustor including CI. They measured the coefficient at the inlet of the lean-premixed combustor with the wave decomposition method. Recently, the result of the CI evaluated with the reflection coefficients measured as a function of frequency has been reported [9].

In this study, a method to determine the reflection coefficient on the boundary plane of a model gas turbine combustor as a function of frequency have been developed by using the two microphone method (TMM) to predict the CI in the gas turbine combustor for power generation. Theory associated with this approach has basically been borrowed from ISO standards [10] that were prepared for the determination of the absorption coefficients of sound absorbing materials. Factors associated with the characteristics of the model gas turbine combustors such as the mean flow and thermo-viscous dissipation effects were also taken into account. The validity of the developed procedure was verified by comparing the results with those of the benchmark experimental data. Further, it has been used to determine the reflection coefficient of a partially-premixed model gas turbine combustor burning SNG [11].

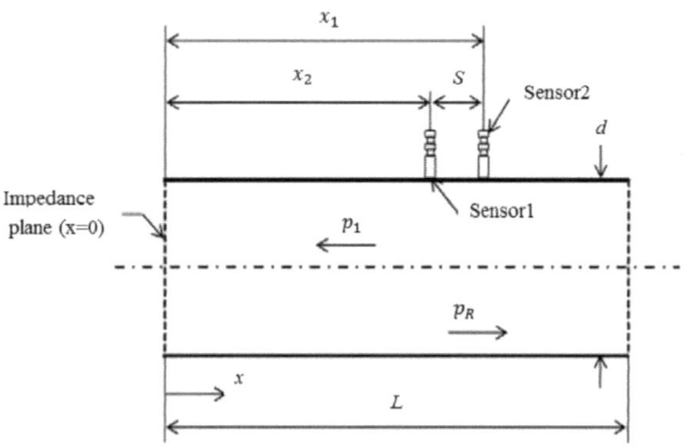

Figure 1. Geometry of two microphones for measuring reflection coefficient at an impedance plane [10].

2. Theory and development of analysis code

The acoustic reflection coefficient R of the plane wave inside the duct of a model gas turbine combustor is determined by the pressure transfer function (PTF) H_{12} that is measured with the two microphones installed in front of the impedance plane (see Figure 1). The theory associated with this method is as follows. The planar incident wave p_I and planar reflected wave p_R in a duct (or in an impedance tube) are expressed as in the following equations (1) and (2), respectively.

$$p_I = \hat{p}_I e^{jk_o x} \tag{1}$$

$$p_R = \hat{p}_R e^{-jk_o x} \tag{2}$$

where $\hat{p}_I$ and $\hat{p}_R$ are the magnitudes of p_I and p_R in the duct, and the constant k_0 is a complex wave number. The sound pressure of p_1 and p_2 at the two microphone positions are as expressed in the following equations (3) and (4), respectively.

$$p_1 = \hat{p}_I e^{jk_o x_1} + \hat{p}_R e^{-jk_o x_1} \tag{3}$$

$$p_2 = \hat{p}_I e^{jk_o x_2} + \hat{p}_R e^{-jk_o x_2} \tag{4}$$

The transfer function for the incident wave H_I can be expressed as in the following equation (5).

$$H_I = \frac{P_{2I}}{P_{1I}} = e^{-jk_o(x_1 - x_2)} = e^{-jk_o S} \tag{5}$$

where s(x_1 - x_2) denotes the distance between the two microphones. Similarly, the transfer function for the reflected wave alone H_R can be expressed as in the following equation (6).

$$H_R = \frac{P_{2R}}{P_{1R}} = e^{jk_o(x_1 - x_2)} = e^{jk_o S} \tag{6}$$

The pressure transfer function H_{12} for the total sound field in the duct can be derived from equations (3) and (4) and the relation of $\hat{p}_R = R\hat{p}_I$ as follows:

$$H_{12} = \frac{P_2}{P_1} = \frac{e^{jk_o x_2} + \mathrm{R}e^{-jk_o x_2}}{e^{jk_o x_1} + \mathrm{R}e^{-jk_o x_1}} \tag{7}$$

Substituting equations (3) and (4) into equation (7) yields the reflection coefficient R at the impedance plane ($x = 0$) as follows:

$$R = \frac{H_{12} - H_I}{H_R - H_{12}} e^{2jk_o x_1} \tag{8}$$

Figure 2 illustrates the flowchart of the procedure for determining the PFT of model gas turbine combustor and thereof the reflection coefficient at the combustor inlet. Sound pressures p_1 and p_2 at the two microphone positions are time-domain data, and, to convert them into frequency domain, the certain data block of finite length is taken to perform fast Fourier Transform (FFT). The original dynamic pressure data were divided into blocks of the certain length, and the degree of overlapping between two consecutive blocks was taken as a parameter. For most of runs to be discussed, there were around 100 blocks. When FFT is performed, the time weighted function (that is, window function) is usually applied to reduce errors attributable to the aperiodicity and discontinuity of the data block. In MATLAB code developed in this study, one of three window functions (i.e., uniform (or boxcar), Hanning, and Hamming windows) can be selectively chosen during the transformation.

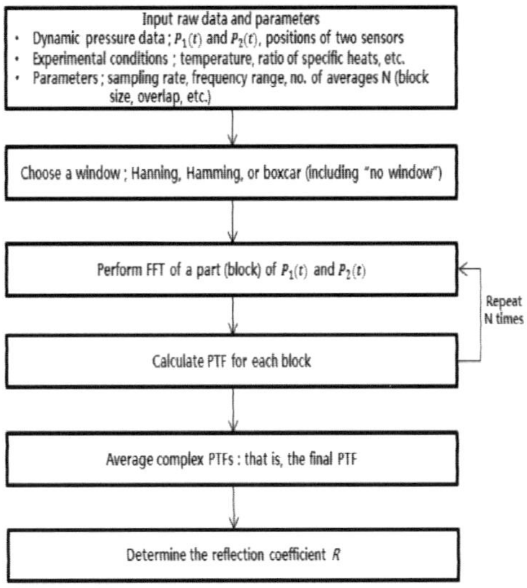

Figure 2. Flowchart for a MATLAB code calculating PTF and thereof reflection coefficient R at the inlet of a model gas turbine combustor.

IOP Conf. Series: Materials Science and Engineering **164** (2017) 012010 doi:10.1088/1757-899X/164/1/012010

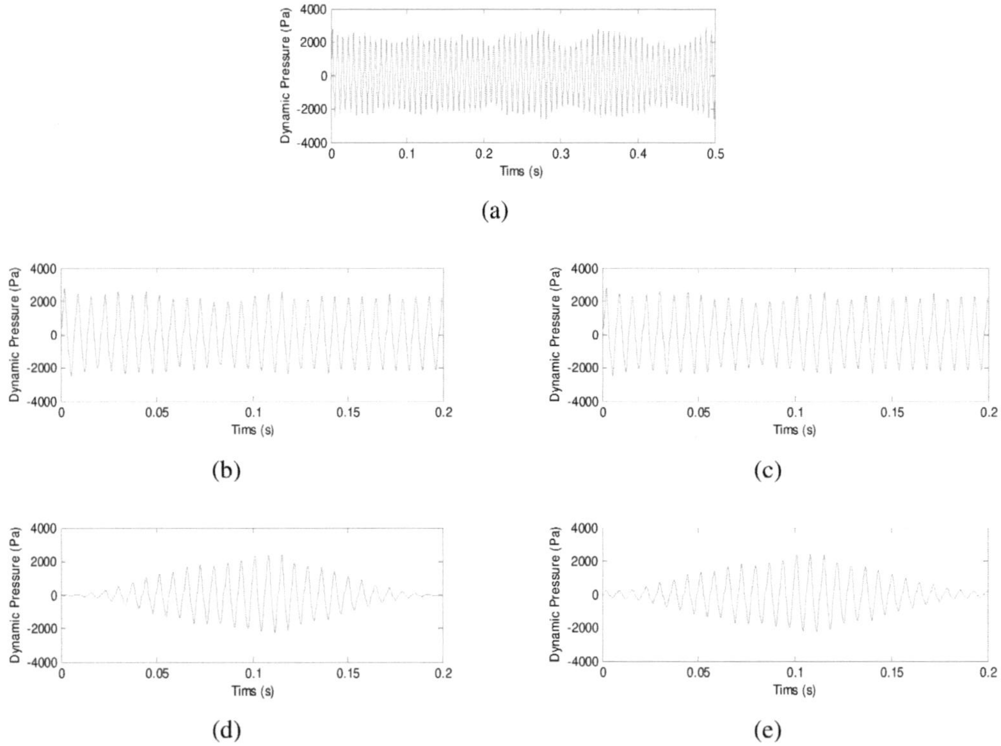

Figure 3. A typical p' data as a time record of finite length: (a) the raw data, (b) the selected block of the data, (c) the modified with a boxcar window, the same as (b), (d) the modified with Hanning window and (e) the modified with Hamming window.

Figure 3 shows a typical p' data as a time trace of finite length and its modifications with various windows. The PTF H_{12} in equation (7) was calculated with the dynamic pressure data p_1 and p_2 taken from the two microphones by utilizing MATLAB built-in function *tfestimate*. The reflection coefficient R was then determined by using equation (8).

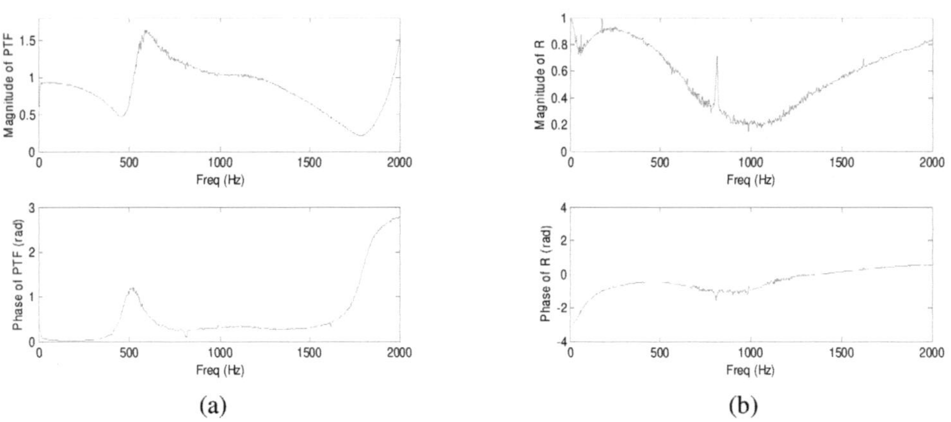

Figure 4. Benchmark data used for verifying the reflection coefficient calculation code developed in this study: (a) the PTF and (b) the corresponding R.

The developed MATLAB code, based on the flowchart in Figure 2, has been tested with a set of benchmark data shown in Figure 4 for its validation. Figure 4(a) depicts a calculated PTF of the TMM data, in terms of its magnitude and phase, taken from an impedance tube mimicking a lean premixed model gas turbine combustor, which is similar to the one in Figure 1. Figure 4(b) shows the calculated reflection coefficient *R*, in terms of the magnitude and phase. In order to verify the developed code in this study, the PTF in Figure 4(a) has been provided into the code, and its result of the calculated reflection coefficient is shown in Figure 5. The nearly identical reflection coefficients in Figure 5 and Figure 4(b) show that the developed code works well.

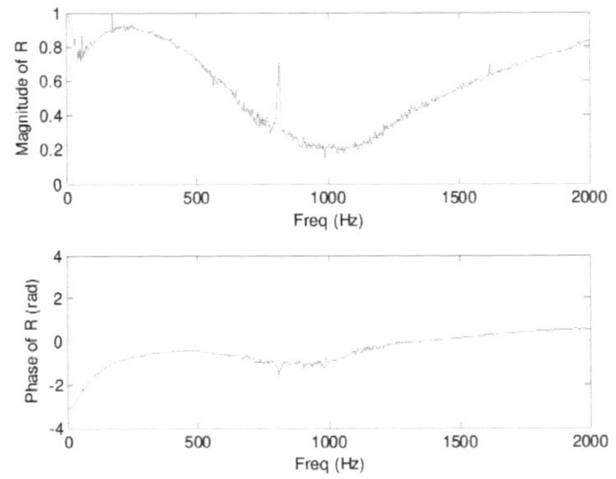

Figure 5. Calculated reflection coefficient in this study with the PTF in Figure 4(a).

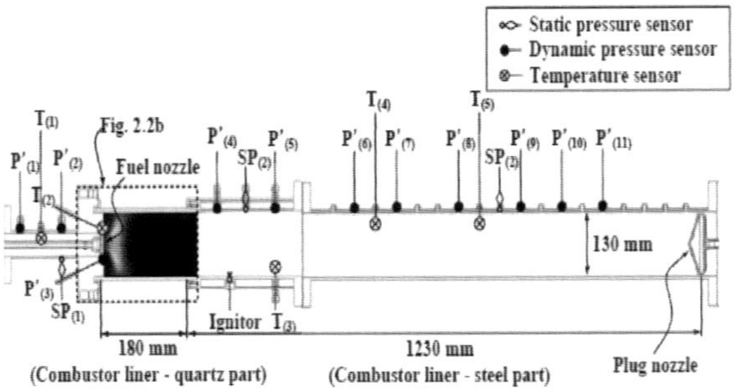

Figure 6. Schematic of a partially-premixed model gas turbine combustor showing locations of TMM dynamic pressure sensors [11].

3. Application to a model gas turbine combustor case

The combustion characteristics of a partially premixed model gas turbine combustor (see Figure 6) were investigated by varying both fuel composition of synthetic natural gas (see Figure 7) and other operation conditions [11]. Dynamic pressure transducers (DP01 - DP11; PCB® model 102M205) installed at 11 locations along the axis of both the partially premixed burner and the combustor, as shown in Figure 6, to identify the spatial distribution of dynamic pressure. The dynamic pressure signals from DP01 and DP02 were used for determining the reflection coefficient. The sampling frequency of the data is 10 kHz. Dynamic pressure signals from DP01 and DP02 for several test cases shown in Figure 7(a) have been input to the MATLAB code to calculate their reflection coefficients at the inlet of the combustor which is just upstream of flange shown in Figure 6. There is a flow straightener made out of the perforated metal disk at the inlet. Figures 8(a), 8(b), 8(c), and 8(d) show the time traces of dynamic pressure, their zoomed-in data, the corresponding pressure transfer function H_{12}, and the calculated reflection coefficient at the inlet of the combustor for the test case of 11 in Figure 7(a). Fuel composition of the test case is 12.5% H_2, 12.5% CO, and 75% CH_4 in volume. Since the fundamental frequency of the 1st longitudinal mode of the combustor would be less than 500 Hz, the result of the reflection coefficient is shown up to the frequency, as shown in Figure 8(d). The inlet of the combustor is frequently considered to be "choked" acoustically, at which the reflection coefficient is 1, if there is no experimental measurement available. As shown in Figure 8(d), the magnitude of the measured reflection coefficients at the inlet is generally in the vicinity of unity within the frequency range and thereby, the acoustic results obtained in this study could be concluded to be reasonable.

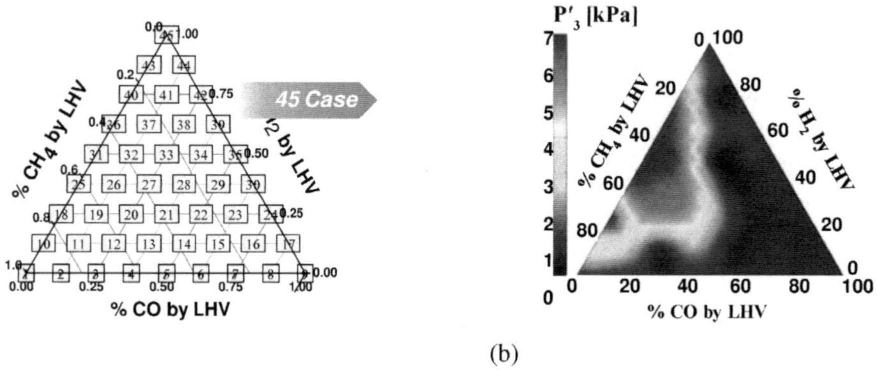

Figure 7. (a) Test cases with different fuel compositions of SNG and (b) their peak dynamic pressures measured at the dump plane of the combustor with DP03 [11].

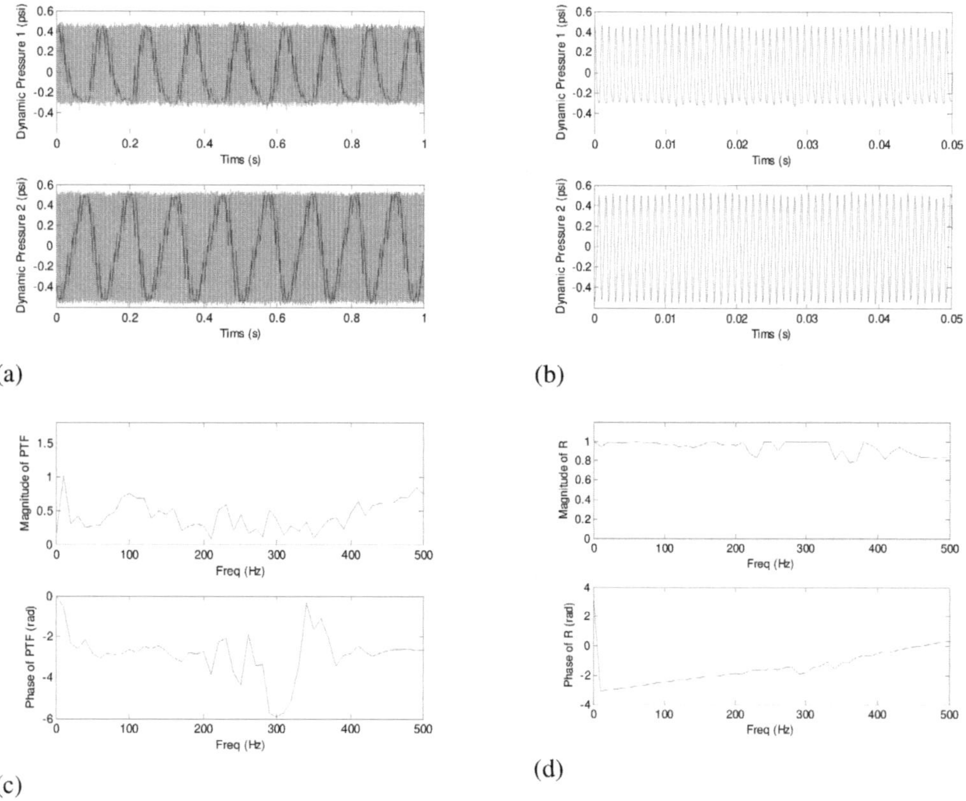

Figure 8. Reflection coefficient measurements at the inlet of the partially premixed model gas turbine combustor: (a) the time trace of TMM signals with DP01 and DP02, (b) the time trace of the TMM signals for the first 0.05 s, (c) the corresponding PTF H_{12}, and (d) the calculated reflection coefficient at the inlet of the combustor for the test case of 11.

Currently, the calculated reflection coefficient is being provided to the CI prediction codes including the finite element method (FEM)-based Helmholtz solver [4] and an open source OSCILOS (open source combustion instability low order simulator) [12] to predict the CI characteristics of the partially-premixed model gas turbine combustor [11]. The occurrence of CI and their dominant frequencies to be predicted are to be compared with those of the FFT analysis of the corresponding experimental dynamic pressure data measured at the dump plane of the combustor, DP03 in Figure 6.

4. Conclusion

A MATLAB code that can determine the acoustic reflection coefficient on the acoustic boundaries of model gas turbine as a function of frequency has been developed. The feasibility of the developed code was verified with the experimental benchmark data. Further, the reflection coefficient at the inlet of partially-premixed model gas turbine combustor has been calculated with dynamic pressure data measured synchronously at two locations of an experimental rig, resulting in quite reasonable values. A prediction of the CI for the model gas turbine combustor is under way with the calculated reflection coefficient at its inlet and other input parameters including the flame transfer functions, and will be reported in the near future.

Acknowledgments

This work was supported by the New & Renewable Energy Core Technology Program of the Korea Institute of Energy Technology Evaluation and Planning (KETEP) granted financial resource from the Ministry of Trade, Industry & Energy, the Republic of Korea (No. 2011951010001C). The authors are also grateful to Dr. Min-cheol Lee at Seoul National University, currently affiliated with Incheon National University for discussion and providing his invaluable experimental data.

References

[1] Lieuwen T C and Yang V 2005 *Combustion Instabilities in Gas Turbine Engines: Operational Operational Experience, Fundamental Mechanisms and Modelling* (Reston, VA: AIAA, Inc.)
[2] International Energy Agency 2012 *World Energy Outlook 2012: Executive Summary*. Available from: http://www.worldenergyoutlook.org/publications/weo-2012/
[3] Cha D J *et al* 2009 *Journal of Mechanical Science and Technology* **23** 1602-1612
[4] Kim S K *et al* 2010 *Journal of the Korean Society of Propulsion Engineers* **38**(5) 445-455
[5] Dowling A P and Stow S R 2003 *Journal of Propulsion and Power* **19**(5) 751-764
[6] Balachandran R *et al* 2005 *Combustion and Flame* **143** 37-55
[7] Palies P *et al* 2010 *Combustion and Flame* **157** 1698-1717
[8] Kim K T *et al* 2010 *Combustion and Flame* **157** 1718-1730
[9] Richecoeur F *et al* 2013 *Comptes Rendus Mecanique* **341** 141-151
[10] ISO 1998 Acoustics–Determination of Sound Absorption Coefficient and Impedance in Impedance Tubes Reference No. ISO 10534-2
[11] Lee M C 2014 An Experimental Study on Combustion Instability and NO_x Emission Characteristics of H_2/CO/CH_4 Syngas in a Gas Turbine Combustor Ph.D. thesis (Seoul: Seoul National University)
[12] http://www.oscilos.com

Research on fracture performance of epoxy asphalt concrete based on double-K fracture criterion

M Zhang, Z D Qian[1] and Y C Xue

Intelligent Transport System Research Center, Southeast University, Nanjing, Jiangsu, China

Email: qianzd@seu.edu.cn

Abstract. After cracks appear on steel bridge deck epoxy asphalt concrete pavement, cracks propagate fast under vehicle load. This paper studied the fracture performance of epoxy asphalt concrete, utilized single edge notched beam (SEB) three-point bending test, measured the load (P) exerted on epoxy asphalt SEB; utilized digital camera to record the fracture process of epoxy asphalt SEB, extracted the images according to the required sampling frequency and utilized Image-Pro Plus to measure the crack mouth opening displacement (CMOD) of epoxy asphalt SEB on the extracted images; calculated the double-K fracture parameters according to the P-CMOD curve. Results indicate that K_{IC}^{ini} of epoxy asphalt concrete is 1.11 MPa$\sqrt{m}$ and K_{IC}^{un} of epoxy asphalt concrete is 2.31 MPa$\sqrt{m}$ at -15℃; K_{IC}^{ini} of epoxy asphalt concrete is 1.02 MPa$\sqrt{m}$ and K_{IC}^{un} of epoxy asphalt concrete is 1.83 MPa$\sqrt{m}$ at -5℃; K_{IC}^{ini} of epoxy asphalt concrete is 0.77 MPa$\sqrt{m}$ and K_{IC}^{un} of epoxy asphalt concrete is 1.82 MPa$\sqrt{m}$ at 5℃. The double-K fracture parameters of epoxy asphalt concrete increase slightly when the temperature decreases at the scope of -15℃ to 5℃. The relation of K_{IC}^{ini}, K_{IC} and K_{IC}^{un} is $K_{IC}^{ini} < K_{IC} < K_{IC}^{un}$.

1. Introduction

Epoxy asphalt concrete is a kind of steel bridge deck pavement materials in China. Fatigue crack is the typical damage of steel bridge deck epoxy asphalt pavement. When steel bridge deck epoxy asphalt pavement have cracks, there is stress concentration at each crack tip of epoxy asphalt pavement under repeated load. It will lead to a rapid crack propagation until steel bridge deck pavement system damages.

Previous studies on fracture behavior of epoxy asphalt concrete were mainly about the usual fracture criterion, such as stress intensity factor criterion, CTOD criterion and J-integral criterion by laboratory test and numerical simulation [1-5]. In recent years, researchers began to focus on the double-K fracture criterion. And the double-K fracture criterion had been successfully used to estimate the fracture behavior of cement concrete [6, 7].

Epoxy asphalt concrete belongs to thermosetting asphalt concrete. The fracture of epoxy asphalt concrete belongs to brittle fracture below 5℃. The fracture of epoxy asphalt concrete belongs to elastoplastic fracture above 5℃ [8]. In this paper, double-K fracture parameters of epoxy asphalt concrete has been tested by laboratory experiment and double-K fracture criterion of epoxy asphalt concrete has been put forward.

2. The double-K fracture criterion of concrete and test method

The crack propagation process of concrete can be divided into 3 stages: initial fracture, stable propagation and unstable propagation [9]. According to double-K fracture criterion, when stress intensity factor (K_I) is smaller than initial critical stress intensity factor (K_{IC}^{ini}), crack cannot propagate; when stress intensity factor (K_I) equals initial critical stress intensity factor (K_{IC}^{ini}), crack begins to propagate; when stress intensity factor (K_I) is bigger than initial fracture critical stress intensity factor (K_{IC}^{ini}) and smaller than unstable fracture critical stress intensity factor (K_{IC}^{un}), crack propagation is stable; when stress intensity factor (K_I) equals unstable fracture critical stress intensity factor (K_{IC}^{un}), unstable propagation of crack begins; when stress intensity factor (K_I) is bigger than unstable fracture critical stress intensity factor (K_{IC}^{un}), the crack is at unstable propagation stage.

Traditional double-K fracture parameters test method of single edge notched beam [10] is adopted in this paper and the detail steps is shown as below.

(1) Plot the force-crack mouth opening displacement curve (P-CMOD curve). Choose a P_i and the corresponding $CMOD_i$ of the linear growth stage of P-CMOD curve. Measure the crack depth (a_0). Calculate the elastic modulus (E) by equation (1) [11]. The parameter "a" represents the crack depth, "S" represents the span of 3-point bending test, "B" represents the width of beam and "W" represents the height of beam.

$$E = \frac{6aSV_1(\alpha)}{C_i BW^2} = \frac{24a}{C_i BW} V_1(\alpha) \tag{1a}$$

$$C_i = CMOD_i/P_i \tag{1b}$$

$$\alpha = a/W \tag{1c}$$

$$V_1(\alpha) = 0.76 - 2.28\alpha + 3.87\alpha^2 - 2.04\alpha^3 + \frac{0.66}{(1-\alpha)^2} \tag{1d}$$

(2) Put the force value of the initial fracture moment (P_{ini}) and initial crack-depth ratio (a_0) into equation (2) and calculate K_{IC}^{ini}.

$$K_I = \frac{3PS}{2BW^2}\sqrt{W}\,f(\alpha) \tag{2a}$$

$$f(\alpha) = \frac{\alpha^{1/2}}{(1-\alpha)^{3/2}(1+3\alpha)}\left\{p_\infty(\alpha) + \frac{4}{\beta}\left[p_4(\alpha) - p_\infty(\alpha)\right]\right\} \tag{2b}$$

$$p_4(\alpha) = 1.9 + 0.41\alpha + 0.51\alpha^2 - 0.17\alpha^3 \tag{2c}$$

$$p_{\infty}(\alpha) = 1.99 + 0.83\alpha - 0.31\alpha^2 + 0.14\alpha^3 \tag{2d}$$

$$\alpha = a/W \tag{2e}$$

$$\beta = S/W \tag{2f}$$

(3) Put the maximum force value (P_{max}), corresponding crack mouth opening displacement ($CMOD_c$) and the calculated modulus (E) into equation (1). Then calculate the critical effective crack depth (a_c).

(4) Put the maximum force value (P_{max}) and critical effective crack depth (a_c) into equation (2) and calculate K_{IC}^{un}.

3. Laboratory test

The mechanical characteristics of 3-point bending test is similar to the mechanical characteristics of steel bridge deck pavement cracks [12]. The 3-point bending test of single edge notched beam is used in this paper so that to test the double-K fracture parameters of epoxy asphalt concrete. The schematic of 3-point bending test device and the specimen size is shown in Figure 1.

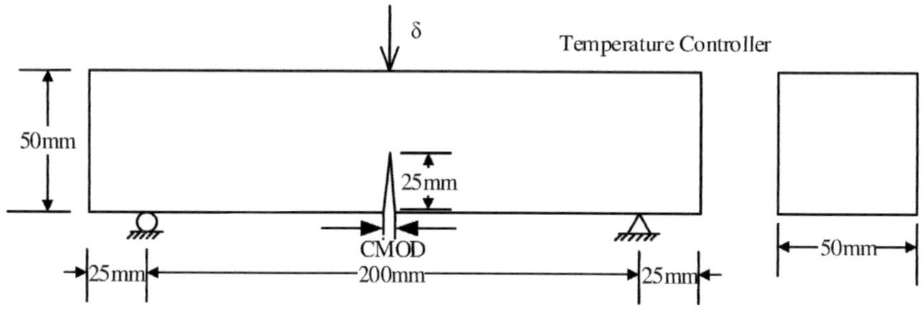

Figure 1. Schematic of 3-point bending test device.

Servo-hydraulic testing machine is used in this paper to load on single edge notched beam. The rate of deflection is 1mm/min. The test device will record the force (P) and the displacement of load location (δ). Clip gauge was usually used to measure the crack mouth opening displacement (CMOD) of single edge notched beam [13]. However, this method to measure the crack mouth opening displacement (CMOD) of single edge notched beam cannot obtain the crack mouth opening displacement (CMOD) at initial fracture moment. Digital camera is used to record the fracture process of single edge notched beam in this paper so that to obtain the crack mouth opening displacement (CMOD) at initial fracture moment and unstable propagation moment. The extracted pictures from video is shown in Figure 2. Image-Pro Plus is used to measure the crack mouth opening displacement (CMOD). Moreover, according to the recorded P and CMOD, the P-CMOD curve can be plotted.

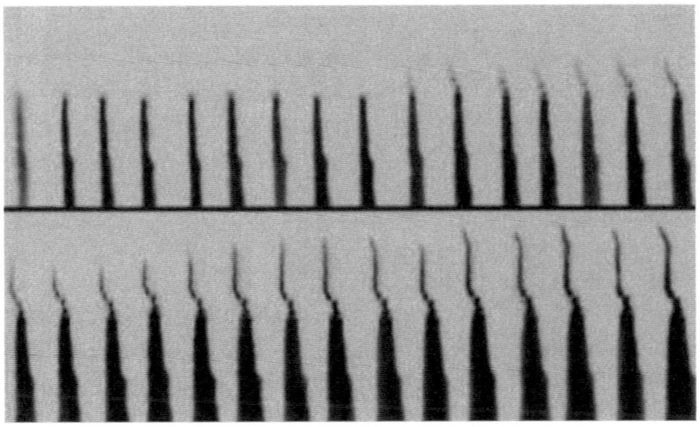

Figure 2. Crack propagation process.

The gradation scope and synthetic gradation curve of epoxy asphalt mixture is shown in Figure 3. Asphalt-aggregate ratio is 6.5%. Single edge notched beams are fabricated according the specimen size which is shown in Figure 1. In order to reinforce the contrast between crack and specimen, white paint was painted on crack edges, as is shown in Figure 4.

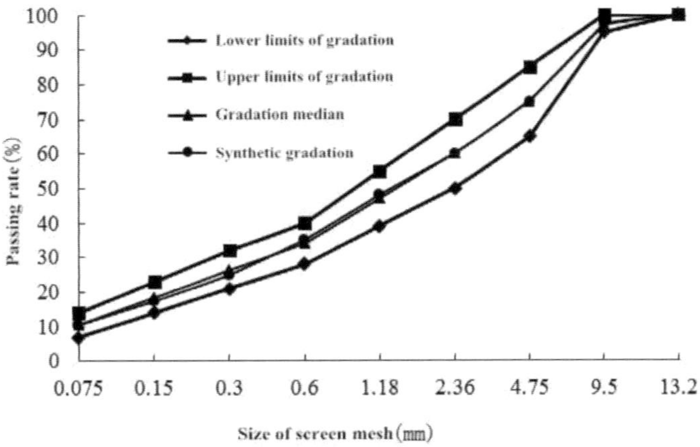

Figure 3. Gradation scope and synthetic gradation curve of epoxy asphalt mixture.

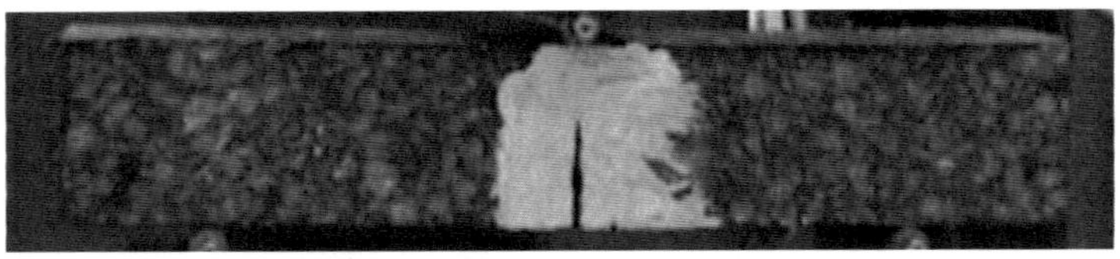

Figure 4. 3-point bending specimen.

Double-K fracture criterion is used to describe the fracture behavior of brittle cement concrete. Epoxy asphalt concrete fracture is brittle fracture below 5℃ while epoxy asphalt concrete fracture is elastoplastic fracture above 5℃ [8]. Double-K fracture parameters of epoxy asphalt concrete is tested at 5℃, -5℃ and -15℃, respectively.

4. Results and discussion

The P-CMOD curves of epoxy asphalt concrete single edge notched beams under the 3 specified temperatures are shown in Figure 5.

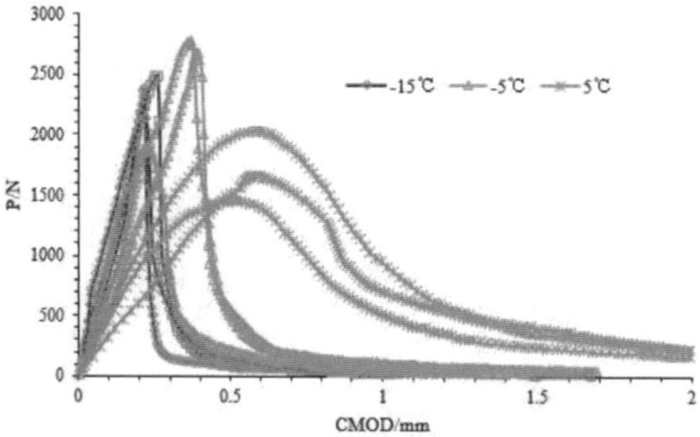

Figure 5. P—CMOD curves under different temperatures.

According to Figure 5, epoxy asphalt concrete fracture is brittle fracture at the temperature of 5℃, -5℃ and -15℃.

According to the recorded video of fracture process, the initial fracture moment is found out and then the load value at initial fracture moment is extracted. The parameters of equation (1) and equation (2) to calculate double-K fracture parameters is shown in Table 1. The span length of 3-point bend test is 20cm and the width of the single edge notched beam is about 5cm. The modulus of epoxy asphalt concrete single edge notched beam are calculated by equation (1).

Double-K fracture parameters of epoxy asphalt concrete at the temperature of 5℃, -5℃ and -15℃ are calculated by equation (2). Then the critical stress intensity factor of epoxy asphalt concrete at the temperature of 5℃, -5℃ and -15℃ are calculated according to ASTM-E399 [14].

Table 1. Calculating parameters.

T(℃)	No.	a_0(cm)	W(cm)	α_0	P_{ini}(N)	E(MPa)	P_{max}(N)	$CMOD_c$(mm)
	-15-1	2.6	5.13	0.507	2151.2	3057.6	2392.1	0.215
-15	-15-2	2.72	5.38	0.506	2348.4	3225.8	2508.6	0.256
	-15-3	2.63	4.98	0.528	2054.6	3211.0	2151.7	0.213
	-5-1	2.22	4.98	0.446	2691.8	1893.1	2782.7	0.365
-5	-5-2	2.37	5.35	0.443	2578.1	1742.9	2687.0	0.389
	-5-3	2.44	4.93	0.495	1828.6	2308.7	1908.5	0.223
	5-1	2.70	5.31	0.508	1607.1	869.8	1670.2	0.568
5	5-2	2.50	5.28	0.473	1398.1	943.6	1456.6	0.532
	5-3	2.58	5.35	0.482	1969.8	1153.5	2040.8	0.583

According to Table 2, the double-k fracture parameters and critical stress intensity factor of epoxy asphalt concrete increase slightly when the temperature decreases at the scope of -15℃ to 5℃. However, the modulus of epoxy asphalt concrete increase much faster than double-k fracture parameters and critical stress intensity factor of epoxy asphalt concrete when the temperature decreases at the scope of -15℃ to 5℃, so the lower the temperature is, the more easily the epoxy asphalt concrete crack will propagate.

Table 2. Double-K fracture parameters of epoxy asphalt concrete.

T(℃)	No.	α_0(cm)	$\bar{\alpha}_0$	K_{IC}^{ini}	$\bar{K}_{IC}^{ini}$	K_{IC}^{un}	$\bar{K}_{IC}^{un}$	K_{IC}	$\bar{K}_{IC}$
	-15-1	0.507	0.	1.07		1.91		2.04	
-15	-15-2	0.506	5	1.14	1.11	2.72	2.31	1.96	2.02
	-15-3	0.528	1	1.11		2.30		2.05	
	-5-1	0.446	0.	1.13		1.93		2.05	
-5	-5-2	0.443	4	1.04	1.02	1.84	1.83	1.76	1.82
	-5-3	0.495	6	0.89		1.73		1.65	
	5-1	0.508	0.	0.79		1.49		1.35	
5	5-2	0.473	4	0.62	0.77	1.75	1.82	1.07	1.31
	5-3	0.482	9	0.89		2.22		1.50	

[a] T: Temperature.

[b] No.: Specimen number.

[c] The units of K_{IC}^{ini},$\bar{K}_{IC}^{ini}$, K_{IC}^{un}, $\bar{K}_{IC}^{un}$,K_{IC} and $\bar{K}_{IC}$ are MPa$\sqrt{m}$.

Comparing the double-k fracture parameters and critical stress intensity factor of epoxy asphalt concrete at the same temperature, the relation of K_{IC}^{ini}, K_{IC} and K_{IC}^{un} is $K_{IC}^{ini} < K_{IC} < K_{IC}^{un}$. And K_{IC}^{un} is similar to K_{IC}, so double-K fracture parameters can reflect the fracture behavior of epoxy asphalt concrete more accurately than critical stress intensity factor.

5. Conclusions

The results of this paper indicate that K_{IC}^{ini} of epoxy asphalt concrete is 1.11 MPa$\sqrt{m}$ and K_{IC}^{un} of epoxy asphalt concrete is 2.31 MPa$\sqrt{m}$ at -15℃; K_{IC}^{ini} of epoxy asphalt concrete is 1.02 MPa$\sqrt{m}$ and K_{IC}^{un} of epoxy asphalt concrete is 1.83 MPa$\sqrt{m}$ at -5℃; K_{IC}^{ini} of epoxy asphalt concrete is 0.77 MPa $\sqrt{m}$ and K_{IC}^{un} of epoxy asphalt concrete is 1.82 MPa$\sqrt{m}$ at 5℃. The double-K fracture parameters of epoxy asphalt concrete increase slightly when the temperature decreases at the scope of -15℃ to 5℃. The relation of K_{IC}^{ini}, K_{IC} and K_{IC}^{un} is $K_{IC}^{ini} < K_{IC} < K_{IC}^{un}$.

Acknowledgements

The authors are very grateful to the Intelligent Transport System Research Center of Southeast University for providing field assistance.

References

[1] Qian Z D, Li Z and Chen C H 2008 Fracture Criterion for Mode I Crack of Epoxy Asphalt Concrete Paving Course of Steel Bridge Pavement *China Journal of Highway and Transport* **21**(5) 33-38

[2] Wang J Y, Qian Z D and Wang L B 2014 Three-dimensional Visualization Discrete Element Modeling of the Crack Development of Epoxy Asphalt Concrete *Journal of Hunan University (Natural Sciences)* **41**(6) 112-119

[3] Chen T J, Huang W and Wang S J 2008 Numerical Analysis of Fatigue Crack Propagation in Asphalt Pavement on Steel Deck *Journal of Traffic and Transportation Engineering* **8**(4)

52-57

[4] Chen C H 2008 Fracture Criteria and Fatigue Life Estimation of Steel Bridge Deck Pavement Based on Composite Beam Southeast University

[5] Qian Z D and Hu J 2012 Fracture Properties Of Epoxy Asphalt Mixture Based on Extended Finite Element Method *Journal of Central South University* **19**(11) 3335-3341

[6] Xu S L and Zhao G F 1992 Double-K Criterion of Concrete Structure Crack Propagation *China Civil Engineering Journal* **25**(2) 32-38

[7] Xu S L and Reinhardt H W 1999 Determination Of Double-K Criterion for Crack Propagation in Quasi-Brittle Fracture, Part II: Analytical Evaluating And Practical Measuring Methods for Three-Point Bending Notched Beams *International Journal of Fracture* **98**(98) 151-177

[8] Wang J Y 2013 Research on Fracture Behavior of Epoxy Asphalt Concrete Used for Steel Bridge Wearing Course under Meso-scale Southeast University

[9] Xu S L 2002 The Calculation Approaches of Double-K Fracture Parameters of Concrete and a Possible Coding Standard Test Method for Determining Them *J of China Tree Gorges Univ. (Natural Science)* **24**(1) 1-8

[10] Xu S L and Reinhardt H W 1999 Determination Of Double-K Criterion for Crack Propagation in Quasi-Brittle Fracture, Part I: Experimental Investigation Of Crack Propagation *International Journal of Fracture* **98**(2) 111-149

[11] Jenq Y S and Shah S P 1985 Two Parameter Fracture Model for Concrete *Journal of Engineering Mechanics* **111**(10) 1227-1241

[12] Li Z 2007 Research on Fracture Parameters and Fracture Energy of Epoxy Asphalt Concrete Southeast University

[13] Wu Z M and Zhao G F 1995 $CTOD_C$ Criterion of Concrete Crack Propagation *Journal of Dalian University of Technology* **35**(5) 699-702

[14] ASTM E1290-99 2002 Standard Test Method for Crack-Tip Opening Displacement (CTOD) Fracture Toughness Measurement

Research on steel bridge deck pavement mechanical response of Urumqi express way

M Zhang, Z D Qian[1], Y C Xue and Q B Huang

Intelligent Transport System Research Center, Southeast University, Nanjing, Jiangsu, China

Email: qianzd@seu.edu.cn

Abstract. The cold climate of Urumqi influences the pavement performance of steel bridge. Mechanical response and suitable pavement system of Urumqi express way steel bridge deck pavement were studied in this paper. ABAQUS was utilized to build a model of steel bridge deck pavement. Maximum tension stress, maximum tension strain, maximum vertical displacement and maximum inter-laminar shearing stress of "double EA" pavement structure and "lower EA+upper SMA" pavement structure under different pavement thicknesses and different ambient temperatures were calculated. Results indicate that when both the pavement materials of upper layer and lower layer of pavement structure are same, the maximum tension stress, the maximum tension strain, the maximum vertical displacement and the maximum inter-laminar shearing stress of "lower 3cm+upper 4cm" pavement structure are smaller than both "lower 2.5cm+upper 3.5cm" pavement structure and "lower 3cm+upper 3.5cm" pavement structure; at the range of -45°C ~ 70°C, when the pavement materials of upper layer and lower layer of pavement structure is different, the increase of temperature will lead to the decrease of maximum tension stress and maximum inter-laminar shearing stress and will also lead to the increase of the maximum tension strain and the maximum vertical displacement; the maximum tension stress of "double EA" pavement structure is larger than "lower EA+upper SMA" pavement structure; the maximum tension strain, the maximum vertical displacement and the maximum inter-laminar shearing stress of "double EA" pavement structure are close to "lower EA+upper SMA" pavement structure; the "lower 3cm EA+upper 4cm SMA" pavement structure is recommended for the steel bridge deck pavement of Urumqi expressway.

Keywords: maximum tension stress, maximum tension strain, maximum vertical displacement, maximum inter-laminar shearing stress.

1. Introduction

Vehicle load will lead to fatigue cracking and rutting of steel bridge deck pavement, and the failure of bonding layer. The mechanical response of steel bridge deck pavement under vehicle load also affects the durability and stability of the bridge structure. The mechanical evaluation indexes should be used to design the steel bridge deck pavement.

Existing research mainly focused on the effects of static load response, moving load and ambient temperature on steel bridge deck pavement. The mono-wheel load was usually adopted to study the mechanical response of steel bridge deck pavement and the modulus at room temperature (30°C) was usually adopted. Qian et al. [1] adopted finite element program to analysis the mechanical response of steel bridge deck pavement, put forward the mechanical evaluation indexes and analyzed the influence of pavement parameters on the mechanical response of steel bridge deck pavement. Cheng [2] built a steel bridge deck pavement model under moving vehicle load and analyzed the mechanical response of pavement under different moving loads. Lu et al. [3] studied the temperature field distribution of steel bridge deck pavement and analyzed the temperature distribution regularity of steel bridge deck pavement. Liu et al. [4] measured the temperature of asphalt pavement of Runyang Bridge when the bridge was paving and the influence of asphalt concrete pavement on the temperature distribution of steel-concrete box girder was obtained. However, there were few studies on the mechanical response on steel bridge deck pavement under different pavement thicknesses, different pavement materials and low temperature.

Urumqi Express Way steel bridge is located in extremely cold areas. The annual average ambient temperature of Urumqi is 6.9°C and the ambient temperature ranges from -45°C to 40°C. Higher performance of steel bridge deck pavement material and structure should be put forward so that to accommodate the extremely cold climate of Urumqi. Relying on the Urumqi Express Way steel bridge project, this paper will build a steel bridge deck pavement system model through numerical simulation. The maximum tensile stress, tensile strain, vertical displacement and inter-laminar shearing stress can be used to evaluate the mechanical performance of steel bridge deck pavement [5-7]. So the maximum tensile stress, tensile strain, vertical displacement and inter-laminar shearing stress of pavement upper surface of "double epoxy asphalt" pavement structure ("double EA" pavement structure) and "lower epoxy asphalt +upper stone matrix asphalt "pavement structure ("lower EA+upper SMA" pavement structure) under different pavement thicknesses and working temperatures were calculated in this paper so that to study the mechanical response of Urumqi Express Way steel bridge deck pavement and put forward a suitable pavement structure of Urumqi Express Way steel bridge.

2. Numerical simulation

2.1. Numerical model of steel bridge deck pavement system

Steel bridge deck epoxy asphalt pavement system model has been built by ABAQUS. There are 4 diaphragm plates in longitudinal direction and 7 stiffening rib in transverse direction, as is shown in Figure 1. According to the designed parameters of Urumqi Express Way steel bridge, the detail parameters of steel bridge deck pavement system model is shown in Table 1.

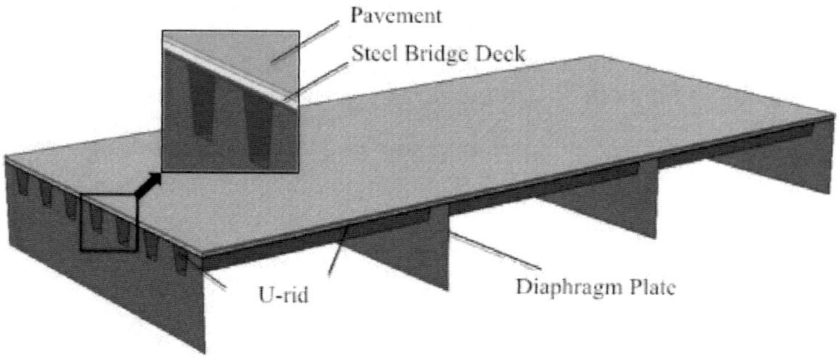

Figure 1. The model of steel bridge deck pavement system.

Table 1. The detail parameters of steel bridge deck pavement system model.

Geometry Size			
Width of U-rid (mm)	Height of U-rid (mm)	Thickness of U-rid (mm)	Spacing of U-rid (mm)
280	280	8	600
Thickness of Diaphragm Plate	Spacing of Diaphragm Plate	Thickness of pavement (mm)	Thickness of steel deck (mm)

(mm)	(m)		
14	3	60	16

2.2. Load position

According to the investigation of Urumqi Express Way steel bridge engineering, the main damages of bridge deck pavement is longitudinal cracks because of the transverse tensile stress [8, 9]. So the key mechanical control index of bridge deck pavement is maximum transverse tensile stress located on the pavement surface above the stiffening ribs. The second important mechanical control index of bridge deck pavement is maximum longitudinal tensile stress located on the pavement surface above diaphragm plates. The dual wheels load (0.7MPa) is adopted in this paper. And the dual wheels load is exerted on the pavement surface above longitudinal stiffening rib and the pavement surface above diaphragm plate, as is shown in Figure 2(a). The dimension of dual wheels load is 200mm×180mm×2 which is shown in Figure 2(b).

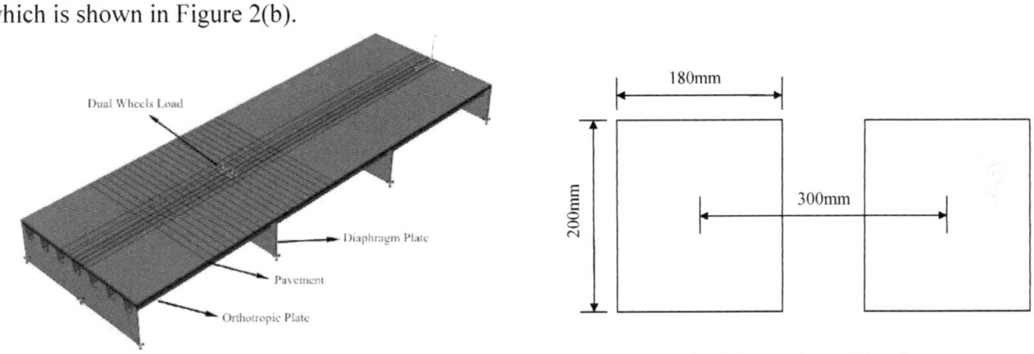

(a) Location of load (b) Dimension of load

Figure 2. The setting of load.

2.3. Output of mechanical control indexes

Fatigue cracking, rutting and the failure of bonding layer is the general damages of steel bridge deck pavement. The tensile stress and strain on pavement surface will lead to fatigue cracking. The insufficient resistance to permanent deformation will lead to rutting. The insufficient shear strength will lead to the failure of bonding later [10]. So the maximum tensile stress, tensile strain, vertical displacement and inter-laminar shearing stress are selected as the mechanical control indexes to evaluate the mechanical performance of steel bridge deck pavement so that to put forward a suitable pavement project of Urumqi Express Way steel bridge.

3. Results and conclusions

3.1. Effects of pavement thickness on the mechanical control indexes

According to the meteorological data of Urumqi, Urumqi is located in extremely cold area and the annual average ambient temperature is 6.9℃. The modulus and Poisson's ratio of epoxy asphalt concrete and SMA at 10℃ is shown in Table 2. The thickness of steel bridge deck asphalt concrete pavement ranges from 50mm to 80mm [11]. The effects of pavement thickness on the mechanical control indexes of steel bridge deck pavement are discussed in this section. Respectively calculate the mechanical control indexes of "double EA" pavement structure and "lower EA+upper SMA" pavement structure. The thickness of pavement layer adopts "lower 3cm+upper 3.5cm", "lower 2.5cm+upper 3.5cm" and "lower 3cm+upper 4cm", respectively. Moreover, according to the results, a suitable pavement thickness project of Urumqi Express Way steel bridge will be put forward. The calculated results are shown in Figure 3.

Table 2. Parameters of steel bridge deck system.

Parameters of Steel Deck		Parameters of Epoxy Asphalt Concrete		Parameters of SMA	
Modulus (MPa)	Poisson's Ratio	Modulus (MPa)	Poisson's Ratio	Modulus (MPa)	Poisson's Ratio
2.1×10^5	0.25	10000	0.25	7500	0.25

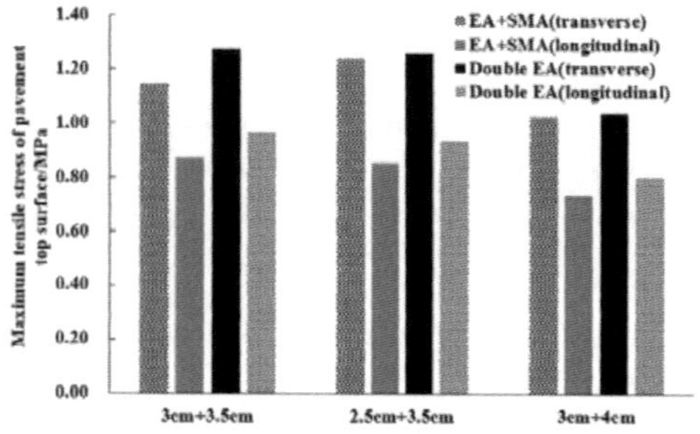

(a) Pavement thickness versus maximum tensile stress of pavement top surface

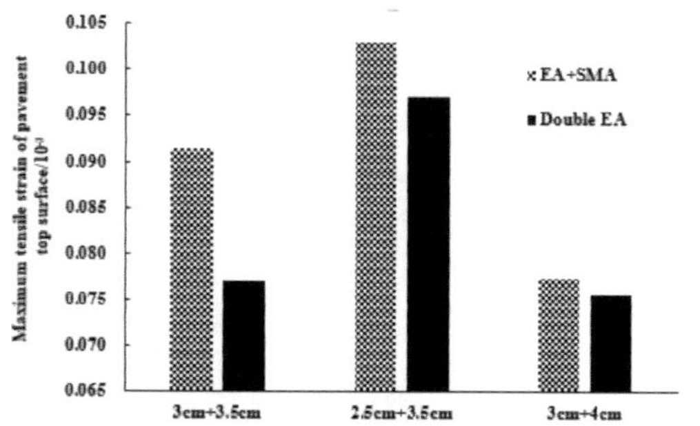

(b) Pavement thickness versus maximum tensile strain of pavement top surface

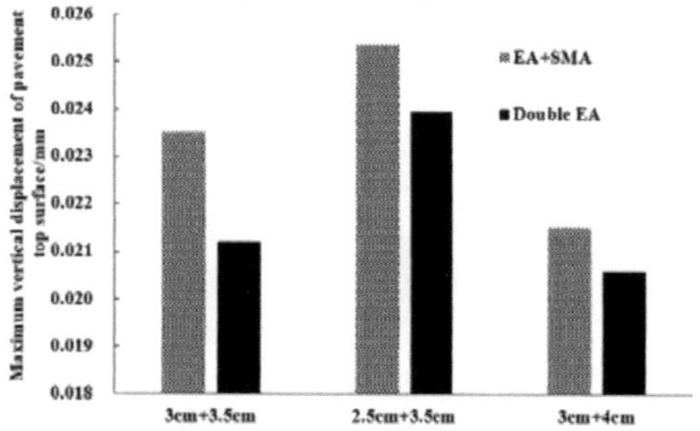

(c) Pavement thickness versus maximum vertical displacement of pavement top surface

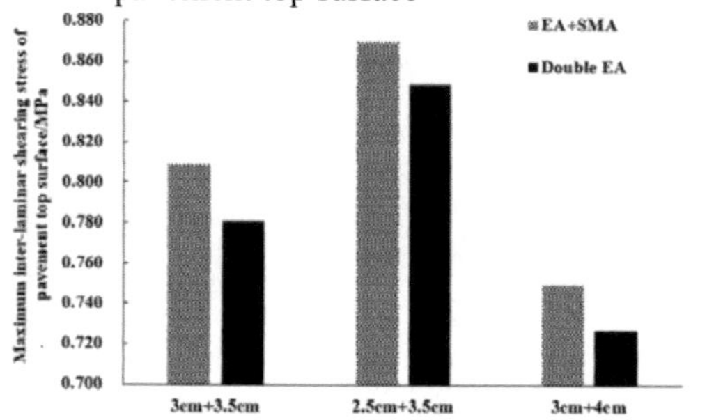

(d) Pavement thickness versus maximum inter-laminar shearing stress of pavement top surface

Figure 3. Effects of pavement thickness on the mechanical control indexes.

As is shown in Figure 3, the maximum tensile stress, tensile strain, vertical displacement and inter-laminar shearing stress of the 3 kinds of pavement thickness projects are compared and the regularity is "lower 2.5cm+upper 3.5cm" pavement structure > "lower 3cm+upper 3.5cm" pavement structure > "lower 3cm+upper 4cm" pavement structure. The "lower 3cm+upper 4cm" pavement structure is recommended to apply to Urumqi Express Way steel bridge.

Then the maximum tensile stress (σ), tensile strain (ε), vertical displacement (U) and inter-laminar shearing stress (τ) of the 3 kinds of pavement material projects of "lower 3cm+upper 4cm" pavement structure are compared and the regularity is $\sigma_{\text{double EA}}>\sigma_{\text{EA+SMA}}$, $\varepsilon_{\text{double EA}}<\varepsilon_{\text{EA+SMA}}$, $U_{\text{double EA}}<U_{\text{EA+SMA}}$ and $\tau_{\text{double EA}}<\tau_{\text{EA+SMA}}$.

The mechanical control indexes of "double EA" pavement structure and "lower EA+upper SMA" pavement structure at 10℃ are calculated in this section. However, temperature is the main factor influencing the material properties of asphalt mixture, so the suitable pavement material project of Urumqi Express Way steel bridge cannot be put forward in this section.

3.2. Effects of temperature on mechanical control indexes of steel bridge deck pavement

Temperature is the key factor influencing the material properties of asphalt mixture [12, 13]. According to the meteorological data of Urumqi, the environment temperature of Urumqi ranges from -45°C to 40°C which belongs to the extremely cold area. The temperature of steel bridge deck is

IOP Conf. Series: Materials Science and Engineering 164(2017) 012012 doi:10.1088/1757-899X/164/1/0120

usually 10°C higher than traditional truss bridge, because the steel box girder is not ventilated and the thermal dissipation of steel box girder is slow in summer [14].

This paper mainly studies the pavement design of Urumqi Express Way steel bridge at the working temperature ranging from -45°C to 70°C, the mechanical control indexes of "double EA" pavement structure and "lower EA+upper SMA" pavement structure at different temperatures are calculated and according to the calculated results, the suitable pavement material project of Urumqi Express Way steel bridge can be put forward. The moduli of asphalt mixture at different temperatures are shown in Table 3 and the calculated results is shown in Figure 4.

Table 3. The moduli of asphalt mixture at different temperature(mpa).

Mixture \ Temperature/℃	-45	-40	-30	-20	-10	0	10
EA	21300	21200	21000	20000	18000	13000	10000
SMA	13100	13100	13200	13000	12000	9000	7500
Mixture \ Temperature/℃	20	30	40	50	60	70	
EA	8000	3200	1800	1200	800	600	
SMA	5500	2500	1100	800	600	550	

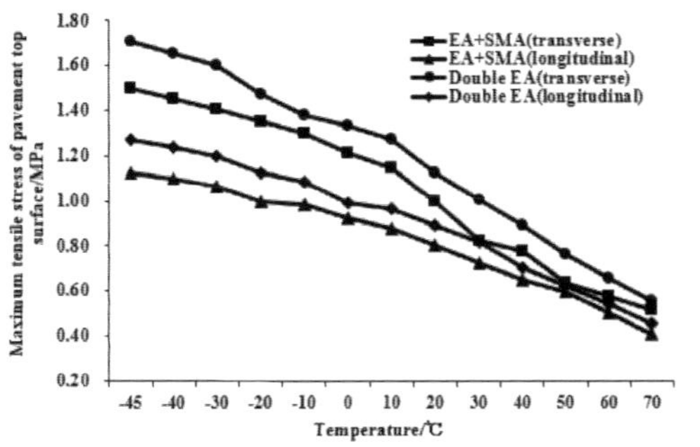

(a) Temperature versus maximum tensile stress of pavement top surface

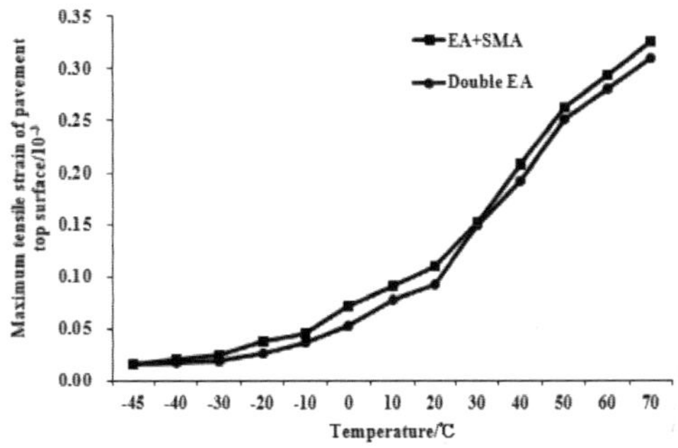

(b) Temperature versus maximum tensile strain of pavement top surface

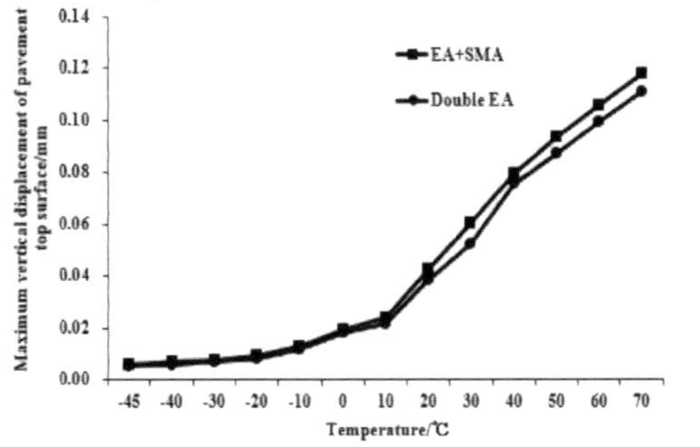

(c) Temperature versus maximum vertical displacement of pavement top surface

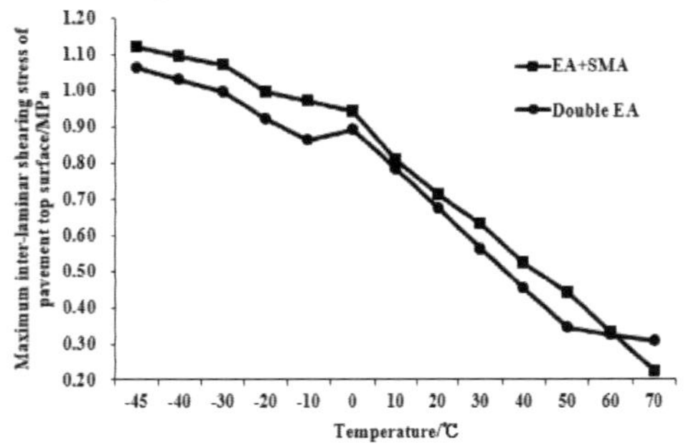

(d) Temperature versus maximum inter-laminar shearing stress of pavement top surface

Figure 4. Effects of temperature on mechanical control indexes.

As is shown in Figure 4, the maximum tensile stress and inter-laminar shearing stress of "double EA" pavement structure and "lower EA+upper SMA" pavement structure increase with the decrease of temperature. The maximum tensile strain and vertical displacement decrease with the decrease of temperature. The maximum tensile stress of "lower EA+upper SMA" pavement structure's top surface is smaller than "double EA" pavement structure at the same temperature. The maximum tensile strain, vertical displacement and inter-laminar shearing stress of "double EA" pavement structure's top surface are similar to "lower EA+upper SMA" pavement structure. The "lower 3cm EA+upper 4cm SMA" pavement structure is recommended to apply to Urumqi Express Way.

4. Conclusions

(1) The maximum tensile stress, tensile strain, vertical displacement and inter-laminar shearing stress of the 3 kinds of pavement thickness projects are compared and the regularity is "lower 2.5cm+upper 3.5cm" structure > "lower 3cm+upper 3.5cm" structure > "lower 3cm+upper 4cm" structure.

(2) The maximum tensile stress and inter-laminar shearing stress of "double EA" structure and "lower EA+upper SMA" structure increase with the decrease of temperature. The maximum tensile strain and vertical displacement decrease with the decrease of temperature. The maximum tensile stress of "lower EA+upper SMA" structure's top surface is smaller than the maximum tensile stress of "double EA" structure's top surface at the same temperature. The maximum tensile strain, vertical displacement and inter-laminar shearing stress of "double EA" structure's top surface is similar to "lower EA+upper SMA" structure's top surface's.

(3) "Lower 3cm EA+upper 4cm SMA" structure is recommended to apply to Urumqi Expressway. This paper has studied the static load response of steel bridge deck pavement, however, the main type of loads on steel bridge deck pavement is dynamic load which is not similar to the load exerted in the model of this paper.

Acknowledgements

The authors are very grateful to the Intelligent Transport System Research Center of Southeast University for providing field assistance.

References

[1] Qian Z, Huang W and Mao Q *et al* 2001 Mechanical Analysis of Nanjing Second Yangtze River Bridge Deck Pavement *Journal of Highway and Transportation Research and Development* **18**(06) 43-46

[2] Cheng F 2004 Further Research on Mechanical Properties of Long-span Steel Bridge Deck Surfacing Southeast University

[3] Lu Y, Chen Y and Sun Z *et al* 2009 Characteristics of Temperature Field Distribution of Steel Bridge Deck Pavement *Journal of South China University of Technology (Natural Science Edition)* **37**(08) 116-121

[4] Liu Q, Ding F and Zhu J *et al* 2006 Temperature Field Caused by Bituminous Deck Pavement of Steel-concrete Composed Box Girder Bridge *Journal of Southeast University (Natural Science Edition)* **36**(4) 572-575

[5] Huang W 2006 Theory and Method of Deck Paving Design for Long-span Bridges (Beijing: China Building Industry Press) pp 42-43

[6] Liu Z, Huang W and Cao D *et al* 2006 Mechanical Characteristics Analysis of Long-span Stable Bridge Steel-deck Surfacing with Longitudinal Clapboard *Journal of Highway and Transportation Research and Development* **23**(03) 44-48

[7] Zi J, Lu Q and Huang X *et al* 2008 Mechanical Analysis of the Asphalt Pavement on Orthotropic Steel Bridge Deck *Journal of Wuhan University of Technology* **30**(02) 90-93

[8] Liu Z 2006 Research on Key Techniques of Long-span Steel Bridge Deck Pavement System Design Southeast University

[9] Qian Z, Luo J and Jing M 2005 Mechanical Analysis of Asphalt Concrete Paving Projects on Steel Bridge Deck *China Journal of Highway and Transport* **18**(2) 61-64

[10] Qian Z and Huang W 2014 Maintenance and Rehabilitation of Asphalt Pavement on Steel Bridge

IOP Conf. Series: Materials Science and Engineering **164** (2017) 012012 doi:10.1088/1757-899X/164/1/012012

(Beijing: Deck China Communication Press) pp 11-12
[11] Liu Z, Huang W and Lin G 2005 Method of Deck Paving Design for Long-span Bridges *Highway* **50**(02) 32-36
[12] Chen X 2005 Research on Fatigue Properties of Bridge Deck Pavement Based on Composite Beam Southeast University
[13] Wang X, Li H and Fang X *et al* 2011 Research on Fracture and Fatigue Performance of SMA *Highway* **56**(09) 196-200
[14] Huang W 2007 Design of Deck Pavement for Long-span Steel Bridges *China Civil Engineering Journal* **40**(09) 65-77

Synthesis of a stable magnetite (magnetic fluid) colloid solution

S I Evdokimov[1] and V S Evdokimov

North Caucasian Institute of Mining and Metallurgy (State Technological University) NCIMM (STU), Republic of North Ossetia-Alania, Vladikavkaz, Nikolaev st., 44, 362021, Russia

Email: eva-ser@mail.ru

Abstract. The article is dedicated to the development and investigation of conditions for obtaining the aggregate stable colloid solutions of magnetite – magnetic fluids. The developed method differs from the known ones and involves new ways of producing and stabilizing magnetic nanoparticles with a narrow particle size distribution.

1. Introduction

Recent decades were marked by a rapid expansion in nanoscience. New magazines, monographs, professional institutes, departments, special laboratories, and even conferences have begun to implement nanoscience in their daily operations [1].

Numerous studies concerning the unique physical properties of nanoparticles have been conducted [2-4]. Many of these studies have focused on the magnetic properties of nanoparticles. These studies have revealed the differences between volumetric (three-dimensional) and nano materials [5-24]. For example, cubic spinel Fe_3O_4 is ferromagnetic at temperatures below 858 K.

Magnetic nanoparticles are prepared via two main approaches by:

(1) Grinding macroscopic materials and
(2) Generating new phases by imposing directional changes on the compositions of original chemical solutions (chemical synthesis) [1, 25].

In chemical synthesis (condensation), particles with specific sizes are obtained by limiting new phase size growth through various methods.

Magnetic fluids are widely used in the extraction of gold from placer (deposit) dressing products [26], medicine, biology, and technology [1].

2. Methodology

Two problems must be solved when synthesizing magnetic fluid. First, single-domain magnetic particles (15,4 nm for Fe, 21 nm for Co, and 40 nm for Ni) must be obtained since an increase in particle size can result in decreased sedimentation stability and magnetic property deterioration, particularly in low magnetic fields. In addition, the structure of the adsorbed stabilizer on the surface of the particles must provide for the stable aggregation of the colloid solution in the carrier liquid.

The most common and efficient method used to extract the nanoparticles of magnetic is chemical condensation via the Elmore reaction [26, 27]:

$$FeCl_2 + 2FeCl_3 + 8NaOH \rightarrow Fe_3O_4 + 8NaCl + 4H_2O \quad (1)$$

Another well-known reaction used to produce nanosized magnetite in analytical chemistry can be written as:

$$FeSO_4 + 2FeCl_3 + 8NH_4OH \rightarrow Fe_3O_4 + 6NH_4Cl + (NH_4)_2SO_4 + 4H_2O \quad (2)$$

In the reactions shown in (1) and (2), aqueous solutions containing ferrous salts and ferric ions are mixed at a temperature of 25-40°C under constant stirring to limit particle growth. Then, an aqueous basic solution is added in sesquialteral (one-and-a-half), or stoichiometric, excess. The formation and growth of new phase nuclei (nucleation) occur simultaneously via homogeneous condensation. In order to obtain a stable ferromagnetic colloid solution (magnetic fluid), the energy expended through the particle interactions must be either eliminated or significantly reduced in order to prevent the formation of microaggregates. Therefore, the surfaces of magnetic nanoparticles should be protected with a chemically adsorbed stabilizer (e.g., oleic acid) at a temperature of 60-90°C and converted into a carrier liquid (water or hydrocarbon) [25, 26].

The parameters that determine the dispersion properties of magnetic nanoparticles can also be altered. According to Uriev et al. [26, 27], the parameters of the condensation mechanisms of new phase formation include the nucleation rate (u1)

$$u_1 = k\frac{c_n - L}{L} \quad (3)$$

and growth rate (u_2)

$$u_2 = \frac{DS}{\delta}\left(c_n - L\right) \quad (4)$$

where k denotes the proportionality coefficient; cп denotes the concentration of the supersaturated solution; L denotes the solubility of the substance; (cп–L) denotes the condensation pressure, which determines the rate at which the solid is released from the solution; D denotes the diffusion coefficient; S denotes the surface of the particle; and δ denotes the thickness of the solution layer, through which diffusion occurs (the concentration increases from cп to L).

A high rate of nucleation u1 and relatively low rate of particle condensation u2 lead to increased nanoparticle formation rather than coarse sediment formation. This was achieved by Uriev et al. [26-28] through the rapid injection of the concentrated solution of one component into the diluted solution of another component with vigorous stirring.

The kinetic model of particle aggregation-fragmentation (disaggregation) in the hydrodynamic shear field of coagulation dispersion was used to develop and study a method of obtaining monodispersed particles [29-32]. In the equilibrium state, when the annexation rate and aggregation detachment rate are equal, the dispersion process can be divided into two phases consisting of large aggregates and small particles (primary particles and small aggregates). Polydisperse system formation is primarily attributed to the differences in the Brownian and gradient growth mechanisms of small and large particles. The small- and large-fraction particles in a model are characterized by continuum mass spectra f(μ) и F(m) (where μ, m denotes the mass of the relevant particle fractions). Each of the functions f and F is denoted by the number of particles in the fraction. In the condensation method of obtaining nanoparticles, the reaction process slows after ~30% of the particles reach the required size. The formation of very small particles occurs slowly via nucleation. These small particles combine into large, unstable aggregates through coagulation with subsequent peptization. The particles that form the aggregates gradually increase in size, eventually forming stable nanoparticles. These stages can be described by the following kinetic equations:

$$\frac{\partial f(\mu)}{\partial t} = \int_0^{\mu/2} \beta(\mu-\mu',\mu') f(\mu-\mu') f(\mu') d\mu' - \\ - \int_0^{\infty} \beta(\mu,\mu') f(\mu) f(\mu') d\mu' - \int_0^{\infty} \beta(m,\mu) F(m) f(\mu) dm + \\ + \int_{\mu}^{\infty} \gamma(m-\mu,\mu) F(m) dm, \tag{5}$$

$$\frac{\partial F(m)}{\partial t} = \int_0^{m/2} \beta(m-\mu,\mu) F(m-\mu) f(\mu) d\mu - \\ - \int_0^{\infty} \beta(m,\mu) F(m) f(\mu) d\mu + \int_0^{\infty} \gamma(m,\mu) F(m+\mu) d\mu - \\ - \int_{\mu}^{m/2} \gamma(m-\mu,\mu) F(m) d\mu \tag{6}$$

where β (x, y) denotes the coagulation nucleus, or the specific rates of particle association x and y in the unit aggregate x+y, and γ (x, y) denotes the collapse nucleus, or the specific unit rate of aggregate collapse x+y into particles x and y. The functional forms of the coagulation and disintegration nuclei can be recorded based on the aggregate growth mechanism using an approach developed in [32]. Models can be used to investigate the kinetics of the concentration of sol particles in terms of size. This process begins with condensation particle growth and a reduction in the number of small particles, resulting in asymmetry. However, as shown in Figure 1, the degree of asymmetry is reduced once monodispersed sol particles are produced [26].

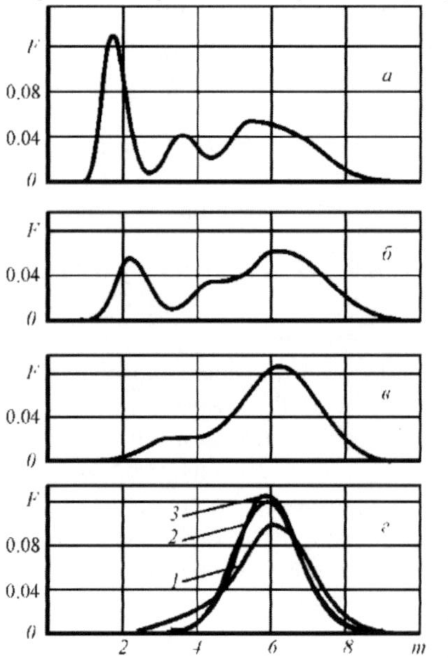

Figure 1. Kinetics of the concentration of large particles per weight.

In a previous study, Erin [8] proposed that magnetite particles should be condensed onto primers through heterogeneous condensation. Using this methodology, Zigmondi developed a method of hydrosol production with a narrow distribution of magnetite particles in terms of size.

First, the primer sol was obtained by pouring concentrated salt solutions of Fe2+ and Fe3+ with 10 % more than the equivalent amount of excess of Fe2+ into boiling ammonia NH4OH. The dispersion phase concentration of the resulting sol was approximately 10-3 g/dm^3.

As shown in Figure 1, the aggregation process began with particle growth via condensation, resulting in a decreased number of small particles and asymmetry. However, the degree of asymmetry gradually decreased over time, at which point particle aggregation played a secondary role to intensive nanoparticle crystallization and fragmentation due to the effects of hydrodynamic stress.

3. Experimental study and analysis

In this study, a method of obtaining magnetite nanoparticles with a narrow particle size distribution through heterogeneous condensation is developed based on the studies of Erin et al. [8, 26, 27, 32]. In the proposed method, condensation is performed on clusters, and <1 nm seeds are received by and adsorbed into the solution.

The seed clusters were prepared by pouring a concentrated salt solution containing Fe^{2+} and Fe^{3+} at a 10% Fe^{2+} salt excess into boiling ammonia NH_4OH. The dispersion phase concentration of the resulting colloid solution was 10^{-3} g/dm^3.

Due to their small size, the particles in seed clusters do not react significantly to magnetic field imposition. As a result, seed cluster particles should be enlarged in order to promote participation in thermal motion (chaotic Brownian oscillation) and the acquisition of explicit magnetic properties. For this reason, basic metallic solutions were poured into the ammonia colloid solution containing seed clusters. After mixing these solutions, new clusters (new phase nuclea) ceased to form and the available clusters increased in size. Monodispersible magnetic nanoparticles with a magnetic phase volume concentration of 1,8-7,8% were obtained through this process.

According to these results, the vibration of the dispersion system hindered the formation of large particles. The excitation energy (low frequency mechanical and ultrasonic frequencies) must exceed the interaction energy in order to achieve aggregation stability.

The obtained particles were washed with water in order to remove electrolytes, which are the co-products of chemical reactions.

The "double additive" method developed by Uriev et al. [26, 32, 34, 35] was used to obtain the stable magnetite colloid solution. This method involves the execution of the following technological operations.

The magnetic fluid was obtained via the method described by Berkovskii et al. [25, 26], but with double (vs. optimum) excess stabilizer, which consists of a mixture of oleic acid and kerosene. Next, the certain volume (υ_1) was prepared by adding the excess stabilizer of the magnetic fluid to the magnetite colloid solution υ_2. The resulting solution was heated to 90°C. Then, the heterogeneous condensation method was applied to the seed clusters. The obtained magnetic fluid was separated from the water with a Buechner funnel and then "pressed out" onto the magnetic separator.

The developed method resulted in two types of particle interactions. An adsorbed layer of stabilizer was noted on the surfaces of the nanoparticles in the magnetic fluid. The multilayer adsorption of the stabilizer was expected due to the adsorption mechanisms of the surface-active agents (SAA) on the salt-forming surfaces. Oleic acid and its sodium salt (sodium oleate), which are strong SAAs, formed the first adsorbed layer in the form of chemically fixed electrovalent molecules. In the next layer, which consisted of inversely oriented SAA molecules, the hydrocarbyls of the adsorbate and adsorbent molecules were connected through hydrophobic interactions, in which the polar groups of the adsorbent molecules were oriented toward the liquid phase. The structure of the adsorbed layer was maintained by the free SAA molecules, which were diffusion-bonded with the chemically and physically adsorbed molecules on the particle surfaces. The number of these particles n_1 was proportional to the volume of the magnetic fluid υ_1. In addition, the number of other particles n_2 was proportional to the volume υ_2 of the colloid magnetite solution. The number of particles that entered the surface of the system n_1 as well as the number of particles free from the surface of the system n_2 were characterized by $\tau = \upsilon_1/\upsilon_2$.

The number of particles N was directly measured with a flow-line ultramicroscope VDK-4.

In the proposed method, τ_1 represented the contact time of the oleic acid and nanodispersed magnetite, while τ_2 represented the contact time of the magnetic fluid and nanodispersed magnetite. Adsorption equilibrium was rapidly achieved in the system after applying the stabilizer to the particle surfaces. When τ_1 = 10-15 min., the reactions between the oleic acid and magnetite nanoparticles resulted in a reduction in the number of particles N at a relatively slow rate. After further growth from τ_1 to 70 min. (in the individual tests, the value of τ_1 exceeded 24 hours) the effects of the stabilizer were almost unobservable. The number of particles N decreased continuously as the contact time τ_2 increased. Therefore, the kinetic factor significantly influenced the adsorption of the oleic acid by the magnetite nanoparticles. The particle concentration N was estimated via direct measurements with a flow-line ultramicroscope VDK-4 (Table 1).

Table 1. Relationship between the particle concentration of the colloid solution N and τ_1 and τ_2 at $\theta = \upsilon_1/\upsilon_2 = 1$.

Process parameter	Time (τ_1 or τ_2), min						
	5	10	15	20	30	50	70
$N \cdot 10^{-7}$ at τ_1 (τ_2 = 30 min)	14,0	1,9	1,7	1,6	1,5	1,4	1,4
$N \cdot 10^{-7}$ at τ_2 (τ_1 = 30 min)	13,7	1,7	1,6	1,5	1,4	1,2	1,1

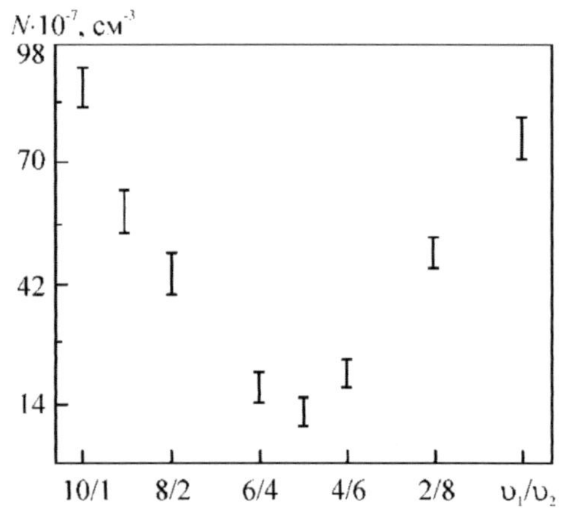

Figure 2. Particle concentration of the colloid solution N as a function of $\theta = \upsilon_1/\upsilon_2$, where the experimental points denote the confidence intervals constructed with the Student-t distribution.

According to the results shown in Figure 2, the minimum nanoparticle concentration of the magnetite colloid solution N was achieved at $\theta = \upsilon_1/\upsilon_2 \approx 1$ – the same number of n_1 and n_2 particles.

The probability that two E particles will bond can be determined using the coagulation probability α (according to the theory of hydrophobic colloid stability DLFO [26, 27]) and flocculation probability φ (the probability of aggregate formation on the surface of a particle upon contact with the stabilizer). If Σ is a part of the surface with the stabilizer, the probability of contact between the surface areas of two particles without the stabilizer is (1-Σ). Therefore, the coagulation and flocculation probabilities can be written as $\alpha\cdot(1-\Sigma)$ and $\varphi\cdot\Sigma^2$, respectively. The probability that a particle will bond via two mechanisms simultaneously can be expressed as:

$$E = 1-(1-\alpha)\cdot(1-\Sigma)^2-(1-\varphi)\cdot\Sigma^2$$

This probability measurement can be used to determine the availability of the minimum value function $N = f(\theta = \upsilon_1/\upsilon_2)$ shown in Figure 2. As shown in Figure 3, the minimum value on the curve $N = f(\upsilon_1/\upsilon_2)$ did not influence the relationship between the magnetic fluid density (ρ, g/cm^3) and the ratio $\gamma = n_1/n_2$, which also passed through the minimum value at $\gamma = n_1/n_2 \approx 1$. Dependence was achieved at nanodispersed magnetite and oleic acid concentrations of $5{,}4\cdot10^{-2}$ g/l and 10^{-3} g-mol/l, respectively.

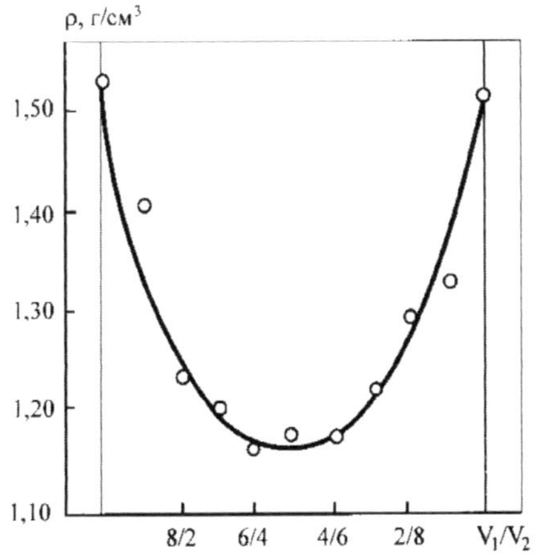

Figure 3. Magnetic fluid density ρ as a function of $\gamma = n_1/n_2$.

The effects of the stabilizer and dispersion phase concentrations on the stability of the aggregated nanoparticles in the magnetic fluid are shown in Figure 4. Dependence was achieved at nanodispersed magnetite concentrations of $4{,}7\cdot10^{-2}$ g/l (curve 1) and $4{,}4\cdot10^{-4}$ g/l (curve 2) when $\theta = \upsilon_1/\upsilon_2 = 1$.

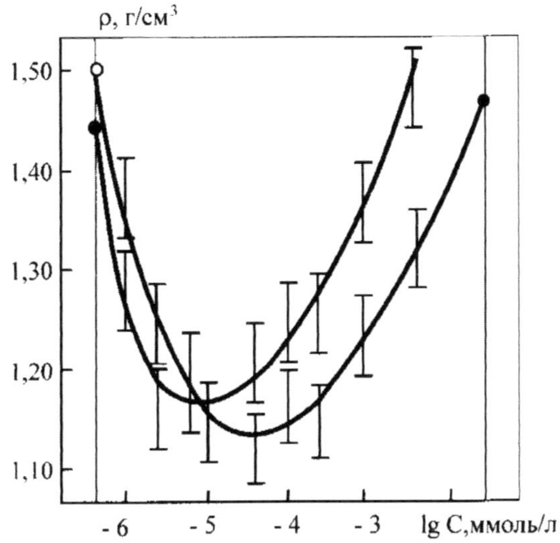

Figure 4. Magnetic fluid density as a logarithmic relationship of the oleic acid concentration, where the experimental points denote the confidence intervals constructed with the Student-t distribution.

The interparticle repulsive force was lowest when $\theta = \upsilon_1/\upsilon_2 = 1$, which is equal to the same number of both particle types. This system state was characterized by decreased aggregate stability resulting from the aggregation of particles in the dispersion phase, the formation of residue, and a reduction in the magnetic fluid density. Any change in the ratio $\theta = \upsilon_1/\upsilon_2 = 1$ increased the aggregate stability of the magnetic fluid. The peptizating effects of solutions of low stabilizer concentrations can be described by the Rehbinder effect [26, 27]. In the Rehbinder effect, the elastic repulsion and mechanical strength of adsorbed SAA layers interfere with the aggregation of particles. High concentrations of oleic acid in the dispersion phase of the system resulted in the stabilization of the colloid solution. Therefore, the amount of excess oleic acid as well as the features of micellar structures should be considered since micelle can form structural and mechanical barriers in SAA solutions. These effects resulted in an increase in the minimum value on curve $\rho = f(lgC)$ in Figure 4. According to the results of this study, the ratio $\theta = \upsilon_1/\upsilon_2$ should be adjusted in order to achieve an oleic acid concentration of $4\cdot10^{-2}$ mol/l when developing a stable magnetite colloid solution.

4. Conclusions

According to the results of this study, the mass crystallization of magnetite colloids can be achieved with a combination of the heterogeneous chemical condensation method developed by Zigmondi and the Weimam Rule. In addition, the method proposed by G.J. Fleer and J. Lyklema, in which portions of a colloid solution are mixed with and without stabilizer, can be used to achieve stabilization in colloid magnetite solutions.

Acknowledgement

This article was prepared with the support of the agreement on granting of subsidies № 14.577.21.0142 (RFMEFI57714X0142).

References

[1] Gubin S P, Koksharov Y A, Khomutov G B and Jurkov G Y 2005 Magnetic nanoparticles: preparation methods, structure and properties *Russian Chemical Reviews* **74**(6) 539-574
[2] Baraton M I 2002 *Synthesis, Functionalization, and Surface Treatment of Nanoparticles* (Los Angeles: American Scientific Publishers) pp 144-157
[3] Krutyakov Y A, Kudrinsky A A, Olenin A Y and Lisichkin G B 2008 Synthesis and properties of silver nanoparticles: advances and prospects *Russian Chemical Reviews* **77**(3) 242-269
[4] Dykman L A and Bogatyrev V A 2007 Gold nanoparticles: preparation, functionalisation and applications in biochemistry and immunochemistry *Russian Chemical Reviews* **76**(2) 199-213
[5] Taketomi S and Tikadzumi S 1993 *Magnetic fluid* (Moscow: World) p 240
[6] Kalaeva S Z 2015 *Directed change in the properties of minerals and rock deposit wastes to obtain magnetic fluids, ensuring the solution of engineering problems of extraction and processing of minerals* PhD. Engineering (Tula: Tula State University) p 32
[7] Mamiya H, Nakatani I and Furubashy T 2000 Phase transitions of iron-nitride magnetic fluids *Phys. Rev. Lett.* **84** 6106-6109
[8] Erin K V 2001 *The study of the kinetics of birefringence in colloidal systems under the influence of external electric and magnetic fields* PhD. Physics and Mathematics (Stavropol: Stavropol State University) p 151
[9] Dikanskii Y I and Nechayev O A 2003 Structural changes in magnetic fluids in electric and magnetic fields *Colloid Journal* **65**(3) 338-342
[10] Bondarenko E A 2001 *Mechanism of formation of multilayer structures in magnetic fluids in the sheath region* PhD. Physics and mathematics (Stavropol: Stavropol State University) p 130
[11] Jennings B R, Xu M and Ridler P J 2001 Structure in magneto-rheological fluids: a theoretical analysis *Journal of Physics D: Appl. Phys.* **34** 1617-1623
[12] Dikanskii Y U, Wegera J G, Suzdalev V N and Smerek Y L 2003 About magnetic fluids with disperse non-magnetic inclusions of various shapes *Proceedings of Universities of North-Kavkaz Region* (Rostov-on-Don: Southern Federal University) pp 37-40
[13] Verolaynen N V, Zhuravlev O E and Voronchihina L I 2012 Investigation of citric acid and its salts as stabilizers of magnetic aqueous fluids *VestnikTvSU Series: Chemistry* **13** 89-98
[14] Brikov E S, Zhuravsky D V, Mikheev V A, Novikov V F and Smirnov I A 2011 The effect of steady magnetic fields on the formation of the mean value of magnetite nanoparticles in aqueous ion exchange reaction with an excess of alkali precipitation *Chemistry, Physics and Technology of Surface* **2**(3) 229-234
[15] Burkova E N 2014 *Spatial segregation of particles in concentrated magnetic fluids: numerical simulation* PhD. Physics and Mathematics (Perm: Perm State University) p 18
[16] Kayal S and Ramanujan R V 2010 Anti-cancer drug loaded iron–gold core–shell nanoparticles (Fe and Au) for magnetic drug targeting *J. Nanosci. Nanotech.* **10** 1-13
[17] Faraji M, Yamini Y and Rezaee M 2010 Magnetic nanoparticles: synthesis, stabilization, functionalization, characterization, and applications *J. Iran. Chem. Soc.* **7**(1) 1-37

[18] Jaganathan H, Gieseck R L and Ivanisevic A 2010 Characterizing proton relaxation times for metallic and magnetic layer-by-layer coated, DNA-templated nanoparticle chains *Nanotechnology* **21** 1-7
[19] Robinson I, Tung L D, Maenosono S, Walti C and Thanh N T K 2010 Synthesis of core-shell gold coated magnetic nanoparticles and their interaction with thiolated DNA *Nanoscale* **2**(12) 2624-2630
[20] Tamer U, Gündoğdu Y, HakkBoyac I and Pekmez K 2010 Synthesis of magnetic core–shell Fe3O4–Au nanoparticle for biomolecule immobilization and detection *J. Nanoparticle Research* **12**(4) 1187-1196
[21] Wang Y, Shen Y, Xie A, Li S, Wang X and Cai Y 2010 A simple method to construct bifunctional Fe3O4/Au hybrid nanostructures and tune their optical properties in the nearinfrared region *J. Phys. Chem. C.* **114**(32) 4297-4301
[22] Kirui D K, Rey D A and Batt C A 2010 Gold hybrid nanoparticles for targeted phototherapy and cancer imaging *Nanotechnology* **21** 1-10
[23] Jelveh S and Chithrani D B 2011 Gold nanostructures as a platform for combinational therapy in future cancer therapeutics *Cancers* **3** 1081-1110
[24] Tang M M 2014 *The elastic properties of the magnetic fluids with an air cavity created and transported by the magnetic field* PhD. Physics and Mathematics (Kursk: Southwestern State University) p 108
[25] Berkovskii B M, Medvedev V F and Krakow M S 1989 *Magnetic fluid* (Moscow: Chemistry) p 240
[26] Uriev N B and Kuchin A V 2006 Modelling of the dynamic state of disperse systems *Russian Chemical Reviews* **75**(1) 36
[27] Evdokimov S I, Panshin A M and Kanashvili M J 2008 Magnetic fluid: new technology *Non-Ferrous Metallurgy* **2** 15-19
[28] Panshin A M, Evdokimov S I, Solodenko A A *Gold: Theory and textile arts* (Vladikavkaz: NPKP Ltd.) p 960
[29] Evdokimov S I, Solodenko A B, Hutuev T Y and Maksimov R N 1995 Obtaining iron oxide hydrosol for the synthesis of magnetic fluid *Non-Ferrous Metallurgy* **3** 22
[30] Evdokimov S I and Solodenko A A 2006 The theory of preparation of magnetic fluids and the movement of mineral particles in separators of deflecting type *Non-Ferrous Metallurgy* **4** 28-34
[31] Evdokimov S I, Panshin A M and Kanashvili M J 2009 Investigating the stability of colloidal solutions of magnetite as a separation medium *Chemistry and Chemical Engineering* (Ivanovo: Ivanovo state University of chemistry and technology) pp 73-77
[32] Protodyakonov I O, Lublinskaya I E and Ryzhkov A E 1987 *Hydrodynamics and Mass Transfer of Liquid-Solid Bodies in Dispersed Systems* (Leningrad: Chemistry) p 336

Research on construction technology for orthotropic steel deck pavement of Haihe River Chunyi Bridge

Y C Xue, Z D Qian[1] and M Zhang

Intelligent Transportation System Research Center, Southeast University, Nanjing, Jiangsu, China

Email: xueyc@seu.edu.cn

Abstract. In order to ensure the good service quality of orthotropic steel deck pavement of Haihe River Chunyi Bridge in Tianjin, and to reduce the occurrence of pavement diseases like lateral and longitudinal cracks, the key working procedures such as steel deck cleaning, anticorrosive coating, bonding layer spraying, seam cutting, epoxy asphalt concrete's mixing, transportation, paving and compaction were studied. The study was based on the main features of epoxy asphalt concrete which is the pavement materials of Haihe River Chunyi Bridge, and combined with the basic characteristics and construction conditions of Haihe River Chunyi Bridge. Furthermore, some processing measures like controlling time and temperature, continuous paving with two pavers, lateral feeding, and improving the compaction method were proposed. The project example shows that the processing measures can effectively solve the technical difficulties in the construction of orthotropic steel deck pavement of Haihe River Chunyi Bridge, can greatly improve the construction speed and quality, and can provide reference for the same kinds of orthotropic steel deck pavement construction.

1. Introduction

Epoxy asphalt concrete (EAC) is an excellent concrete with several decades of application as orthotropic steel deck pavement material. Generally, the aggregate used in EAC is highly processed basalt aggregate with a dense gradation. Epoxy asphalt is consisted of petroleum asphalt and thermosetting acid epoxy [1]. The epoxy asphalt is premixed before being combined with the heated aggregate through epoxy asphalt static mixer.

Previous studies [2-5] have clearly demonstrated that the EAC is a sustainable material with long service life, and has superior waterproofness, fatigue performance, moisture susceptibility and high temperature stability, can apply to serious service conditions in high-cold and high-latitude areas. EAC was first applied to pave the steel deck of the San Mateo–Hayward Bridge in the mid-1960s, which has exhibited superior performance for many years [6]. From then on, EAC has been widely used as surface material on numerous orthotropic steel decks around the world [7, 8].

With the distinct advantage of superior performance, using EAC does have some downsides. The main one is the difficult in construction of EAC: Once the two separate components of epoxy asphalt are mixed, curing agent and epoxy resin begin an irreversible chemical reaction that increases the strength and stiffness of the EAC. If we don't finish the paving and compaction of EAC in construction allowable time and specific temperature range, the EAC will become hard to compact and pave and its pavement performance will become very poor [9-11]. Therefore, the mixing temperature and the construction allowable time need to be carefully controlled.

Haihe River Chunyi Bridge using orthotropic steel deck lies in the southeast of Tianjin, China as shown in Figure 1. It is located between the Haijin Bridge and Jizhao Bridge. The span and width of the bridge are 663m and 40m, respectively. The pavement structure of orthotropic steel deck on main bridge is double-EAC. Besides, Haihe River Chunyi Bridge is the final bridge in the upstream of Haihe River since 2002 after overall development of Haihe River. It can cross the Haihe River and ease traffic pressure of Tianjin, and is very important to transportation planning and economic development of Tianjin. Hence the construction of Haihe River Chunyi Bridge got sufficient attention. This paper presents a research on the construction technology for orthotropic steel deck pavement of Haihe River Chunyi Bridge.

Figure 1. Aerial view of Haihe River Chunyi Bridge.

2. Materials

The technical specifications of the epoxy asphalt used in the paper are listed in Table 1, Table 2 and Table 3. The gradation of the basalt aggregate is presented in Table 4 and the nominal maximum size is 9.5 mm. The basalt aggregate has a compressive strength of 140 MPa and crushing value of 4.2%. The optimum binder content of EAC with was 6.5% determined based on the standard Marshall Mix design procedure.

Table 1. Technical specifications of epoxy resin (component A).

Technical specification	Test results	Criteria	Test Method
Viscosity (23°C, Pa·s)	13	11~15	ASTM D 445 [12]
Epoxide equivalent weight	188	185~292	ASTM D 1652 [13]
Color (Gardner)	3	≤4	ASTM D 1544 [14]
Moisture content (%)	0.03	≤0.05	ASTM D 1744 [15]
Flash point, Cleveland open cup (°C)	247	≥200	ASTM D 92 [16]
Specific gravity at 23°C	1.165	1.16~1.17	ASTM D 1475 [17]
Appearance	Transparent amber	Transparent amber	Visual

Table 2. Technical specifications of mixture of curing agent and petroleum asphalt (component B).

Technical specification	Test results	Criteria	Test Method
Viscosity at 100°C (Pa·s)	0.18	≥0.14	ASTM D 2983 [18]
Acid Value (mg, KOH/g)	52	40~60	ASTM D 664 [19]
Flash point, Cleveland open cup (°C)	255	≥200	ASTM D 92 [16]
Specific gravity at 23°C	1.00	0.98~1.02	ASTM D 1475 [17]
Color	Black	Black	Visual

Table 3. Technical specifications of epoxy asphalt (mixture of component A and component B).

Technical specification	Test results	Criteria	Test Method
Time (Viscosity from 0 to 1000 mPa·s, min)	51	≥50	ASTM D 4402 [20]
Tensile strength (23°C, MPa)	2.22	≥1.52	ASTM D 638 [21]
Fracture elongation (23°C, %)	462	≥190	ASTM D 638 [21]

Table 4. Grading limits and design value.

Sieve size (mm)		0.075	0.15	0.3	0.6	1.18	2.36	4.75	9.5	13.2
Percentage passing (%)	Upper limit	14	23	32	40	55	70	85	100	100
	Lower limit	7	14	21	28	39	50	65	95	100
	Design value	10.8	16.7	23.8	35.9	48.4	59.7	75.2	96.9	100

3. Construction

3.1. Steel deck cleaning
The oil, salt and other dirt in the deck were cleaned through detergent. The appearance of the steel bridge deck was checked to ensure that there was not splash, pinhole and burr in the deck.

3.2. Anticorrosive coating
The deck was descaled through using a portable automatic dust-free sand machine, the grinding material in machine was steel shot and grit, and the mix ratio of steel shot and grit was 7:3. The roughness of deck must reach the requirement of 50-100μm before coating anticorrosive and the

relative air humidity must be less than 85% when coating anticorrosive. Epoxy zinc-rich paint widely used as an anticorrosive material was coated in the deck through high-pressure airless spraying method. The thickness of epoxy zinc-rich paint was 60-80μm.

3.3. Bonding layer spraying

As the middle layer in composite pavement structure, the bonding layer should play an important role in connecting EAC and orthotropic steel deck, that is to say, the bonding layer should have a great adhesive performance in maintaining the continuity of each layer under the action from the external load. Hence the procedure of spraying bonding layer is very important to the integral pavement performance of orthotropic steel deck. The bonding layer was sprayed through special spreader combined with artificial coating as shown in Figure 2. The spraying volume is about 0.5 L/m^2.

Figure 2. Bonding layer spraying.

3.4. EAC mixing

The procedures of EAC mixing were as follows: Firstly, the epoxy asphalt was prepared by mixing component A and component B at a fixed ratio of 1:2.45, when component A was heated to 87℃ and component B to 133℃, then blended the epoxy asphalt with basalt aggregate when basalt aggregate was heated to 120-125℃. The final temperature of EAC should be controlled at 110-130℃. A faster chemical reaction of epoxy asphalt caused by a higher temperature would shorten the construction allowable time of EAC, while a lower temperature cannot ensure sufficient drying of aggregates. The properties of EAC are presented in Table 5.

Table 5. Properties of EAC.

Technical specification	Test results	Criteria	Test Method
Density (g/cm^3)	2.566	-	JTG E20-2011 T0705
Air void (%)	2.0	1.5~3.0	JTG E20-2011 T0705
Voids in the mineral aggregate (%)	17.5	≥17	JTG E20-2011 T0705

Stability (kN)	60.18	≥40	JTG E20-2011 T0709
Flow value (0.1mm)	38.2	20~60	JTG E20-2011 T0709

To guarantee the favorableness and quality of construction, the method of controlling time and temperature was proposed as shown in Table 6. The construction allowable time was determined by marshall molding state and strength developing rule measured in a mass of Marshall Stability tests. The time from mixing epoxy asphalt to the beginning of paving EAC must be more than the shortest construction allowable time, while the time from mixing epoxy asphalt to the finish of compacting EAC must be less than the longest.

Table 6. Construction allowable time of EAC at different temperatures.

Temperature (°C)	Shortest Construction Allowable Time (min)	Longest Construction Allowable Time (min)	Construction Allowable Time Range (min)
110	45	96	51
111	43	91	48
112	41	86	45
113	39	82	43
114	38	79	41
115	37	77	40
116	36	74	38
117	35	71	36
118	35	69	34
119	35	70	35
120	34	69	35
121	34	68	34
122	34	67	33
123	34	66	32
124	34	66	32
125	33	64	31
126	33	63	30
127	33	63	30
128	33	62	29
129	33	62	29
130	33	61	28

3.5. EAC transportation

In order to reduce the time spent for transportation of EAC and guarantee the temperature, some measures were taken, including covering canvas and quilt on the top of lorry, formulating the shortest travel path, arranging dispatchers for lorries.

3.6. EAC paving

To prevent the pollution of lorries on bonding layer which had not paved, the lateral feeding machine was used in paving EAC as shown in Figure 3. Continuous paving with two pavers could be adopted to avoid the longitudinal seam as shown in Figure 4. The distance and speed of two pavers was about 10 m and 2 m/min, respectively.

Figure 3. Lateral feeding machine.

Figure 4. Continuous paving with two pavers.

3.7. EAC compaction

The compaction method of EAC was improved as shown in Table 7. Vegetable oil was smeared constantly on the wheels of rollers when compacting. Moreover, Diversion, head back and emergency brake were forbidden. Artificial hammer and handheld vibration rammer were adopted to compact the pavement where the roller could not compact.

Table 7. Compaction method.

Compaction stages	Preliminary	Next	Final	
Roller kinds	Oller	Steel wheel	Oller	Steel wheel
Compaction times	4	4	4	4
Compaction temperature (°C)	≥82	≥72	≥65	

3.8. Seam cutting

The lateral construction seam was cut at angle of 90°, its position should be the middle of adjacent diaphragm plates.

4. Conclusions

There were no any diseases like cracks occurred on the orthotropic steel deck pavement of Haihe River Chunyi Bridge after in service for three years. The excellent operations quality shows that the EAC with construction technology above has a long service life with superior low temperature crack resistance, high temperature stability, moisture susceptibility and fatigue performance.

The project example shows that the processing measures can effectively solve the technical difficulties in the construction of orthotropic steel deck pavement of Haihe River Chunyi Bridge, can greatly improve the construction speed and quality, and can provide reference for the same kinds of orthotropic steel deck pavement construction.

Acknowledgments

This study was sponsored by the National Natural Science Foundation of China (No. 51178114, 51378122). This funding is greatly appreciated.

References

[1] Li Y, Wu G and Shao L 2005 Application of epoxy asphalt concrete to pavement of long span steel bridge *Central South Highway Engineering* **3** 046-050

[2] Hu J, Qian Z D and Xue Y C *et al* 2016 Investigation on fracture performance of lightweight epoxy asphalt concrete based on microstructure characteristics *Journal of Materials in Civil Engineering* **28** 1-8

[3] Xue Y C and Qian Z D 2016 Development and performance evaluation of epoxy asphalt concrete modified with mineral fiber *Construction and Building Materials* **102** 378-383

[4] Li Z and Chen S 2001 Research and development of steel deck pavement technology *Highway* **1** 2-3

[5] Huang W and Qian Z D 2003 Epoxy asphalt concrete paving on the deck of long-span steel bridges *Chinese Science Bulletin* **48** 2391-2394

[6] Gaul R 1996 Epoxy asphalt concrete: a polymer concrete with 25 years' experience *American Concrete Institute Technical Session Conference* (Minneapolis Minnesota: Traffic Institute Press) **166** 233-252

[7] Cubuk M and Guru M 2009 Improvement of bitumen performance with epoxy resin *Fuel* **8** 1324-1328

[8] Bocci E and Canestrari F 2012 Analysis of structural compatibility at interface between asphalt concrete pavements and orthotropic steel deck surfaces *In Transportation Research Record: Journal of the Transportation Research Board, Transportation Research Board of the National Academies* **2293** 1-7

IOP Conf. Series: Materials Science and Engineering **164** (2017) 012014 doi:10.1088/1757-899X/164/1/012014

[9] Chen L L and Qian Z D 2010 Research on key construction technique of domestic epoxy asphalt concrete *Construction Technology* **39** 74-76
[10] Lu Q and Bors J 2015 Alternate uses of epoxy asphalt on bridge decks and roadways *Construction and Building Materials* **78** 18-25
[11] Wang J W and Shen J L 2009 Construction controlling research of domestic epoxy asphalt mixture *Journal of Southeast University (Natural Science Edition)* **39** 1226-1230
[12] ASTM D445-14a 2014 Standard Test Method for Kinematic Viscosity of Transparent and Opaque Liquids and Calculation of Dynamic Viscosity
[13] ASTM D1652-04 2004 Standard Test Method for Epoxy Content of Epoxy Resins
[14] ASTM D1544-04 2010 Standard Test Method for Color of Transparent Liquids (Gardner Color Scale)
[15] ASTM D1744-13 2016 Standard Test Method for Determination of Water in Liquid Petroleum Products by Karl Fischer Reagent
[16] ASTM D 92-01 2001 Standard Test Method for Flash and Fire Points by Cleveland Open Cup
[17] ASTM D 1475-13 2013 Standard Test Method for Density of Liquid Coatings, Inks, and Related Products
[18] ASTM D 2983-15 2015 Standard Test Method for Low-Temperature Viscosity of Lubricants Measured by Brookfield Viscometer
[19] ASTM D 664-11 2011 Standard Test Method for Acid Number of Petroleum Products by Potentiometric Titration
[20] ASTM D 4402-06 2006 Standard Test Method for Viscosity Determination of Asphalt at Elevated Temperatures Using a Rotational Viscometer
[21] ASTM D 638-10 2010 Standard Test Method for Tensile Properties of Plastics

Numerical simulation of leakage rates of labyrinth seal in reciprocating compressor

H N Tang[1,3], H Yao[2], S J Wang[1], X S Meng[1], H T Qiao[1] and J H Qiao[1]

[1] School of Mechanical Engineering, Shenyang University of Technology, Shenyang, China
[2] AVIC Aerodynamic Research Institute, Shenyang, China

Email: 28775147@qq.com

Abstract. The influence of labyrinth seal structure on leakage behaviour in a reciprocating compressor was addressed in this paper. The effects of the main labyrinth seal parameters, such as tooth angle, sealing clearance, and cavity depth, were compared using FLUENT software and iterative calculation results. Simulations of the sealing process with the influence of internal structure size of labyrinth seal performance in different structures were conducted to explore the characteristics of fluid flow. By comparing the simulations of leakage of fluid-structure interaction and experience formula calculations, the results revealed the validity of the fluid-structure interaction analysis method. The CFD analysis method for fluid-structure coupling was adopted to verify the theory of labyrinth seals and for the design of a labyrinth structure.

1. Introduction

As a kind of non-contact dynamic seal, labyrinth seals are composed of a series of throttling backlash and expansion cavities. A labyrinth seal reduces the leakage rate by increasing the local kinetic energy loss of fluid flow to improve the sealing performance. Widely used in centrifugal compressors, turbine coolers, turbine expansion engines, and other mechanical applications, labyrinth seals do not need lubrication, offer stable and reliable performance, and are convenient [1-4]. Through the variation of cavity expansion, local kinetic energy of the fluid flow is incrementally lost in a labyrinth seal. Currently, due to the strict requirements of the materials, difficulty of manufacturing, high standards of mechanical processing, and assembly accuracy, labyrinth seals are produced by only a small number of companies, such as Swiss Sulzer, throughout the world. The Sulzer company has produced labyrinth compressors since 1935, accounting for about ninety percent of the world market production. Japan Steel Works, Germany's Linde, and only a few other companies produce labyrinth compressors, and their products are mainly used to satisfy the internal demands of their enterprise groups. How to effectively reduce leakage is one of the key technical problems afflicting labyrinth sealing technology. To date, many research projects have made progress on design optimization of labyrinth compressors, but the mechanism of labyrinth seals, which play a decisive role in the production capacity of labyrinth compressors, has not yet been perfected.

Presently, calculating leakage rates is conducted by utilizing FLUENT software to analyze flow fields independently [5-7]. Research methods mainly involve CFD numerical analysis, thermodynamic theory analysis, leakage measurement and flow visualization, and the analysis method combining numerical simulation and experimental analysis. Toff adopted the FDM numerical analysis method to perform numerical simulations on incompressible flow fields with a straight-through labyrinth seal structure, and Versteeg et al. used the same approach [8, 9]. Rhode and Bobolink first simulated [10].

2. Selection of leakage rate algorithm

There are several methods for calculating leakage rates of labyrinth seals including the Martin, Sodalist, Egli, Dumbarton, and Vermes calculation methods. These algorithms, based on specific types of labyrinth seals, are suitable for different applications. Presently, geometric analysis of labyrinth seals is primarily based on two- or three-dimensional flow motion equations, which simplifies the calculation model of seal flow fields. As such, it is not possible to accurately obtain the internal flow field of a seal and pressure field distribution under the interplay of an actual flow field and solid wall. There is interaction between the piston reciprocating movement and gas, which means that the piston reciprocating movements affect the pressure distribution of the gas flow field. Meanwhile, the internal flow field structure is also changed by the gas forces on the piston. Consequently, the pressure distribution of the gas is influenced. Because there is the potential for significant errors if these aspects are studied separately, combining the two problems in fluid-solid coupling can obtain more accurate results.

The flow discipline of leakage fluid inside the seal cavity follows the mass, energy, and momentum conservation formulas. The following general form can be derived by the general variable ϕ [8]:

$$\frac{\partial(\mathrm{p}\phi)}{\partial \mathrm{t}}+\operatorname{div}(\rho \mathrm{u}\phi)=\operatorname{div}(\varphi \operatorname{grad}\phi)+s \tag{1}$$

The formula can be expanded as follows:

$$\frac{\partial(\rho\phi)}{\partial \mathrm{t}}+\frac{\partial(\rho \mathrm{u}\phi)}{\partial \mathrm{x}}+\frac{\partial(\rho v\phi)}{\partial y}+\frac{\partial(\rho w\phi)}{\partial z} =\frac{\partial}{\partial x}\left(\varphi\frac{\partial\phi}{\partial x}\right)+\frac{\partial}{\partial y}\left(\varphi\frac{\partial\phi}{\partial y}\right)+\frac{\partial}{\partial z}\left(\varphi\frac{\partial\phi}{\partial z}\right)+s \tag{2}$$

where ϕ, φ, and S are the general variable, generalized diffusion coefficient, and generalized source term, respectively.

In light of the general equation, the following iterative calculation approach was applied to obtain an equation for theoretical leakage rates of labyrinth seals [9]:

$$G_{\mathrm{i}}=\gamma_{\mathrm{i}}A\sqrt{\frac{2\mathrm{k}}{\mathrm{k}-1}}\frac{P_{\mathrm{i}-1}}{\sqrt{RT_0}}g(\lambda_{\mathrm{i}}) \tag{3}$$

In Eq. (3), i= 1,2,…,n.

Expressly, $\mathrm{g}(\lambda_i)=-3.0495+7.599\lambda_i+(-7.014)\lambda_i^{\,2}+2.945\lambda_i^{\,3}+(-0.462)\lambda_i^{\,4}$ (4)

$$\mathrm{g}(\lambda_i)=-3.0495+7.599\lambda_i-7.014\lambda_i^{\,2}+2.945\lambda_i^{\,3}-0.462\lambda_i^{\,3} \tag{5}$$

$$\gamma_{\mathrm{i}}=\sqrt{\frac{n}{(1-J)n+1}} \tag{6}$$

$$J=1-\left(1+16.6\frac{h}{L}\right)^{-2} \tag{7}$$

$$\lambda_{\mathrm{i}}=\frac{P_{i-1}}{P_i} \tag{8}$$

where A is the cross-sectional area, K is the gas adiabatic index, n is the sealing teeth number; h is the gap width of the sealing tooth mouth, L is the gap of the two adjacent teeth, and λ_{i} is the pressure ratio.

Because the kinetic energy overload effect of a staggered labyrinth seal is smaller, r_{i} was set as 1 directly.

In order to determine which $k-\varepsilon$ turbulence model to use for this research, the calculation results of three $k-\varepsilon$ turbulence models from literature were compared [11]. The results of the standard $k-\varepsilon$ model were determined to be closer to the actual experiment. Therefore, in this paper, the standard $k-\varepsilon$ model was used as the turbulence model.

In the standard $k-\varepsilon$ model, the eddy viscosity model of Reynolds stress was as follows [12]:

$$\tau_{ij} = -\rho \overline{uu_j} = 2u_t\left(S_{ij} - \frac{S_{nn}\delta_{ij}}{3}\right) - \frac{2\rho k\delta_{ij}}{3} \tag{9}$$

$$u_t = c_u f_u \rho k^2 / \varepsilon \tag{10}$$

$$\varepsilon = \frac{\mu}{\rho}\overline{\left(\frac{\partial u_1'}{\partial x_k}\right)\left(\frac{\partial u_1'}{\partial x_k}\right)} \tag{11}$$

where u_t is eddy viscosity, S_{ij} is the strain rate tensor of average velocity, δ_{ij} is the kronecker operator, and K is the turbulent kinetic energy.

In the rectangular coordinate system, the governing equation of the standard $k-\varepsilon$ model is as follows for K equation (turbulent kinetic energy equation):

$$\frac{\partial \rho k}{\partial t} + \frac{\partial(\rho u_j k)}{\partial x_j} = \frac{\partial}{\partial x_j}\left[\left(\mu + \frac{u_t}{\sigma_k}\right)\frac{\partial k}{\partial x_j}\right] + G_k + G_b - \rho\varepsilon - Y_M + S_k \tag{12}$$

The ε equation (turbulent dissipation rate equation) is as follows:

$$\frac{\partial \rho \varepsilon}{\partial t} + \frac{\partial(\rho u_j \varepsilon)}{\partial x_j} = \frac{\partial}{\partial x_j}\left[\left(\mu + \frac{u_t}{\sigma_\varepsilon}\right)\frac{\partial \varepsilon}{\partial x_j}\right] + G_{\varepsilon 1}\frac{\varepsilon}{k}(G_k + c_{\varepsilon 3}G_b) - c_{\varepsilon 2}\rho\frac{\varepsilon^2}{k} + S_\varepsilon \tag{13}$$

3. Analysis of geometrical models

Because of the smaller cavity volume and bigger restriction clearance, the staggered labyrinth seal structure was selected as the study object. In the manufacturing process, the common seal structure materials include metal (such as aluminium or stainless steel), non-metallic materials (such as rubber, ceramic, or graphite), composite materials (such as rubber-asbestos or aerogel felt-polyurethane), and the most widely used material, rubber-like elastic material. The reliability of numerical simulation results depends on the correct application of key technologies. For example, the choice of mathematical model, numerical solution method, boundary conditions, and grid generation are all factors in obtaining accurate results. Five different sizes of tooth profile angles for staggered labyrinth seal structures, including 15°, 30°, 45°, 60°, and 75°, were studied. The piston velocity was set as v = 4.25 mm/s, seal clearance was H, and space widths were B_1 and B_2 while the cavity depths T_1 and T_2 were set as constants. The geometric parameter model and parameter sizes are listed in Figure 1 and Table 1, respectively. The solid model and 3D solid domain mesh of a piston are shown in Figure 2, and Table 2 lists the cylinder boundary conditions.

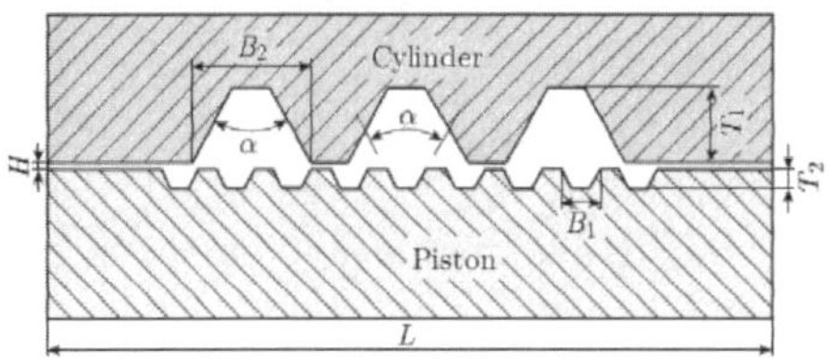

Figure 1. Geometric parameters of the model.

The boundary conditions of the solid surface were set without the slippage or permeation and acted as the adiabatic wall. The kinetic fluid field adopted the spring-based smoothing and reconstruction model of the local grid to renovate the dynamic mesh. Gambit 2.0 was used to generate the unstructured grids in the labyrinth channels. However, the local grid reconstruction model can only be used for a tetrahedron grid or triangular grid, so the former was used for the 3-D fluid domain and solid structure grid.

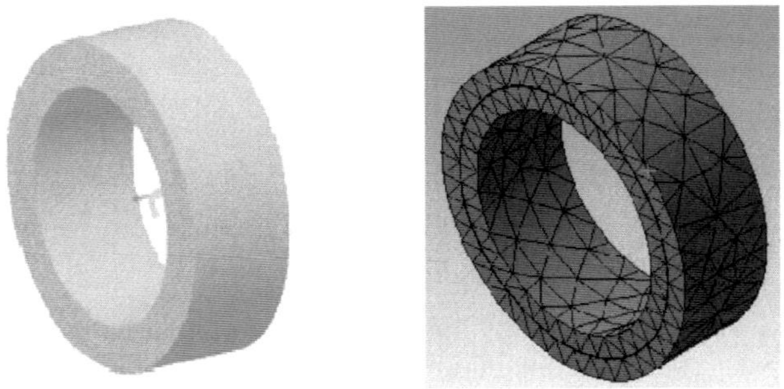

a. Solid model b. 3D solid domain mesh

Figure 2. Models of piston and cylinder.

Table 1. Geometrical parameters of the sealed cavity.

Cylinder teeth $/n_1$	Piston teeth $/n_2$	Seal clearance $/H$ (mm)	Cylinder cavity depth $/T_1$ (mm)	Piston cavity depth $/T_2$ (mm)	Sealing length $/L$ (mm)	Piston cavity width $/B_1$ (mm)	Cylinder cavity width $/B_2$ (mm)	Piston radius $/R$ (mm)
3	9	0.4	4	1	40	2.15	6.5	50

Table 2. Boundary conditions of the model.

Working medium	Specific heat at constant pressure J/(kg·K)	Thermal conductivity W/(m·K)	Inlet of model	Outlet of model	Temperature (K)	Behavior of gas
Ideal gas	1006.43	0.0242	Given total pressure	Backpressure environment	300	Compressible gas

4. Analysis of calculation results

4.1. Tooth angle influence on leakage rate

Tooth shape is an important parameter affecting labyrinth seal structure, and the most common cavity tooth form of labyrinth seals in engineering are the triangle labyrinth seal, trapezoidal seal tooth, rectangular seal tooth, and stepped seal tooth. The trapezoidal labyrinth seal tooth, offering outstanding performance and simple machine processing, has been widely applied in engineering fields. Focused on the isosceles trapezoid tooth, the five different sizes of labyrinth seal structure teeth of 15°, 30°, 45°, 60°, and 75° were respectively discussed. With simulation analyses, Figures 3-7 illustrate the turbulence kinetic energy and total displacement nephogram for each tooth angle of the 3-D flow field. It can be seen that the amount of structure deformation showed an initial decline, followed by an increase with the degree of tooth angle. When the configuration size was fixed, the formation of the cavity eddy current was facilitated with the degree of the tooth angle. In the shear at the junction between the jet flow and backflow, the turbulence kinetic energy also showed an increasing trend, and the energy conversion was subsequently enhanced. It was observed that the strongest whirl vortex appeared in the expansion cavity at the seal tooth profile angle of 60°. The energy dissipated completely, and the seal leakage rate was the lowest. At the extreme tooth profile angle, the conversion between the kinetic energy and heat energy in the expansion cavity was relatively weak, and the leakage rate increased significantly at that moment. In conclusion, both the energy dissipation of the subject vortex in the expansion cavity and energy exchange at the throttling gear had significant effects on the improvement of the seal leakage rate, and the former was more apparent. Different tooth profile angles of labyrinth seal leakage are listed in Figure 8 along with simulation and iterative calculations.

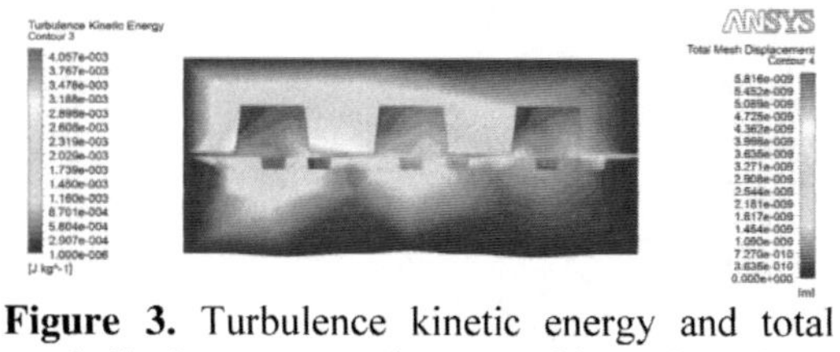

Figure 3. Turbulence kinetic energy and total mesh displacement nephogram with tooth angle of 15°.

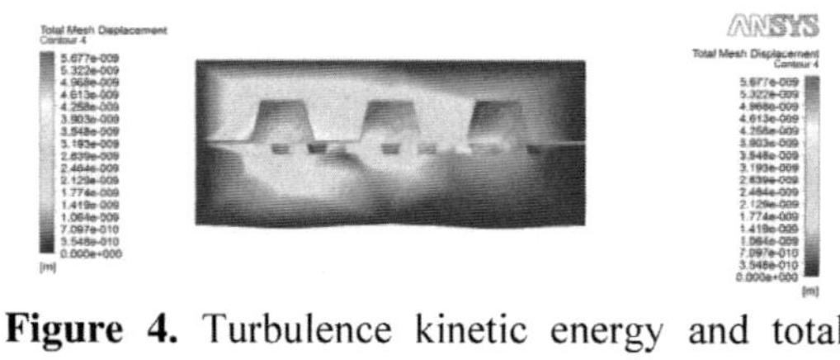

Figure 4. Turbulence kinetic energy and total mesh displacement nephogram with tooth angle of 30°.

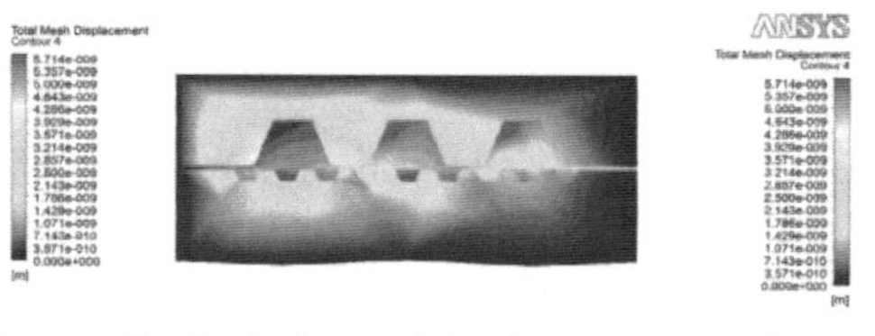

Figure 5. Turbulence kinetic energy and total mesh displacement nephogram with tooth angle of 45°.

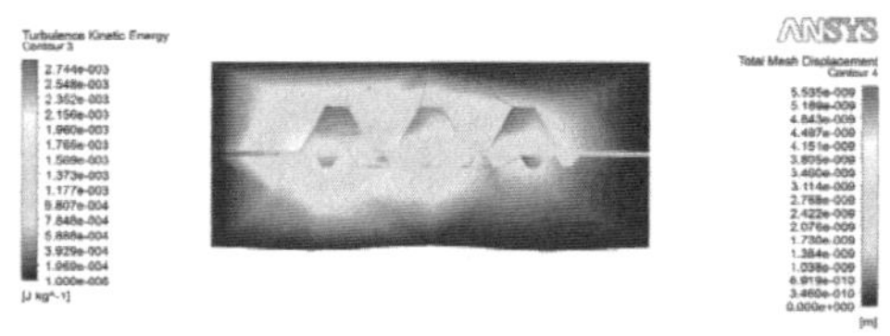

Figure 6. Turbulence kinetic energy and total mesh displacement nephogram with tooth angle of 60°.

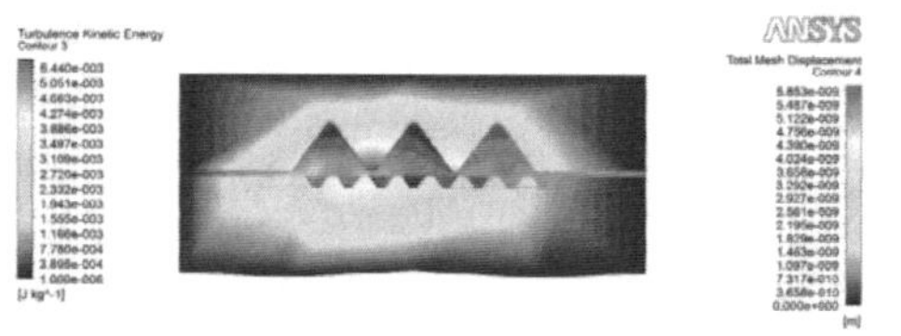

Figure 7. Turbulence kinetic energy and total mesh displacement nephogram with tooth angle of 75°.

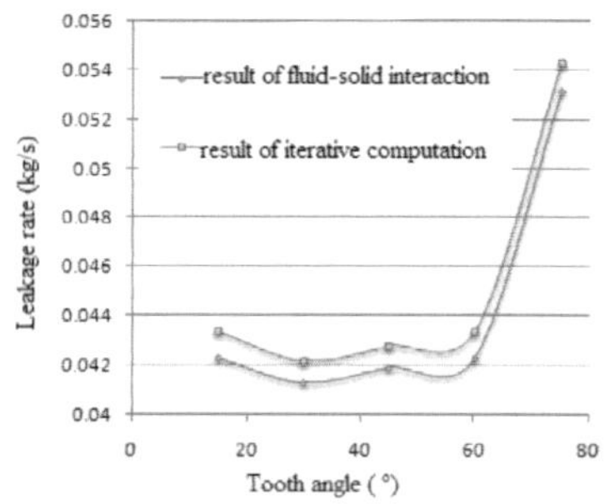

Figure 8. Leakage rate of labyrinth seal with the degree of tooth angles.

When the tooth angle was between 15° and 30°, the leakage decreased gradually, and between 30° and 60°, the leakage increased gradually. However, from 60° to 70°, the leakage increased sharply. The leakage was the lowest at the tooth profile angle of 30°. The major forms of fluid flow within the seal chamber were the main vortex and the jet, which were important factors resulting in energy loss. Energy loss was also related to the formation of small local vortexes in the shear layer of the strong jet and the merging primary vortex, jet impingement against the sealed wall, changes of jet direction, and other factors. When the jet flowed through the orifice, deflection flows directly impacted on the cavity wall, which caused kinetic energy loss; jet branches were thus produced. At the appropriate tooth profile angles, the jet velocity in the seal clearance decreased gradually with the degree of tooth angle, and most of the fluid energy dissipated completely after the long seal clearance. If the tooth angle was too wide, the flow velocity was accelerated into the cavity and the corner position, so the leakage rate decreased.

4.2. Comparison of leakage rate with the sealing clearance

The size of the sealing clearance has significant effects on labyrinth seals. For labyrinth compressors, too-wide seal clearance increases leakage, and thus reduces the efficiency of the compressor. Too-narrow seal clearance gives rise to difficulty in the manufacturing process, and the material performance requirements of labyrinth pistons are also strict. Figures 9-13 illustrate the turbulence kinetic energy and total mesh displacement nephogram for different sealing clearances.

As the figures demonstrate, labyrinth seal leakage increased with the widening of sealing clearance. Because the wider sealing clearance affected the resistance of the labyrinth seal, the ability of gas to convert pressure energy to kinetic energy was weakened, and the speed at which it entered the seal cavity in jet form dropped. With the turbulent kinetic energy of the gas in sealed cavity weakened, the energy dissipation of the gas decreased, and consequently, the leakage rate increased.

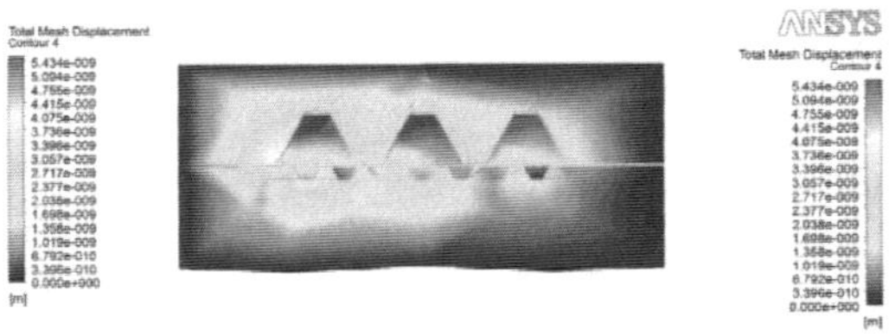

Figure 9. Turbulence kinetic energy and total mesh displacement nephogram (gap of 0.3 mm).

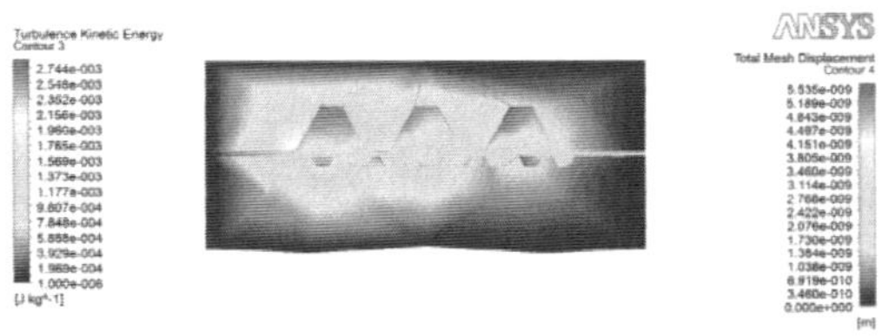

Figure 10. Turbulence kinetic energy and total mesh displacement nephogram (gap of 0.4 mm).

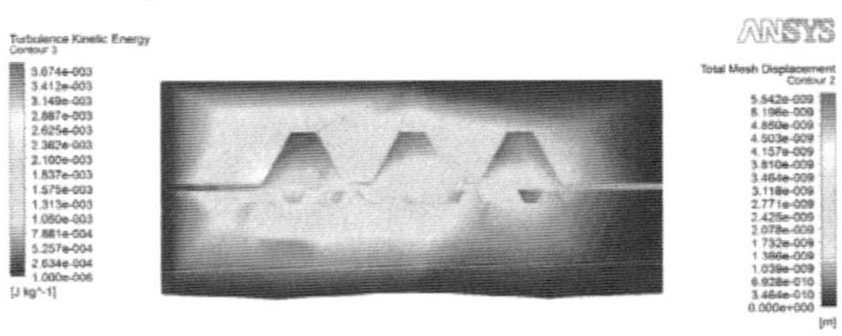

Figure 11. Turbulence kinetic energy and total mesh displacement nephogram (gap of 0.5 mm).

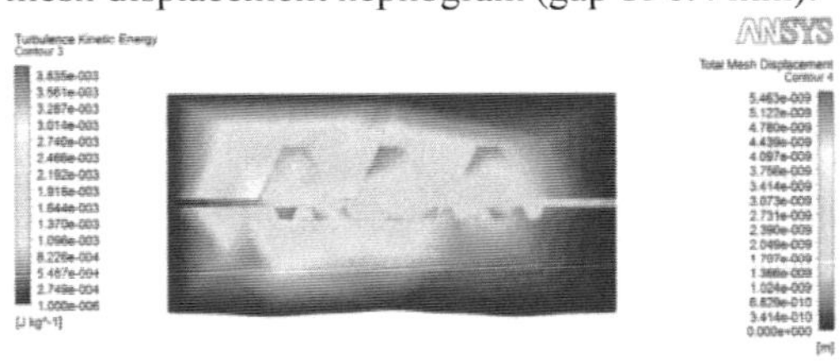

Figure 12. Turbulence kinetic energy and total mesh displacement nephogram (gap of 0.6 mm).

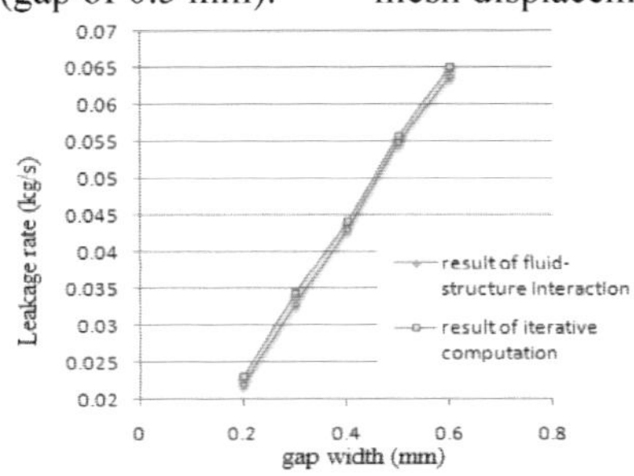

Figure 13. Labyrinth seal leakage for different sealing clearances.

4.3. Effects of leakage rate on the cavity depth

The configuration of fluid load can be affected by cavity depth, which changes the fluid leakage rate and the sealing effect. A smaller depth not only causes difficulty in the manufacturing process, but the performance requirement of labyrinth piston material is becoming more strict. Figures 14-19 illustrate pressure and von mises stress nephogram with different cavity depths. With a gradual increase of cavity depth, the wall deformation of the seal increased at first, then decreased to a certain point, and lastly, demonstrated a gradual increasing trend. It can be seen that the minimum leakage rate occurred at the cavity depth of 3.5-4 mm, during which time optimal sealing effect was obtained. When the airflow through the throttling cavity, mainly by direct jet, and the flow with a certain angle entered the sealed cavity, a vortex inside the sealed cavity was generated, exacerbating the conversion from kinetic energy into heat energy, and causing the energy to dissipate. For a smaller cavity depth, most of the fluid was in the form of a jet flow, which did not easily form a vortex, and as a result, the cavity fluid was rendered nearly motionless.

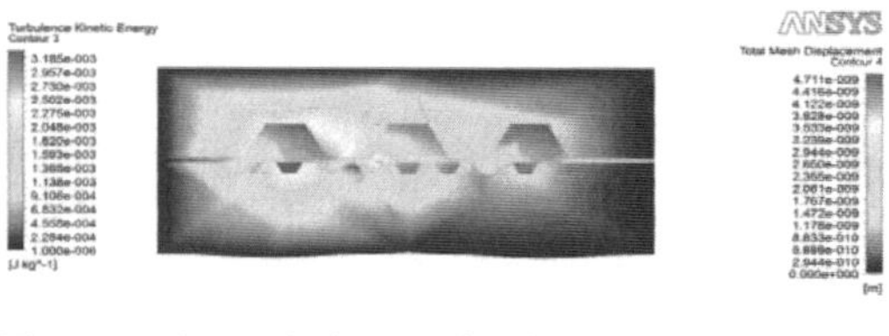

Figure 14. Turbulence kinetic energy and total mesh displacement nephogram (cavity depth of 3 mm).

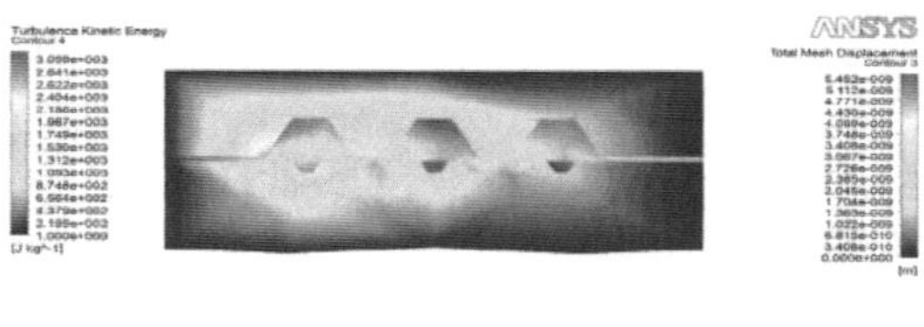

Figure 15. Turbulence kinetic energy and total mesh displacement nephogram (cavity depth of 3.5 mm).

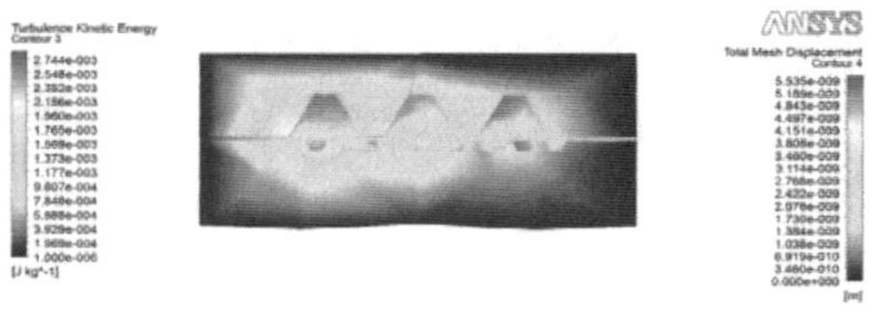

Figure 16. Turbulence kinetic energy and total mesh displacement nephogram (cavity depth of 4 mm).

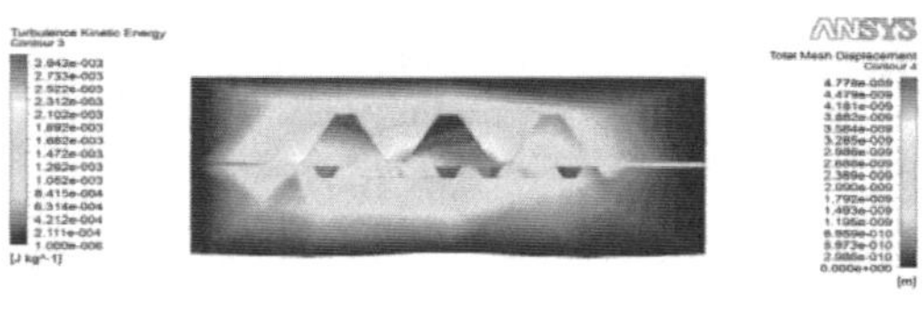

Figure 17. Turbulence kinetic energy and total mesh displacement nephogram (cavity depth of 4.5 mm).

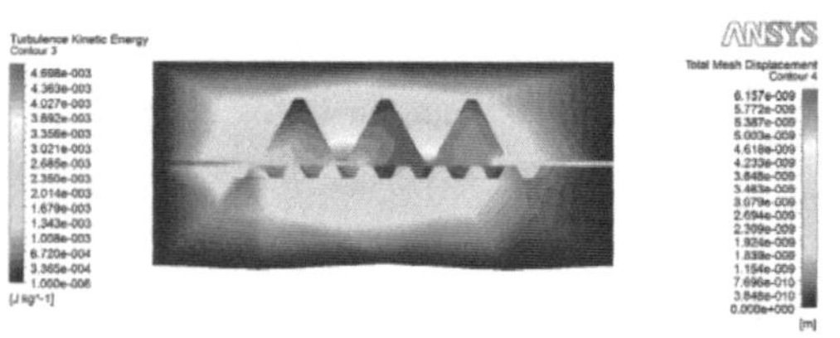

Figure 18. Turbulence kinetic energy and total mesh displacement nephogram (cavity depth of 5 mm).

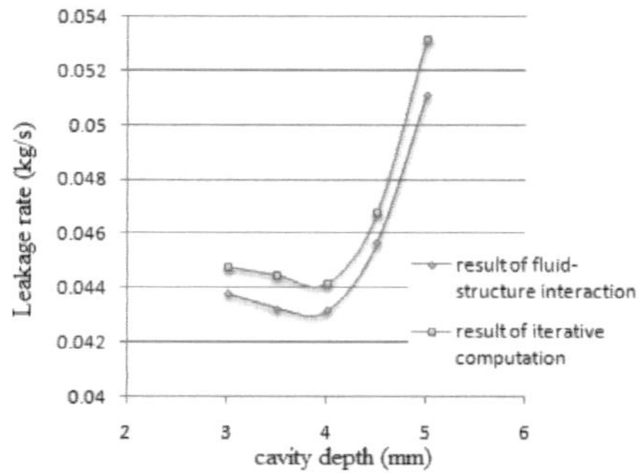

Figure 19. Labyrinth seal leakage for different cavity depths.

5. Conclusion

A numerical simulation model was established in this paper using the CFD analysis method, and the effects of the main labyrinth seal parameters, including tooth angle, sealing clearance, and cavity depth, were researched. After comparing the leakage rates between the fluid-structure interaction simulation and the experience formula calculations, the results revealed the validity of the fluid-structure interaction analysis method for labyrinth seals. In the future, the fluid-structure coupling can be adopted to verify the theory of labyrinth seals and to design labyrinth structures. The conclusions of the present study were drawn as follows:

(1) With the degree of tooth angle, the value of the leakage rate had a tendency to decrease at first, and then increase, according to fluid-solid coupling analysis. The lowest seal leakage rate occurred with a tooth profile angle of 60°.

(2) Labyrinth seal leakage increased with the increase of sealing clearance width. A smaller sealing clearance width should be adopted, but at the same time, reducing the difficulty of the manufacturing process should also be considered.

(3) For this type of reciprocating compressor, the minimum leakage was obtained at the cavity depth of 3.5 to 4 mm, when the sealing effect was the best.

Acknowledgement
This study was supported by Program for Collaborative Innovation Center of Major Machine Manufacturing in Liaoning, National Natural Science Foundation of China (Grant No. 61573249), and National Natural Science Foundation of China (Grant No. 51505298).

References

[1] Larjola J, Honkatukia J and Sallinen P *et al* 2010 Fluid dynamic modeling of a free piston engine with labyrinth seals *Journal of Thermal Science* **19**(2) 141-147
[2] Wu D Y 2000 Analysis and comparison of two commonly used labyrinth seal and critical features *Journal of Aerospace Power* **12**(4) 397-400
[3] Hu D X, Jia L and Yang L X 2014 Dimensional analysis on resistance characteristics of labyrinth seals *Journal of Thermal Science* **23**(6) 516-522
[4] Wang S F, Lv H F and Xia D Y 2004 Numerical analysis and experimental research of seal labyrinth structure features *Journal of Nanjing University of Aeronautics and Astronautics* **36**(6) 732-735
[5] Huang S L, Wang Z Y and Ma D W 2000 Flow analysis influence of cavity inclination on labyrinth leakage *Journal of Propulsion Technology* **16**(2) 40-45
[6] Liu Y J 2002 Study on sealing mechanism of a crenellated radial labyrinth seal *Lubrication Engineering* **154**(6) 20-21
[7] Li L and Liu W H 2007 Study of mechanism of labyrinth seal based on FLUENT *Chinese Mechanical Engineering* **18**(9) 2183-2186
[8] Versteeg H K and Malalasekera W 1995 *An Introduction to Computational Fluid Dynamics: The Finite Volume Method* (New York: Wily) pp 180-182
[9] Demko J A, Morrison G L and Rhode D L 1989 The prediction and measurement of compressible flow in a labyrinth seal *ASME Journal of Engineering Gas Turbines Power* **111**(4) 697-702
[10] Rhode D L and Sobolik S R 1986 Simulation of subsonic flow through a generic labyrinth seal *ASME Journal of Engineering Gas Turbines Power* **108** 674-680
[11] Zhu G T and Liu W H 2006 Analysis of calculational methods on leakage for labyrinth seals *Lubrication Engineering* **31**(4) 123-126
[12] Lin L 2007 Theory Study of the seal mechanism of labyrinth piston compressor Nanjing University of Aeronautics and Astronautics

IOP Conf. Series: Materials Science and Engineering **164** (2017) 012016 doi:10.1088/1757-899X/164/1/012016

Gas metal arc welding in refurbishment of cobalt base superalloys

M S Shahriary[1], Y Miladi Gorji and A M Kolagar

Repair & Product Engineering Department, Mavadkaran Eng. Co., 7th Km Malard Road, MAPNA Blvd. Karaj, Iran

Email: Shahriari.saeed@mapnamk.com

Abstract. Refurbishments of superalloys which are used in manufacturing gas turbine hot components usually consists of removing cracks and other defects by blending and then repair welding in order to reconstruct damaged area. In this study, the effects of welding parameters on repair of FSX-414 superalloy, as the most applicable cobalt base superalloy in order to manufacture gas turbine nozzles, by use of Gas Metal Arc Welding (GMAW) technic were investigated. Results then were compared by Gas Tungsten Arc Welding (GTAW). Metallographic and SEM studies of the microstructure of the weld and HAZ showed that there are no noticeable defects in the microstructure by use of GMAW. Also, chemical analysis and morphologies of carbide in both methods are similar. Hardness profile of the GMAW structure then also compared with GTAW and no noticeable difference was observed between the profiles. Also, proper tensile properties, compared with GTAW, can be achieved by use of optimum parameters that can be obtained by examining the current and welding speed. Tensile properties of optimized condition of the GMAW then were compared with GTAW. It was seen that the room and high temperature tensile properties of the GMAW structure is very similar and results confirmed that changing the technic did not have any significant influence on the properties.

1. Introduction

After developing some tradition kind of cobalt-based superalloys such as X-40, X-45 and Mar-M 509, an advanced type of them was patented by General Electric (GE) Company in 1968-69, which is called FSX-414 that always used for manufacturing of many types of gas turbine nozzles [1]. Because of its chemical composition, corrosion resistance of FSX-414 superalloy is so high that is used without protective coatings in manufacturing of gas turbine component. Since FSX-414 alloy has good welding capability and this advantage give us to correct its defects like casting defects or service conditions defects by welding [2]. Gas turbine nozzles, similar to other parts of turbine, because of severe condition of operation may degrade and many types of defects can be occurred on them. Therefore, they should be repaired. Welding is one of the most common methods for repair of the turbine components. Welding is used for both defect repair and manufacturing. Among many methods of welding, Gas Tungsten Arc Welding (GTAW) is the most important technic. A pre-treatment should be performed before welding to prepare the structure of superalloy. Solution treatment is usually done at temperatures around 1200°C for some hours. In addition to GTAW, Gas Metal Arc Welding (GMAW) is an advanced technic that is widely used for repairing of ferrous and non-ferrous industrial components and materials [3]. Widespread usage of the GMAW is related to high flexibility of the process and the potential for performing automatic and robotic [4]. In recent years, Using of GMAW in gas turbine industry is increasing. The reason is to easy use of GMAW in all positions, being multipurpose and applicable. Weld quality depends to some different parameters such as current, feeding rate, welding speed, protective gas and also welding pool that can be affected by deposition rate. Heat transfer and fluid current in GMAW weld pool is similar to GTAW and in recent decades is widely studied. In GTAW process, metal deposition in weld pool can achieve a flat surface. In contrast with GTAW, in GMAW process, collision of droplets to pool can create free surface and deformed pool affects the distribution of arc current and heat transfer from arc to pool [5].

In previous studies on GMAW, it was seen that in addition to welding parameters, like current, voltage, welding speed and filler material, there are some other parameters that can influence on the welding quality: composition of protective gas and welding technic. High rate of deposition in GMAW can accelerate the repair rate of damaged components. Since rapidity of parts delivery to power plant is one the most important factors of repair, it may be interesting for repair service companies that use GMAW as repair welding technic instead of or in addition to GTAW.

Therefore, in this study, optimization of GMAW method in repair welding of gas turbine nozzles, which is made of FSX-414 cobalt base superalloy, was evaluated and the results was compared with GTAW.

2. Experimental

As mentioned before, the aim of this study is to optimize the GMAW process specification and also comparison of the result with GTAW on the FSX-414 cobalt base superalloy with nominal composition of 53Co 29Cr 10Ni 7W 1Fe 0.25C 0.01B. Because in most cases of repair the surfaces should be reconstructed, an empty pool, as shown in Figure 1, was designed to assure that the volume of repair in the experiments are equal. To fill this, GTAW was performed according to its usual welding process specification, WPS, (Table 1) by use of FSX-414 as filler material with diameter of 1.2mm. GMAW parameters that were studied in this study were current and welding speed. The design of experiments is shown in Table 2.

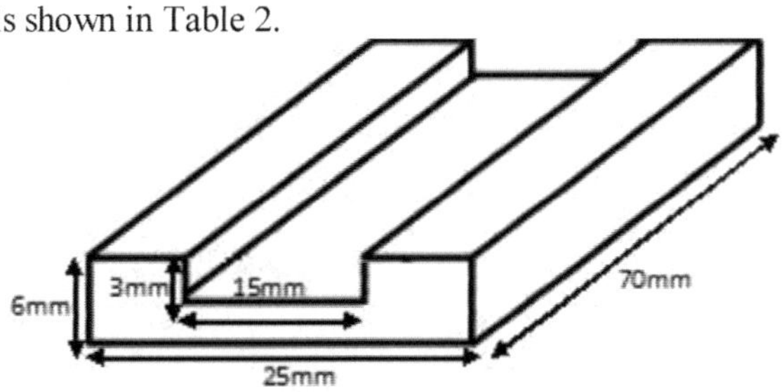

Figure 1. Schematic illustration of weld pool.

Table 1. WPS of GTAW used in this study.

Current (A)	Intermediate Temp. (°C)	Welding Speed (mm/min)	Voltage (V)	Sample Code
80-90	50-70	7-11	10-13	T1

Table 2. Design of experiments for GMAW.

Current (A)	Welding Speed (mm/min)	Code
85	14-16	M1
100		M2
85	24-26	M3
100		M4

Samples then were subjected to optical microscopy, and hardness test. Hardness test was performed according to ASTM E92 standard. SEM observations to analysis of the microstructure were performed using Philips-Tescan model. Finally, to study the mechanical properties of the samples, room and high temperature tensile test were done on the samples. Sub size samples that are proposed in an MSc thesis [6] was used to make samples. To achieve tensile samples, slabs with two similar pools, which are shown in Figure 2, were made and standard samples were taken off (Figure 3). Tensile test were performed in all weld condition. The thickness of the samples was around 1mm [6].

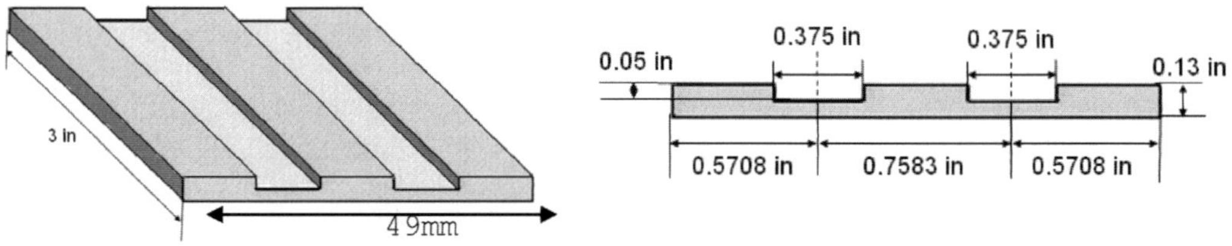

Figure 2. Dimensions of weld pool for tensile test.

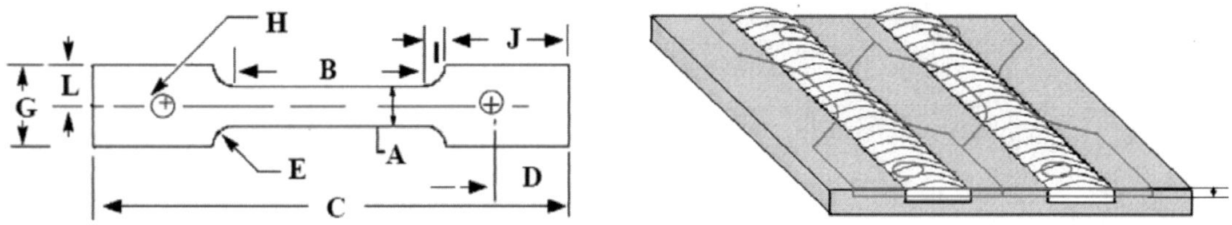

Figure 3. All weld tensile samples used in this study.

3. Results and discussion

3.1. Metallographic examinations

Figure 4 shows the optical microstructure of the weld and its interface with base material after GTAW. Also, because of good weld ability of the alloy, heat affected zone (HAZ) and crack was not observed.

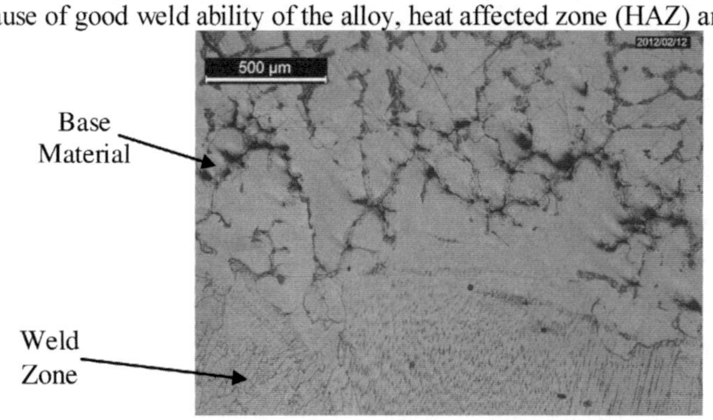

Figure 4. Microstructure of the FSX-414 GTAW.

Figure 5 (a) to (d) shows the microstructure of GMAW of the alloy in different conditions according to Table 2. A good condition in microstructure in terms of cracks, HAZ and also grain structure confirmed that high deposition rate and changing welding technic does not affect the microstructure, when comparing with GTAW (Figure 4).

As can be seen, GMAW welding parameters can affect the microstructure characteristics, but features that can reject the welding are not observed. Welding zone microstructure contains carbide precipitates and by use of cobalt base filler, microstructural characters of cobalt alloys, uniform dendritic structure and precipitates distribution, without any major defects in comparison with GTAW microstructure can be achieved.

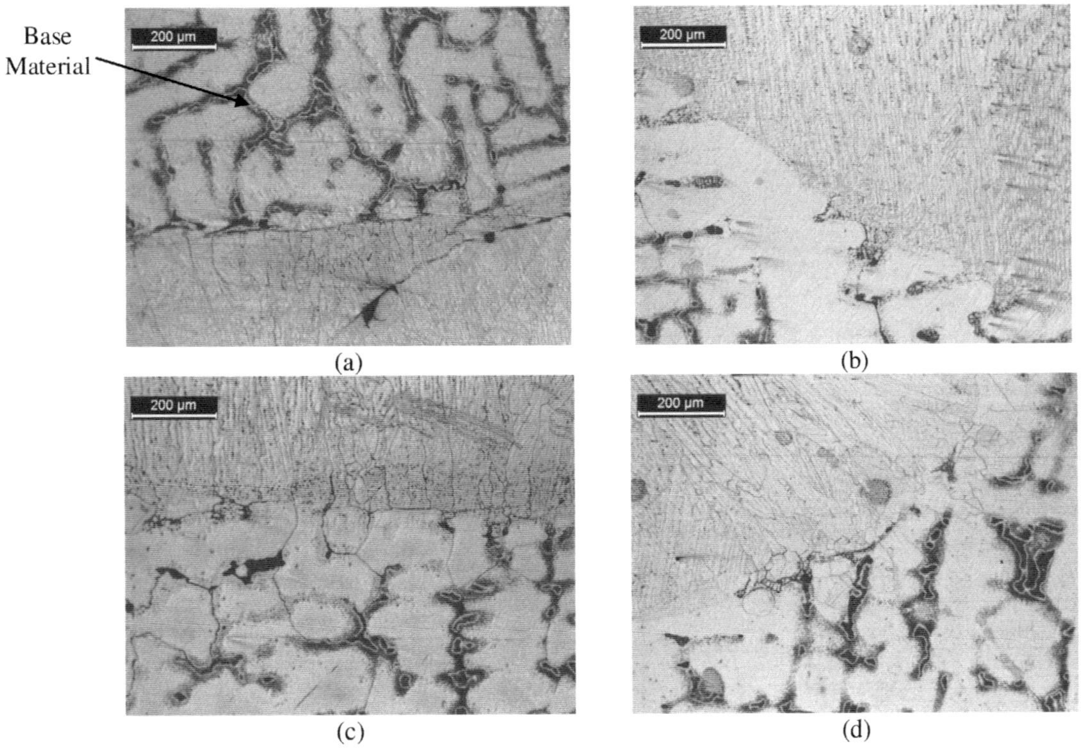

Figure 5. Microstructure of the GMAW of (a) M1, (b) M2, (c) M3 and (d) M4.

3.2. Hardness profile

Figure 6 shows the hardness profile of the GTAW sample (T1) as Vickers hardness measurement versus distance from weld center line. As seen, because of similarity of filler and base material, significant difference between them was not detected. Also, it can be seen that in some distances, hardness of the filler is higher than base alloy. This was referred in previous studies [7].

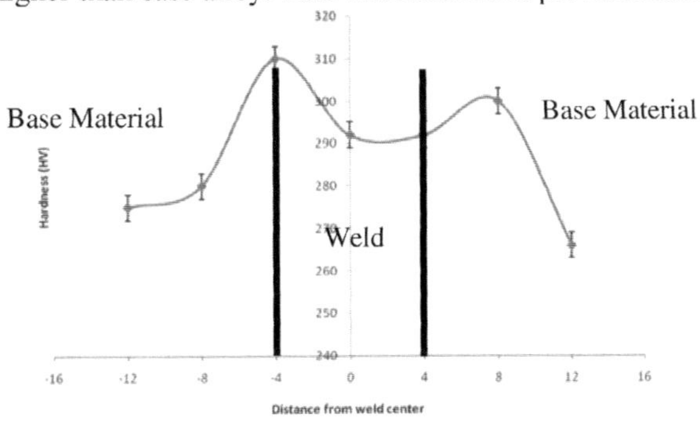

Figure 6. Hardness profile of GTAW microstructure.

Figure 7 illustrates the hardness profile of the GMAW microstructure in different welding conditions. Lines in graphs separate the weld zone and base material. In comparison with center line of the weld and adjacent zones, this is clear that decrement in hardness properties is not significant and this again confirms that replacement of GTAW with GMAW does not affect the weld mechanical properties. It can be observed that by use of high amount of current (samples M2 and M4), hardness profile of the weld and base microstructure has lower difference and increment of welding speed can improve the properties. Also, high welding speed can reduce heat input to the parts and high amount of current can form suitable fusion without any significant microstructural defects in GMAW. According to the above explanations, one can optimize the GMAW technic by use high amount of current and welding speed (M4 sample). In next part, tensile properties of T1 sample (GTAW) and M4 sample (optimized GMAW) will be compared.

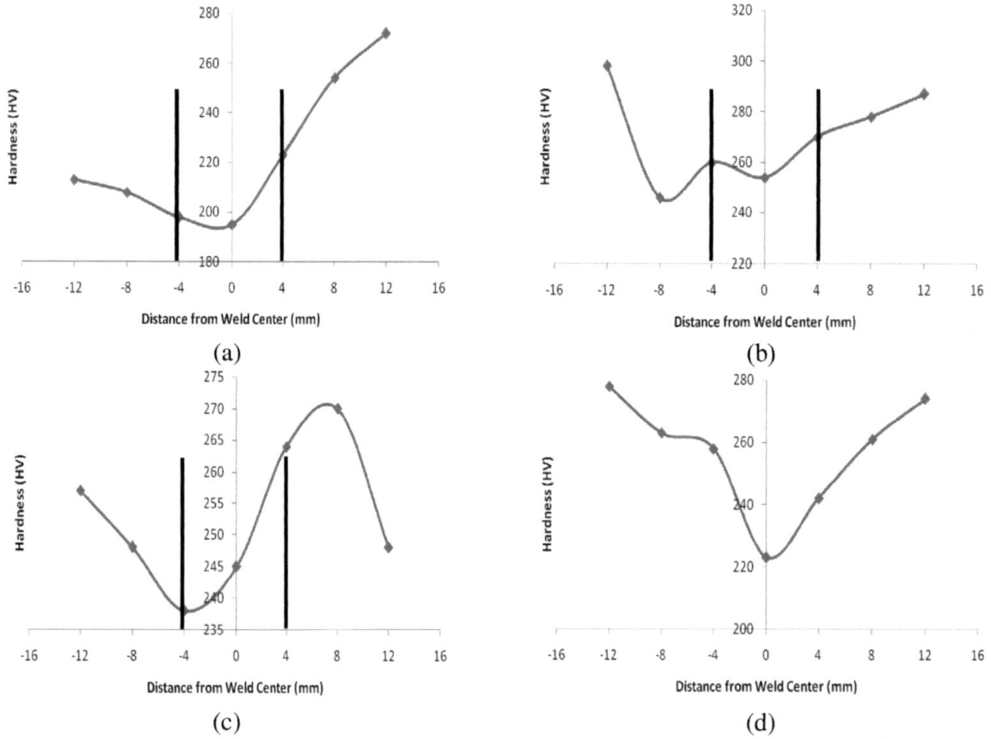

Figure 7. Hardness profile of GMAW of (a) M1, (b) M2, (c) M3 and (d) M4.

3.3. Tensile properties

Figures 8 and 9 show the room and high temperature (600°C) tensile properties of the T1 and M4 samples. The results of mean value of three samples were also summarized in Tables 3 and 4. As mentioned before, mechanical properties of the samples are comparable and there is no significant difference between their properties. Suitable joint between filler and base material, absence of cracks in the microstructure and also margin size of HAZ in the microstructure can form proper mechanical properties. It is noteworthy that high temperature tensile properties of GMAW sample are much better than the GTAW sample.

Table 3. Room temperature mechanical properties of GTAW and GMAW samples.

Code	YS (MPa)	UTS (MPa)	Elongation (%)
T1	496	666	13.2
M4	488	648	16.1

Table 4. High temperature mechanical properties of GTAW and GMAW samples.

Code	YS (MPa)	UTS (MPa)	Elongation (%)
T1	250	350	22.2
M4	275	525	45.5

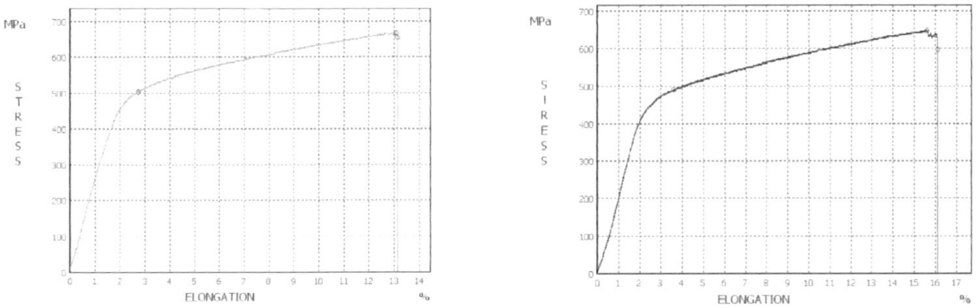

(a) (b)

Figure 8. Room temperature tensile properties of (a) T1 and (b) M4 samples.

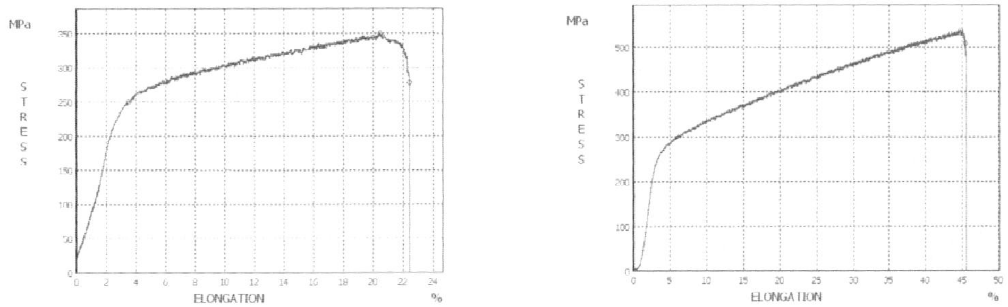

(a) (b)

Figure 9. High temperature tensile properties of (a) T1 and (b) M4 samples.

3.4. SEM observations

Figure 10 shows the SEM microstructure of weld and HAZ of the T1 sample. As seen in Figure 4, there is no noticeable defect in the microstructure. Carbides are created and distributed normally in the structure. Figure 11 shows the EDS analysis of the carbide in the weld zone. Arrow in Figure 10 shows the position of analysis. Also, Table 5 shows EDS analysis results of carbide. The results confirmed that $Cr_{23}C_6$ carbides are formed in the microstructure. Figure 12 shows the microstructure of the optimized GMAW sample (M4) and again no significant defects were observed in the microstructure. Similarly, Arrow in Figure 12 shows the position of EDS analysis. EDS analysis of the M4 sample confirmed that changing the technic and deposition rate do not influence on the formation of carbide's type (Figure 13).

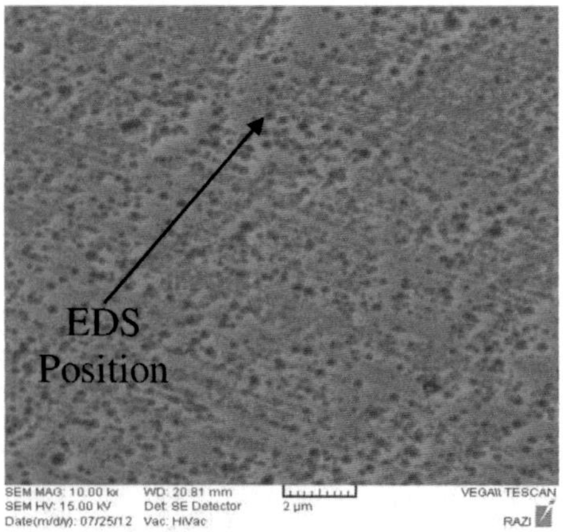

Figure 10. SEM micrograph of the GTAW sample (T1) MAG 10000.

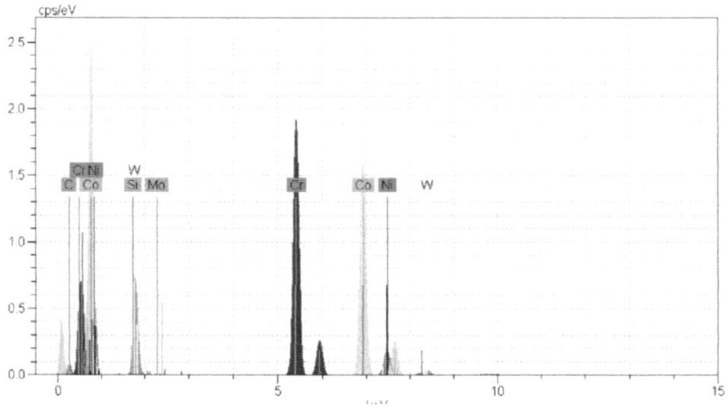

Figure 11. EDS analysis of the carbide in the weld zone of T1 sample.

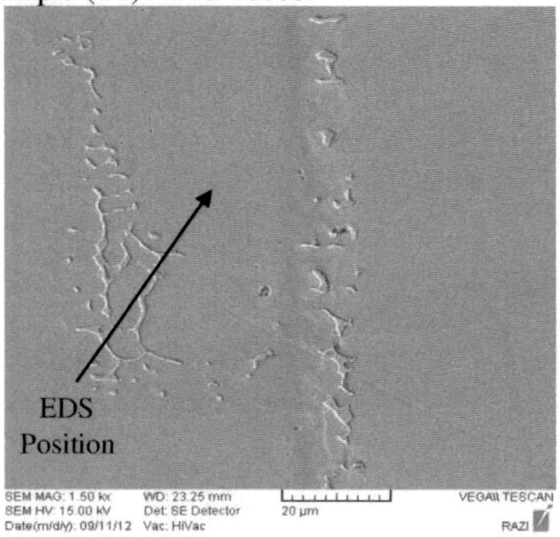

Figure 12. SEM micrograph of the GMAW sample (M4) MAG 1500.

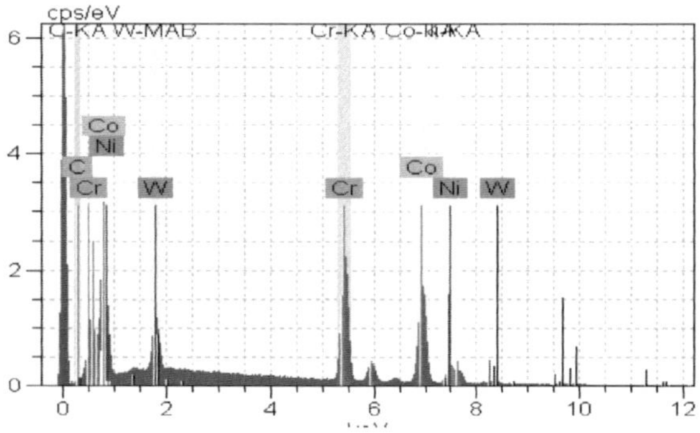

Figure 13. EDS analysis of the carbide in the weld zone of M4 sample.

Table 5. EDS analysis results of carbide in T1 sample.

Element	Series	Atomic Percent
Carbon	K	3.69
Chromium	K	32.94
Cobalt	K	54.01
Nickel	K	7.57
Tungsten	L	1.80
Total		100

4. Conclusions

Weld pools (samples) of FSX-414 cobalt base superalloy, with normal GTAW and four different cases of GMAW method are built up. By comparing the results of GMAW results, first process was optimized. The results were then compared with GTAW, as the most common technic of repair welding of gas turbine nozzles. The Results were summarized below:

1. Optical and SEM observations showed no significant defects like cracks, wide HAZ, etc. in the GTAW and GMAW microstructures.

2. Hardness profile of the GMAW samples showed that increasing the current can create more uniform profile it can improve the microstructural properties. So, welded sample with higher current and welding speed was selected as optimized GMAW sample.

3. Comparison of room temperature mechanical properties of the GTAW and GMAW samples showed that tensile properties of the samples are very close and also high temperature tensile properties of the GMAW sample is better.

4. These results confirm that GMAW with higher deposition rate can be a good method to replace GTAW in repair of superalloys without any significant decrease in properties.

5. Because of higher deposition rate in GMAW, one can improve the rapidity of repair cycle of the superalloys and reduce the defects.

Acknowledgements
The authors would like to show their appreciation for MAPNA Group's support in this research.

References

[1] Mighan M, Samadi A and Hosseinian H 2006 *Proc. Int. Conf. on Power System* (Tehran: Tavanir Company) p 118
[2] Balsone S J 2006 *Buckets and Nozzles, The Gas Turbine Handbook* Section 4.4.1 (U.S.: Department of Energy, National Energy Technology Library) p 411
[3] Parveen P, Yarlagadda P K D V and Kang M J 2005 *Mater. Proc. Tech.* **164-165** 1113
[4] Quintino L and Miranda R M 2007 *Mater. Des.* **28** 1623
[5] Hu J and Tsai H L 2007 *Intl. J. Heat Mass Trans.* **50** 808
[6] Hasselberg T P 2009 MSc Thesis of Rensselaer Polytechnic Institute, Hartford, Connecticut
[7] Jahangiri M R and Fallah A A 2011 *Mater. Des.* **32** 424

An investigation on the behaviour and stiffness of reinforced concrete slabs subjected to torsion

M C T Nguyen[1,2] **and P T Pham**[1, 3]

[1] Faculty of Civil Engineering, Hanoi Architectural University, Km 10 Nguyen Trai Street, Thanh Xuan District, Hanoi, Vietnam
[2] Faculty of Engineering and Technology, Quy Nhon University, 170 An Duong Vuong Street, Quy Nhon, Vietnam

Email: phamphutinh@yahoo.com

Abstract. This paper presents an investigation on RC slab under torsion, by both experiment and finite element analysis. The torsion tests were done on three similar square RC slabs with dimensions of 1900×1900×150 mm. The behaviour of slabs at pre-cracking and post-cracking of concrete phases were investigated, via *Load-displacement*, *twisting moment-curvature* relationships, and torsional stiffness of slabs. The experimental results are compared with the FEA and the results in literatures. The torsional stiffness of slab at the phase of concrete cracked and steel yield is about 1/25 of the stiffness at the pre-cracking phase.

Keywords: Torsional stiffness, RC slab, FEA.

1. Introduction

Reinforced concrete slabs are widely used in the concrete constructions. In structural analysis, the torsional stiffness of slab is ignored in common. When this stiffness is taken into account, the exact theory of bending of elastic plates shows that the twisting moment relieve the bending moments about 25 percent [1, 2].

Among the researchers on torsional stiffness of RC slabs, there were Nielsen [3], Marti and Kong [4]. Their study showed the analytical solution for torsional stiffness of RC slab at pre-cracking and post-cracking of concrete phases. Gudmand-Høyer [5] has investigated the torsional and bending stiffness of RC slab under flexural and membrane effect. All of these researches presented the formulae, which consists of elastic modulus of concrete and steel, thickness of slab, and steel ratio. At pre-cracking phase, torsional stiffness of RC slab are quite close to torsional stiffness of elastic and homogeneous slab. Huber introduced the torsional stiffness of RC slab based on bending stiffness, but his formula is only suitable for pre-cracking phase, this formula was cited in Timoshenko [1].

By experimental study, Marti et al. [6] had tested nine RC slabs under pure torsional moment in the reinforcement directions. Torsion was introduced by two couples of equal and opposite forces, applied at four corners of specimens. Extensive strain on the specimen surfaces and displacement at slab corners were measured. Crack spacings, crack widths, magnitudes of strains, collapsed mechanisms, nonlinear responses of load-deformation, moment-average strains in reinforcement were shown. The ultimate moments from Marti's tests were between 5-46% below yield-line theory predictions, between 8-117% greater than American code predictions, 3-41% greater than Canadian code predictions, these codes are applied to wide beams.

Lopes et al. [7] based on the test model of Marti with some modifications. Torsion was introduced by one force, applied at one corner of specimen, where other corners were restrained from displacement. Displacements and forces (reactions) at four corners were measured. *Load-displacement, twisting moment-curvature* relationships, torsional stiffness of slab at two stages, pre-cracking and post-cracking of concrete had been presented. This research showed that torsional stiffness of RC slabs in the cracked phase is about 1/17-1/15 of the stiffness in the elastic phase.

In this work, three similar RC slabs were tested. The test model was based on Lopes's one, with some modifications. The torsion was introduced as the same manner. Displacements at four points on top surface of each specimen and strain of reinforcing steel were measured. Test results and test interpretations were presented in section 4, and validated by finite element analysis.

2. Experimental program

2.1. Purpose

To study the behaviour of slab under torsion, include occurring and developing of cracks in concrete, stress developing in steel bars, the limit load at some states: start of cracking in concrete, start of yielding in steel, ultimate state, *load-displacement, twisting moment-curvature* relationship, torsional stiffness of slab at pre-cracking and post-cracking stages.

2.2. Specimens

Three similar square slabs, named S1, S2 and S3, with the dimensions of 1900×1900×150 mm, are tested. The sample slabs are reinforced by two identical reinforcing steel meshes, at top and bottom surface. Within each mesh, the steel bars are arranged isotropically, means they are the same in two directions. To study the stress developing in steel bars, there are eight strain gages were glued to the steel bars, see Figure 1. The locations of strain gages are determined with the help of FEA. The mechanical properties of concrete, by cubic test results at 28 days are: mean value of compressive strength f_{cm} is 38 MPa, modulus of elasticity E_c is 28 GPa, and of steel are: modulus of elasticity E_s is 194 GPa, the yield strength f_y is 468 MPa, the ultimate strength f_u is 562 MPa. Details of the samples are shown in Table 1.

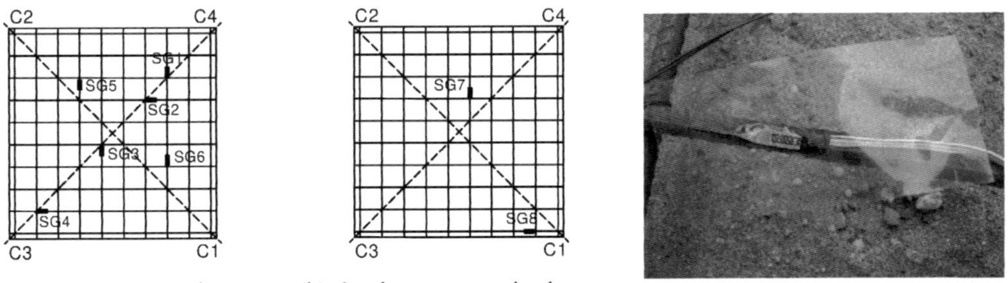

(a) Strain gages at the top mesh of reinforcement

(b) Strain gages at the bottom mesh of reinforcement

(c) A picture for more detail

Figure 1. The location of Strain gages in three specimens.

Table 1. The characteristics of the test specimens.

Name	Thickness (cm)	Steel Mesh	Age at test (days)	f_{cm} (MPa)
S1	15.4	$\phi 10a200(X,Y)$	96	40.21
S2	15.5	$\phi 10a200(X,Y)$	103	41.76
S3	15.2	$\phi 10a200(X,Y)$	107	40.83

2.3. Experimental procedure

The vertical downward displacement at corners *C3* and *C4*, and the vertical upward displacement at corner *C2* are restrained, the test load was applied on corner *C1*, see Figure 2.

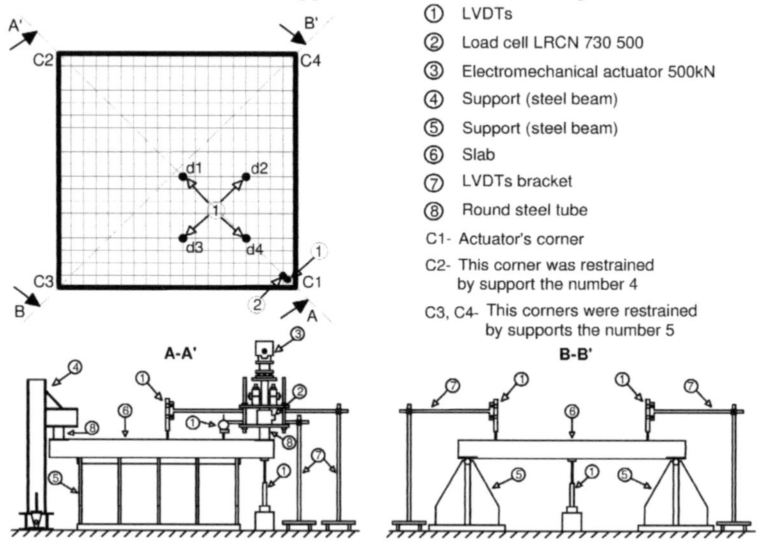

Figure 2. General scheme of the test.

The slabs were applied a concentrated force on corner *C1*, the test load is increased from zero until slab collapsed or the electromechanical actuator reaches its maximum displacement, which was set up 75 mm. The capacity of electromechanical actuator is 500 kN. The loading is controlled by displacement, with the velocity is 0.05 mm per second.

Load cell with a maximum capacity of 500 kN and a LVDT was used to measure the force and displacement at the corner *C1*. Four LVDTs are used to measure displacements at points *d1*, *d2*, *d3*, *d4* on the top surface of slab. Distance between them are 500 mm. All LVDTs, Load cell and Strain gages were connected with a DRA-30A Data Logger to read and record the data from the instruments, see Figure 3. The test results are shown in Figure 7 to 10, in Table 2 and Table 3, in the section 4 of this paper.

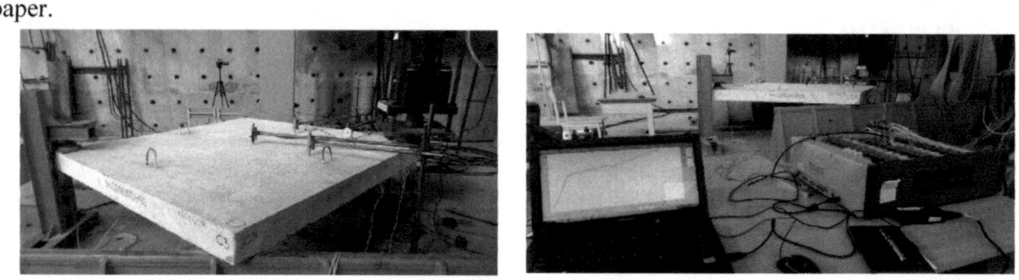

Figure 3. General view of the test.

3. Finite element analysis

3.1. Element, mesh and boundary conditions

The numerical simulations have been done for the slab by the help of ANSYS, V15.0. The slab was discreted by finite elements, where element SOLID65 was applied for concrete, SOLID185 was applied for steel plates at the corners, and LINK180 was applied for reinforcing bars, and the "discrete" model was used.

The FE mesh is shown in Figure 4, where, by the thickness, the slab was discreted into four elements, by two other sides, it was discreted into 22 elements. This mesh was checked to give the stable results, means the mesh refinement is good enough.

The static and kinematic boundary conditions are the same with the experimental conditions. That means, the nodes on steel plate at *C3* and *C4* corners were restrained from downward displacement, the nodes on steel plate at *C2* corner were restrained from upward displacement, and the load was applied on nodes at corner *C1*.

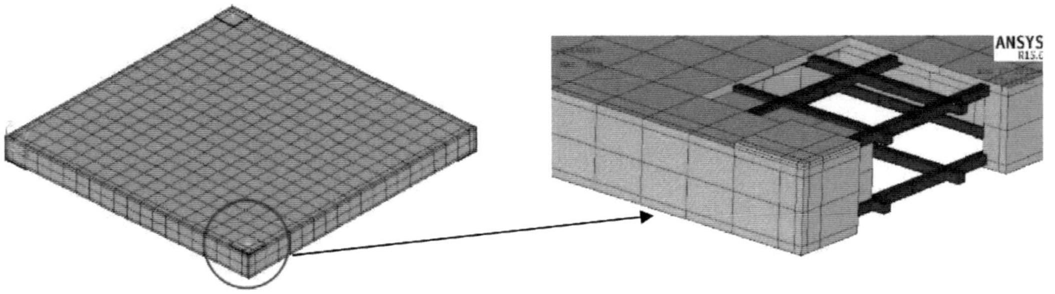

Figure 4. FE mesh.

3.2. Material laws

The Kachlakev law was applied for compression concrete [8], expressed by equations from (1), (2) and (3) [9, 10], and drawn in Figure 5a, and for concrete in tension, the material law was available in ANSYS [11], see Figure 5b. The elastic perfectly-plastic law was applied for reinforcing steel, Figure 6. Elastic modulus and modulus of rupture of concrete was calculated from equation (4) and (5) [2, 12, 13].

$$f = \frac{E_c \varepsilon}{1+\left(\frac{\varepsilon}{\varepsilon_0}\right)^2} \tag{1}$$

$$\varepsilon_0 = \frac{2 f_c^{'}}{E_c} \tag{2}$$

$$E_c = \frac{f}{\varepsilon} \tag{3}$$

$$E_c = 0.043 \rho^{1.5} \sqrt{f_c^{'}} \tag{4}$$

$$f_r = K \sqrt{f_c^{'}} \tag{5}$$

where K =8÷12 as suggestion in [2], and K = 9÷13 as suggestion in [13]. In this study, we take K =9.5. f is the compression stress in concrete at strain ε ; ε_0, $f_c^{'}$, E_c are the ultimate strain, the compression strength, and the elastic modulus of concrete respectively, f_r is rupture modulus, ρ is the concrete density.

The material properties were collected from our material tests. Since, slab S1, S2, S3 are different age at tested date, but their concrete strength are nearly equal, see Table 1, then the average value of concrete strength was calculated.

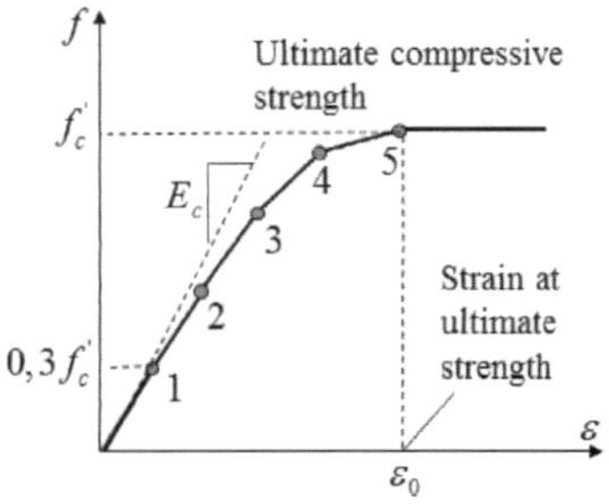

Figure 5a. Concrete law in uniaxial compression [8].

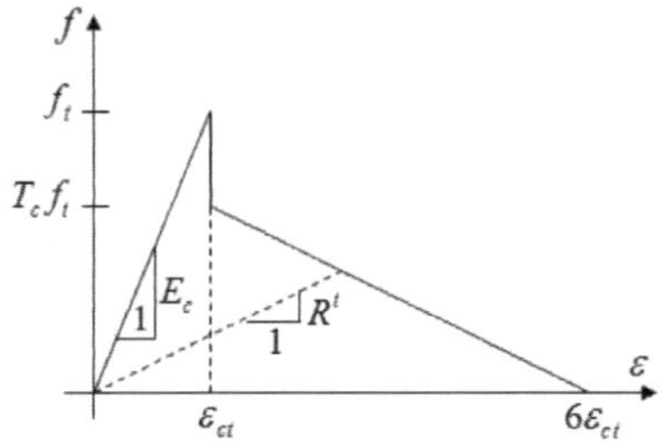

Figure 5b. Concrete law in uniaxial tension [11].

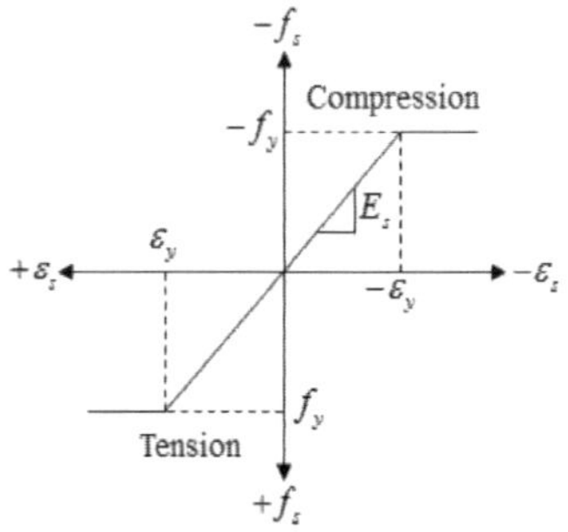

Figure 6. Reinforcing steel law in compression and tension [13].

By FEA, position and the moment when steel bars start to yield are easily investigated. The positions where the steel stress reaches yield stress are the same positions from the experiment, where the strain gages were glued. The FEA results are very close to the experiment results, and have been shown in Figure 9 and 10.

4. Results

4.1. Crack, deformed shape, failure mode of the slabs, and stress in the reinforcement

Crack patterns and failure mode from the present experiment and FEA, as well as from Marti's experiment are very close to each other.

Cracks developed at approximate 45^0 to the edges of the slab. At the top surface, the first crack occurred at corners *C3* and *C4*, started from the edge of slab, and propagated into slab centre. Sequent cracks are also started from the edge of slab, parallel to the first one, and propagated in direction 2, see Figure 7a, 7b.

At the bottom surface, the occurring and the propagating of cracks are the same manner to that at the top surface, but starting from corners *C1* and *C2*, instead of *C3* and *C4*, see Figure 7c, 7d.

The slab was collapsed by crushing of compression concrete at the top and bottom surfaces, at corner *C3* or *C4*, Figure 8. By the FEA results, slab was collapsed by crushing of compression concrete at corners *C3* and *C4* simultaneously, but in this tests, the slabs were collapsed at either corner *C3* or *C4*, since the imperfection and not absolutely symmetry of the specimens.

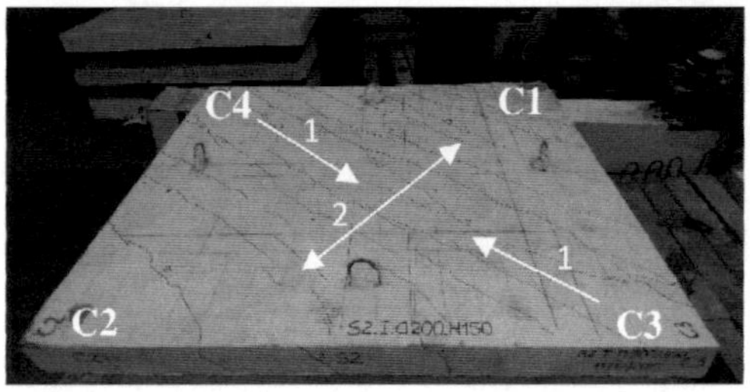

(a) Cracks at the top surface of slab S2

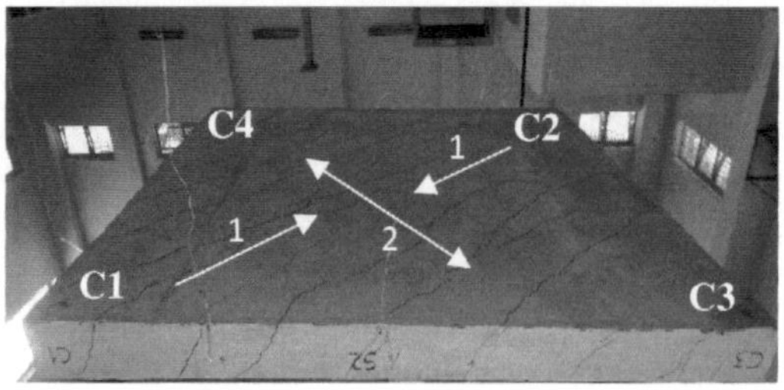

(c) Cracks at the bottom surface of slab S2

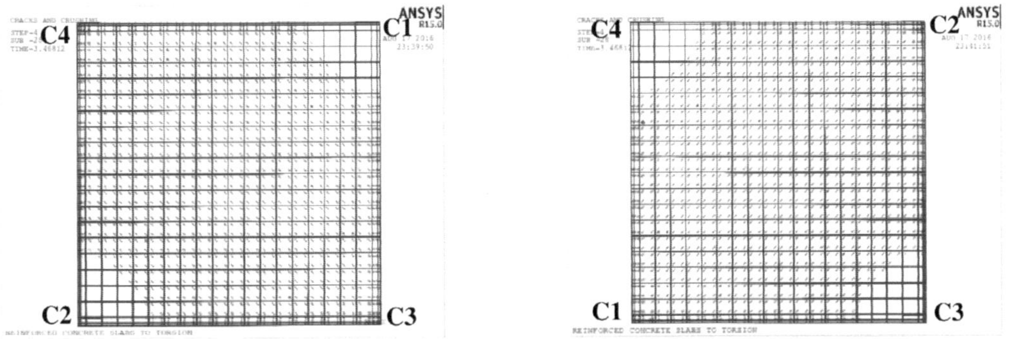

(b) Cracks at the top surface of slab S_FEA (d) Cracks at the bottom surface of slab S_FEA

Figure 7. Crack pattern on the slab surfaces.

Figure 8. Deformed shape and collapsed mechanism of the test slab.

The stresses in reinforcements were calculated from the strain values, which were collected from strain gages, at two states: $\sigma_{s,cr}$ at the point that concrete start to crack, and $\sigma_{s,\max}$ at ultimate load. These stresses are presented in Table 2. In the table, sign (-) means the strain gage did not work. At ultimate load, all reinforcement bars where strain gages were glued, reached yield, except strain gage number 7 in specimen S2.

Table 2. Stress in reinforcements.

Specimen	S1		S2		S3	
Strain gage	$\sigma_{s,cr}$ (MPa)	$\sigma_{s,max}$ (MPa)	$\sigma_{s,cr}$ (MPa)	$\sigma_{s,max}$ (MPa)	$\sigma_{s,cr}$ (MPa)	$\sigma_{s,max}$ (MPa)
1	30.9	492.3	52.0	533.5	35.4	474.1
2	18.1	472.1	3.5	474.6	-	-
3	11.6	471.3	23.8	473.4	6.1	470.5
4	5.9	474.6	3.6	473.2	1.5	471.8
5	5.0	472.4	-	-	6.0	472.3
6	5.1	471.5	4.5	471.7	5.9	472.8
7	23.1	485.1	25.8	372.2	36.3	472.5
8	9.0	472.3	2.7	474.3	-	-

4.2. Load-displacement and twisting moment-curvature diagram

From the load value at corner *C1* and displacements at four points *d1*, *d2*, *d3*, *d4*, twisting moment, m_{xy}, curvature κ_{xy}, κ_{yx} will be calculated by equations (6), (7) hereafter

$$m_{xy} = \left(\frac{P}{2} + \frac{W}{8}\right)\frac{l}{b} \tag{6}$$

$$\kappa_{xy} = \kappa_{yx} = \frac{\partial \varphi_x}{\partial y} \tag{7}$$

where, m_{xy} is twisting moment per unit length; κ_{xy}, κ_{yx} are twisting curvatures in the direction x-y, and y-x respectively; P, W is the applied load at corner *C1* and self-weight of slab respectively, $l = 1.75$ m, $b = 1.8$ m are the spans of slab, with the consideration for concrete cover.

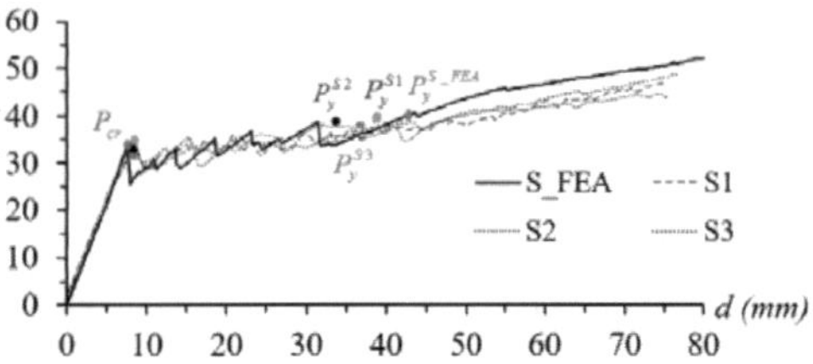

Figure 9. *P-d* relationships (EXP and FEA results).

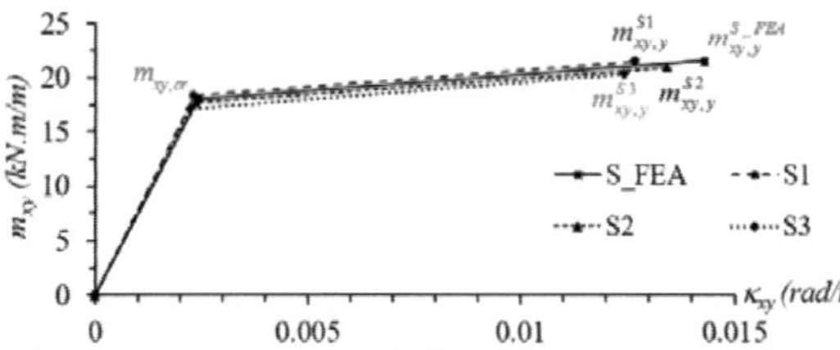

Figure 10. m_{xy}-κ_{xy} relationships (EXP and FEA results).

From Figure 9 and Figure 10, it can be seen that the test curves of three slabs, are very convergent, and in good comparison with FEA results.

4.3. Torsional stiffness of the slabs

The torsional stiffness of slab in the stage I, $D_{xy,I}$, and in the stage II, $D_{xy,II}$, were calculated by equation (8) and (9)

$$D_{xy,I} = \frac{m_{xy,cr}}{\kappa_{xy,cr}} \tag{8}$$

$$D_{xy,II} = \frac{m_{xy,y} - m_{xy,cr}}{\kappa_{xy,y} - \kappa_{xy,cr}} \tag{9}$$

where, $m_{xy,cr}$ and $\kappa_{xy,cr}$ is twisting moment and twisting curvature respectively, at the beginning of concrete crack, $m_{xy,y}$ and $\kappa_{xy,y}$ is twisting moment and twisting curvature respectively, at the moment of reinforcing steel start to yield.

The torsional stiffness of slab also can be calculated from formulae introduced by Nielsen. The formulae are rewritten in equations (10) and (11), as below. The more detail can be seen in [3].

- *Uncracked slab:*

$$D_{xy,I} = \frac{1}{12}E_c h^3 \tag{10}$$

- *Cracked slab:*

$$D_{xy,II} = E_c h^3\left(\frac{2}{3}(n\phi)^3 + (n\phi)^2 + \frac{1}{4}n\phi - \frac{2}{3}(n\phi)^2\sqrt{(n\phi)^2 + 2n\phi} - \frac{2}{3}n\phi\sqrt{(n\phi)^2 + 2n\phi}\right) \tag{11}$$

where $\phi = \dfrac{A_s}{bh}$ and $n = \dfrac{E_s}{E_c}$.

All results are shown in Table 3.

Table 3. Torsional stiffness of the slab.

Slab	Present, experiment			Present, FEA			Nielsen [3]		
	$D_{xy,I}$ (kNm^2)	$D_{xy,II}$ (kNm^2)	$\frac{D_{xy,I}}{D_{xy,II}}$	$D_{xy,I}$ (kNm^2)	$D_{xy,II}$ (kNm^2)	$\frac{D_{xy,I}}{D_{xy,II}}$	$D_{xy,I}$ (kNm^2)	$D_{xy,II}$ (kNm^2)	$\frac{D_{xy,I}}{D_{xy,II}}$
S1	7891	309	25.6				8231	273	30.1
S2	7470	288	26.0	7455	297	25.1	8388	275	30.5
S3	7718	326	23.7				8294	274	30.3

Table 3 shows that the present results from experiment, FEA, are very closed with the Nielsen's solutions.

5. Conclusion

The paper presented an investigation on the behaviour and torsional stiffness of reinforced concrete slab under torsion, by both experiment and finite element analysis. The test results, FEA results and results in literature are close to each other.

Cracks occur firstly on the top surface, then on the bottom surface. Cracks on the top surface and cracks on the bottom surface are orthogonal.

At ultimate load, all reinforcement bars where strain gages were glued, reached yield.

Torsional stiffness of RC slabs in the cracked phase is about 1/25 of the stiffness in the elastic phase for the steel ratio is 0.0032.

References

[1] Timoshenko S and Woinowsky-Krieger S 1959 *Theory of Plates and Shells* 2nd Ed. (New York: McGraw-Hill Companies) pp 118-120

[2] Nilson A H, Darwin D and Dolan C W 2010 *Design of Concrete Structures* 14th Ed. (New York: McGraw-Hill Companies) pp 432-434

[3] Nielsen N J 1920 *Beregninger af Spændinger i Plader* (København: Doktorafhandling ved Den Polytekniske Læreanstalt)

[4] Matri P and Kong K 1987 Response of reinforced concrete slab elements to torsion *Journal of Structural Engineering* **113**(5) 976-993

[5] Gudmand-Høyer T 2004 Stiffness of Concrete Slabs Ph.D Thesis Technical University of Denmark 4 pp 1-57

[6] Marti P, Leesti P and Khalifa W U 1987 Torsion test on reinforced concrete slab elements *Journal of Structural Engineering* **113**(5) 994-1010

[7] Lopes A V, Lopes S M R and Carmo R N F D 2014 Stiffness of reinforced concrete slabs subjected to torsion *Materials and Structures* **47** 227-238

[8] Kachlakev D, Miller T, Yim S, Chansawat K and Potisuk T 2001 Finite element modeling of reinforced concrete structures strengthened with frp laminates *Research Project Work Plan SPR 316* (Washington: Oregon Department of Transportation) pp 11-12

[9] Desayi P and Krishnan S 1964 Equation for the stress-strain curve of concrete *Journal of the American Concrete Institute* **61**(3) 345-350

[10] Gere J M 2004 *Mechanics of Materials* 6th Ed. (New York: McGraw-Hill) pp 23-24

[11] ANSYS 2013 *Inc Theory reference* Release 15.0 Documentation for ANSYS

[12] AS 3600-2009 2009 *Australian Standard: Concrete Structures* (Sydney: The Council of Standards Australia) p 37

[13] Park R and Paulay T 1974 *Reinforced Concrete Structures* (New York: Wiley-Interscience Publication) p 16

The characteristic analysis of the solar energy photovoltaic power generation system

B Liu[1], K Li[1], D D Niu[2,3], Y A Jin[2] and Y Liu[2]

[1]Jilin Province Electric Research Institute Co. LTD, Changchun, 130021, China
[2]College of Automotive Engineering, Jilin University, Changchun, 130025, China

Email: 1941708406@qq.com

Abstract. Solar energy is an inexhaustible, clean, renewable energy source. Photovoltaic cells are a key component in solar power generation, so thorough research on output characteristics is of far-reaching importance. In this paper, an illumination model and a photovoltaic power station output power model were established, and simulation analysis was conducted using Matlab and other software. The analysis evaluated the condition of solar energy resources in the Baicheng region in the western part of Jilin province, China. The characteristic curve of the power output from a photovoltaic power station was obtained by simulation calculation. It was shown that the monthly average output power of the photovoltaic power station is affected by seasonal changes; the output power is higher in summer and autumn, and lower in spring and winter.

Keywords: solar energy, photovoltaic cells, simulation analysis.

1. Introduction

Abundant, clean and environmentally friendly, solar power has become an important renewable resource for energy production all over the world. The degree to which this technology has been utilized, the available economic resources and the prospects for future development vary widely. Technological advances in solar power generation have the potential to impact a variety of stakeholders, such as power plants, power companies, equipment manufacturers and investors [1].

Solar photovoltaic power generation is widely recognized as an important renewable energy technology. The advantages of solar energy include inexhaustible reserves and clean production [2].In rural areas, islands and remote areas where transportation is inconvenient, solar energy is the ideal alternative energy source that is currently available. In addition, as a kind of distributed power system, photovoltaic power generation technology can improve and enhance the stability of the whole power system, especially given its potential to withstand natural disasters [3].

2. Modeling

2.1. The modeling and simulation of the external characteristics of photovoltaic cells

The photovoltaic array is a key component of a photovoltaic power generation system. Its current-voltage characteristic is the nonlinear function of sunshine intensity, environmental temperature and photovoltaic module parameters. Matlab software was selected to establish a simulation model of the photovoltaic array based on its physical mechanism [4]. In the design of a photovoltaic power generation system, the manufacturer of the photovoltaic panels usually provides the parameters of the photovoltaic array, including the open circuit voltage, short circuit current, peak voltage, peak current and maximum power. Applying these parameters to the corresponding mathematical model can directly yield the operation parameters of the photovoltaic array. The characteristic equation for the output of the photovoltaic battery component is:

$$I_L = I_{ph} - I_0 \left\{ \exp\left[\frac{q(U_{oc} + I_L R_S)}{36AkT} \right] - 1 \right\} \tag{1}$$

In equation (1), IL is the current flow through the load; Iph is the photoproduction current; Io is the reverse saturation current of the photovoltaic cell; Rs is the equivalent series resistance; q is the electron charge; Uoc is the open circuit voltage; A is the P-N junction ideal factor; K is the Boltzmann constant; T is the absolute temperature. The unknown variables in equation (1) are Iph, Io, and Rs., which were modeled using SIMULINK software.

Given the Isc, Uoc, Im and Um of the given model photovoltaic cells under standard conditions, we can obtain the current–voltage curve and power-voltage curve, which were used as the photovoltaic battery output characteristic. The current–voltage characteristic curve of the photovoltaic cells shows that a photovoltaic cell is a kind of nonlinear direct-current power supply, and it does not consistently provide the maximum power output. The power-voltage characteristic curve of photovoltaic cells is a single-peak curve with the maximum power point as its extreme value. To obtain the maximum benefit from the photovoltaic arrays, the system must be optimized to function near the maximum power point.

2.2. Acquisition of the solar radiation intensity sequence

The output power of a photovoltaic power station is related not only to the features of the photovoltaic battery, but also to the intensity of solar radiation that reaches the surface of the photovoltaic cells. The illumination model describes the intensity of solar radiation over a certain period of time. The intensity of naturally occurring solar radiation depends primarily on the geographical position of the photovoltaic power station. Using the latitude, longitude and altitude of the given location, the intensity of solar radiation in the region can be roughly determined.

$$R_T = f(\alpha, \beta, h) \tag{2}$$

In equation (2): R_t is the actual solar radiation intensity, kW/m2; α is the longitude of the photovoltaic power station site; β is the latitude; h is the altitude, m.

HOMER (Hybrid Optimization of Multiple Energy Resources) renewable energy microgrid software incorporates region-specific data such as the monthly average radiation and clear-sky index from the NREL (the United States government's National Renewable Energy Laboratory). Given the latitude, longitude and time zone of the photovoltaic power station, HOMER was used to calculate the solar radiation intensity sequence over one year (8,760 hours).

2.3. *Photovoltaic power station output power modeling based on macroscopic analysis*

By comparing local conditions with standard test conditions, especially considering the impact of the local solar radiation intensity and temperature on the photovoltaic battery power generation features, the following photovoltaic power station output power model can be derived:

$$P_{PV} = NR_T S_A \eta \tag{3}$$

$$\eta = \eta_{STC}\left[1+\alpha_P\left(T_C - T_{STC}\right)\right] = \frac{U_L I_L}{R_{STC} S_A}\left[1+\alpha_P\left(T_C - T_{STC}\right)\right] \tag{4}$$

In equation (3) and equation (4): PPV is photovoltaic power station output power, MW; N is the number of photovoltaic cells in the photovoltaic power station; SA is the area of a single photovoltaic cell, m2; RT is the actual intensity of solar radiation, kW/m2; η_{STC} is photovoltaic cell conversion efficiency under standard conditions; η is the actual conversion efficiency of a photovoltaic cell; α_P is the photovoltaic battery power temperature coefficient; TSTC is the temperature of the photovoltaic cells under standard test conditions, ℃; TC is the actual temperature of the photovoltaic cells, ℃; RSTC is solar radiation intensity under standard test conditions, kW/m2; IL is the current flowing through the load; UL is the voltage flowing through the load.

This model takes into account the influence of solar radiation intensity, temperature, photovoltaic conversion efficiency, the surface area of the photovoltaic array and other factors on the output power of a photovoltaic power station. The model can be used for macroscopic planning and quantitative analysis of the scale of photovoltaic power generation in the designated area.

3. The analysis of photovoltaic power station power output characteristics in west Jilin province

3.1. *Distribution and feature analysis of solar energy resources in Jilin province*

According to The Solar Energy Resources Evaluation Method, evaluating solar energy resources is the basis and theoretical foundation for the development and utilization of solar energy. HOMER simulation software evaluates the solar energy resources of a given area, providing information about the effects of seasonal changes on photovoltaic power generation.

The Baicheng area of Jilin province in China has rich solar energy resources. Based on its total annual radiation of 5,570 MJ/m^2, the Baicheng region is the most appropriate area in Jilin province to construct a large-scale photovoltaic power station. The annual temperature of Baicheng City is 276.5 ℃, the average annual sunshine hours is 2915.3 hours, and solar-thermal resources per capita is the highest in Jilin Province [5]. Analysis of solar energy resources in the Baicheng area shows that total monthly solar radiation varies throughout the year between 183.5 MJ/m2 and 710.5 MJ/m2; the minimum appears in December, while the maximum occurs in May. When evaluating the solar resources of a region, just as important as the total annual radiation is the stability of the solar resources, which can be assessed by measuring the sunshine hours per day. The monthly distribution change trend of the number of sunshine hours is approximately the same as the change trend of the total radiation in the region. Sunshine hours are relatively high in the month with the highest total radiation, and significant differences are seen between seasons. During each season, the distribution of sunshine hours is stable for months without sharp fluctuations. A radiation intensity distribution nephogram was obtained using HOMER software by entering the coordinates for the Baicheng region.

On the whole, the solar energy resources in the Baicheng area are relatively rich and stable. Favorable weather and other necessary resources exist to make the development of large-scale solar power generation feasible [6].

3.2. The analysis of the photovoltaic power station output characteristic in west Jilin province

Due to the high level of solar energy resources in the Baicheng region, the "Twelfth Five-Year Plan" for Jilin province includes plans to build a 100 MW photovoltaic power station in Zhenlai County in the Baicheng area. In order to model this proposed power plant, a basic photovoltaic power generation unit with 500 kW capacity will be assumed, using the parameters of GDM-235PE03 (polysilicon) photovoltaic cells. The cells are installed in series of 20 components. A basic power unit consists of 108 parallel series, which totals 2,160 components. Therefore, a photovoltaic power station with 100 MW rated capacity would need 432,000 solar panel components, which would cover an area of 705,672 m2. The annual output curve combined with the intensity of solar radiation and temperature data can be obtained by equation (3), as shown in Figure 1.

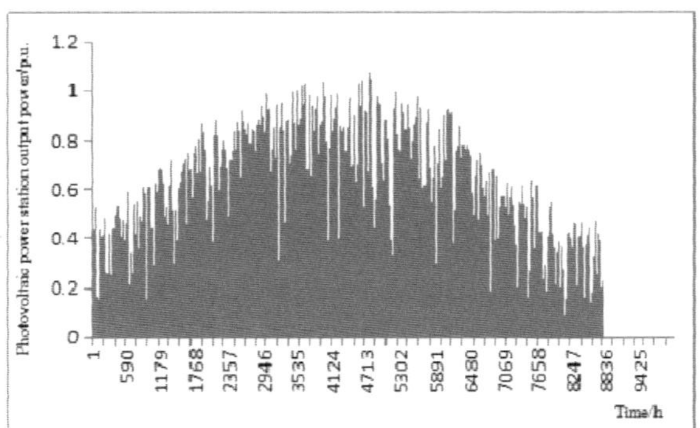

Figure 1. 100 MW photovoltaic power station annual output power.

IOP Conf. Series: Materials Science and Engineering **164** (2017) 012018 doi:10.1088/1757-899X/164/1/012018

The figure shows that the maximum output power is 1.086 p.u., occurring at 4,236 hours, and the number of equivalent full hours are 1,468. Photovoltaic power station annual output power distribution is almost the same as the local solar radiation intensity annual distribution. In order to reflect the fluctuation characteristics of the photovoltaic power station output power in more detail, the output power annual distribution curve was plotted separately for each month, taking a typical sunny day of each month.

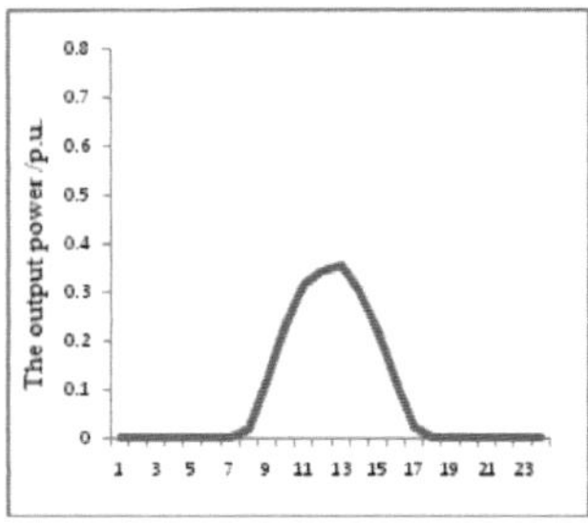

A typical day in January

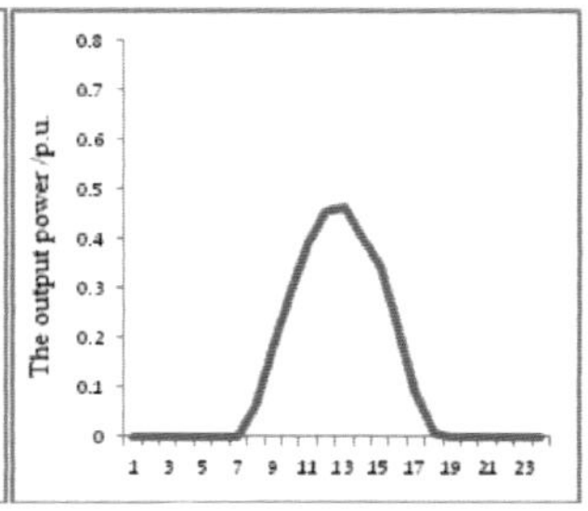

A typical day in February

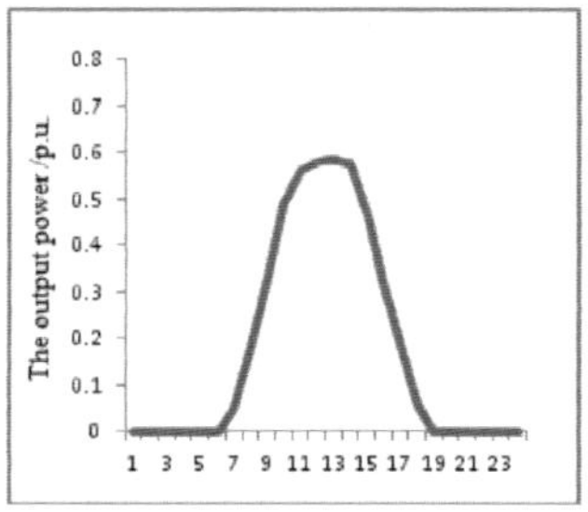

A typical day in March

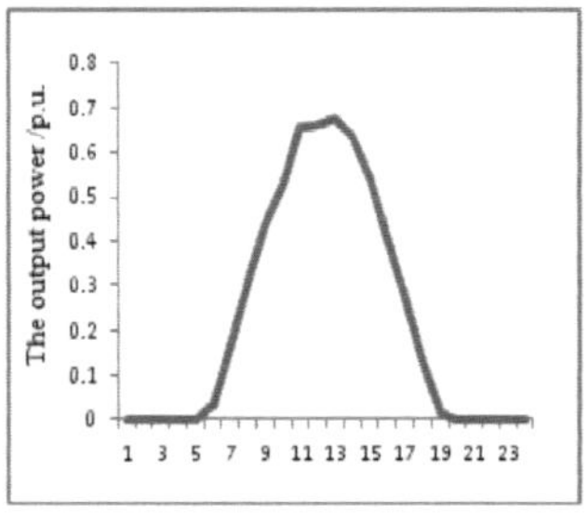

A typical day in April

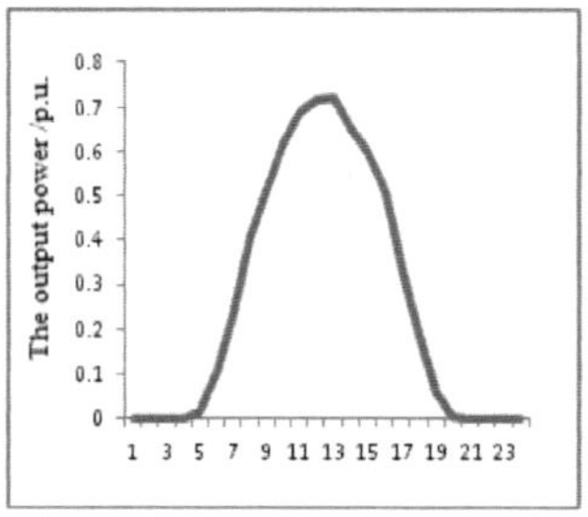

A typical day in May

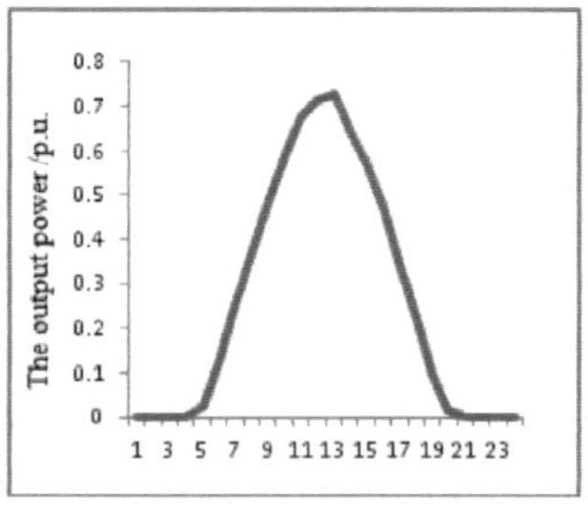

A typical day in June

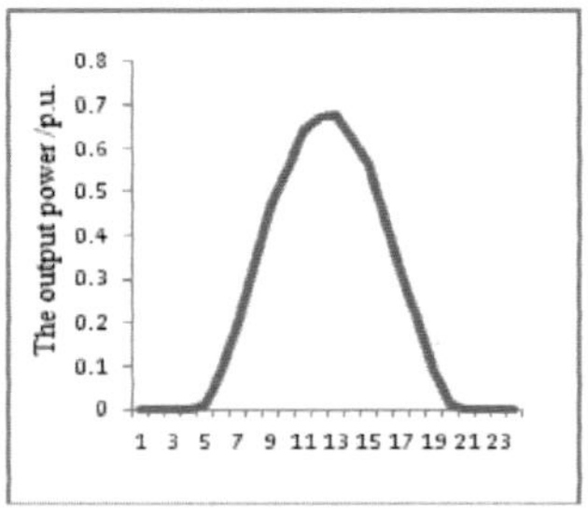

A typical day in July

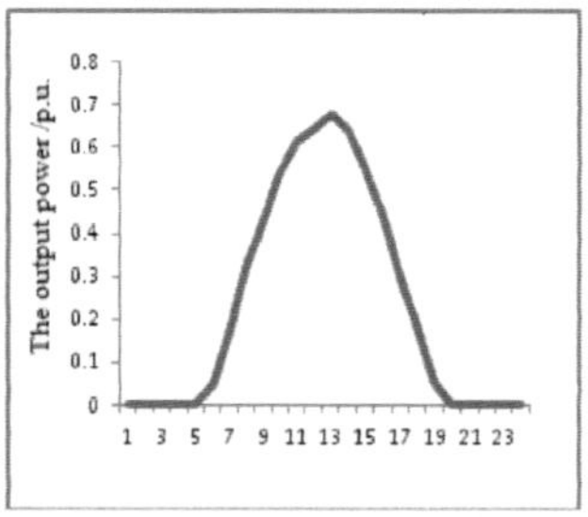

A typical day in August

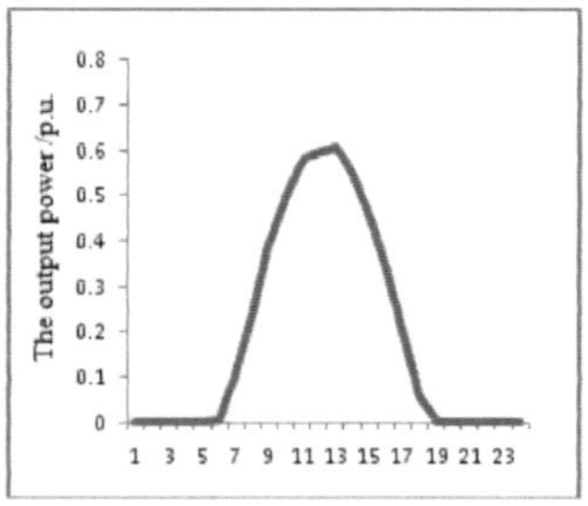

A typical day in September

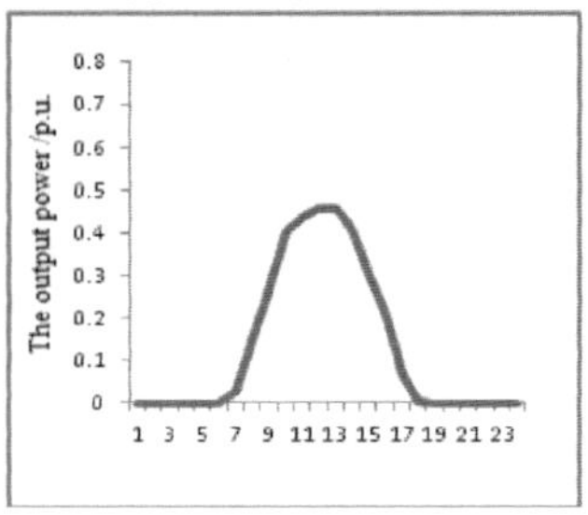

A typical day in October

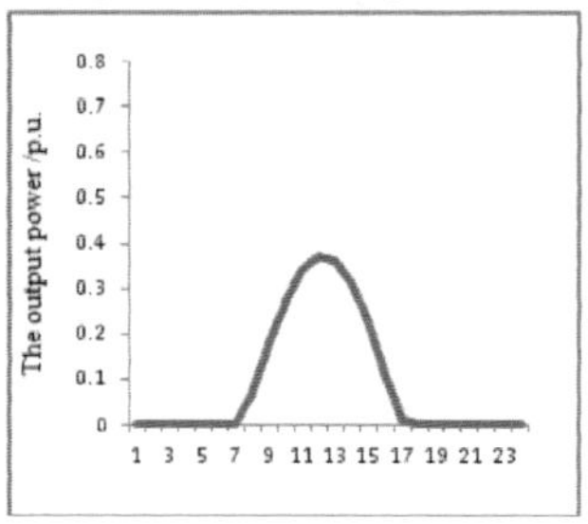

A typical day in November

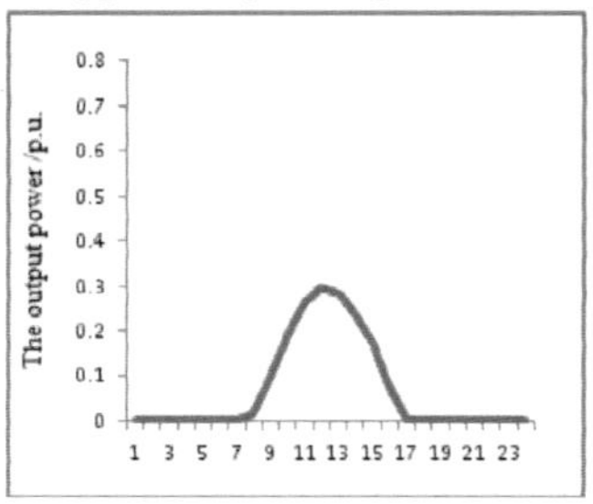

A typical day in December

Figure 2. The month distribution of the output power of a 100 MW photovoltaic power station.

The Figure 2 shows that the monthly average output power of the photovoltaic power station is affected by seasonal changes, starting at a low value in January, increasing each month to reach its peak in May, and then decreasing each month, reaching its minimum value in December. The fact that output power is higher in summer and autumn and lower in spring and winter is consistent with the distribution of radiation intensity in the area. Photovoltaic power station output was limited to the hours of 5:00 to 20:00 every day. Daily maximum output power occurred during the period from 12:00 to 14:00; there is no output at night. The typical daily output was consistently less than 0.8 times the rated capacity of photovoltaic power plants. Most of the time, there is a large gap between the rated capacity and the actual output.

4. Conclusion

This paper analyzes the characteristics of photovoltaic battery power, establishes an illumination model, and builds a model for photovoltaic power station output power that accounts for the effects of solar radiation intensity and temperature change. Based on the condition of solar energy resources in the Baicheng region in the west of Jilin province, China, the photovoltaic power station power output characteristic curve was obtained using the established model. The results show that the monthly average output power of the photovoltaic power station is affected by seasonal changes; the output power is higher in summer and autumn and lower in spring and winter.

Acknowledgement

The authors would like to thank the Project by Clean Development Mechanism (CDM) of China (1113050) and the Science and Technology Project of Jilin Province Development and Reform Commission (2013-729).

References

[1] Xin P Y 2015 *Research on Solar Power Technology Economic Evaluation and Application Prospect* North China Electric Power University pp 1-5

[2] Wang Z H 2015 *Study on Optimizing Configuration of Integrated Power Transmission Capacity of the Clustering Photovoltaic-wind Joint Power Generation* Northeast Electric Power University pp 18-21

[3] Zhang H H 2013 Design for Photovoltaic Power Generation System Based on MATLAB *Control Engineering* **S1** 26-29

[4] Wu X J 2012 *The Simulation of Concentration Solar Cells in Matlab\Simulink* Wuhan University of Technology p 15

[5] Sun Z S 2012 *The Study of Grid-connected Photovoltaic Generation System Based on MATLAB Simulation* Nanchang Aviation University p 1

[6] Cheng L M 2012 The Present Development Situation, Problems and Countermeasures of Solar Industry in Jilin Province *The Regional Economic* **3** 3-4

Economic and safety analysis of unconventional peak regulation on power unit of peak shifting start-stop

X Cao[1], J F Zhao[1], X Q Duan[2, 3] and Y A Jin[2]

[1]Jilin Province Electric Research Institute Co. LTD, Changchun, 130021, China
[2]College of Automotive Engineering, Jilin University, Changchun, 130025, China

Email: 314626370@qq.com

Abstract. Tthe capacity difference of peak regulation between the power gird and the actual demand has become a serious problem considering the growth in the difference between electricity supply and demand. Therefore, peak regulation of power grid needs to be deeply studied. Unconventional peak regulation on unit of peak shifting start-stop is a way that can broaden the range of power regulation, as well as .benefit safe operation of the power grid. However, it requires frequent and fast unit start-stop, complex operation, and more staff labor. By carrying out unconventional thermal power unit load test, the start-stop mode of peak auxiliary equipment is studied in this paper, indicating that it has a positive effect on safety and economic of load-peaking operation. The best working conditions of the peak units is found by analysing consumption cost, safety specifications, and life lost of the start-stop peak regulation mode.

Keywords: unconventional peak regulation tests, the start-stop mode, economic and security analysis.

1. Introduction

The capacity of heating unit accounted for more than 80% of the capacity of thermal power units in Jilin Province by the end of 2011 [1]. Unit-load adjustment range is limited during the heating phase due to the characteristics of heating units. In addition to this, it is indispensable for extra thermal power unit's capacity to maintain power balance with the growth of wind power capacity [2].

For large units in Jilin Province, the safety and economy of the peak load regulation is an important issue in electric power production [3]. Unconventional start-stop peaking regulation research and application has been carried out. In order to achieve the purpose of energy saving, the paper analyzes the economic and safety for the unconventional start-stop equipment and provides reference for the same type of unit in the economic regulation

2. Unit overview

In Shuangliao power plant, No.2 and No.3 steam turbine is subcritical steam turbine produced by Harbin steam turbine plant.
The model is N300-16.7/537/537.
The rated power is 314.4MW.
The main steam flow rate is 963.91t/h.
The main steam pressure is 16.67MPa.
The main steam temperature is 537 ℃.
The No.2 and No.3 boiler with 6 sets of fan mill pulverizing system are produced by Harbin boiler factory production. The main design parameters are as follows.
The superheated steam flow is 1025t/h.
The superheated steam outlet pressure (gauge) is 17.5MPa.
The superheated steam outlet temperature is 540 ℃.
The reheat steam flow rate of is 835 t/h.

3. Operation safety analysis of peak regulation

3.1. Water circulation reliability

Water circulation reliability of natural circulation drum problems attracts attention when the thermal power unit is at a low load peaking operation [4]. Due to the large circulation rate in Shuangliao power plant, it can be estimated by using the circulation rate of K when the boiler is in the low load operation.

$$k = \frac{k_e}{0.15 + 0.85D/D_e}$$

k_e —The circulation ratio of K at rated load

D_e —The main steam flow at rated load, t/h

D —The main steam flow under low load, t/h

The mass flow rate of the water-cooled wall is about 1000 ~ 1200kg/ ($m^2 \cdot s$) when the 300MW unit is in 160WM load. The boiler is 30% under the rated load condition with the minimum flow rate being larger than the limit value of 0.4m/s. The circulation ratio of K is 5.8, which is within the recommended value rage of 4-6.

3.2. Safety of thick wall parts of the drum

The main risk associated with the boiler operation process is the changing stresses of the boiler drum during the repeated starting processes [5]. The boiler drum will bear more mechanical and thermal stress due to the periodical change of the pressure, temperature of the working medium, and the thickness of the wall of the boiler drum. If the control of start and stop is unsuitable, it will lead to larger stress change and reduce the drum's life. The temperature differences of inner and outer walls are the primary parameters of drum stress. Straight through type feed water-piping structure in the interior of the drum is adopted in order to reduce the temperature differences between the drum of water and steam and improve upon the boiler's ability to adapt to the rapidly changing load. When appearing in the middle of the drum, the drum wall has a temperature difference of 10℃.

It is noticed in Figure 1 that the temperature difference between the inside and outside wall of the drum is within the range of 14-20 ℃, with the maximum changing rate of the temperature being 1.52 ℃ /min. When units work in the range of 300-160MW, 9769 is the useful service life of the drum.

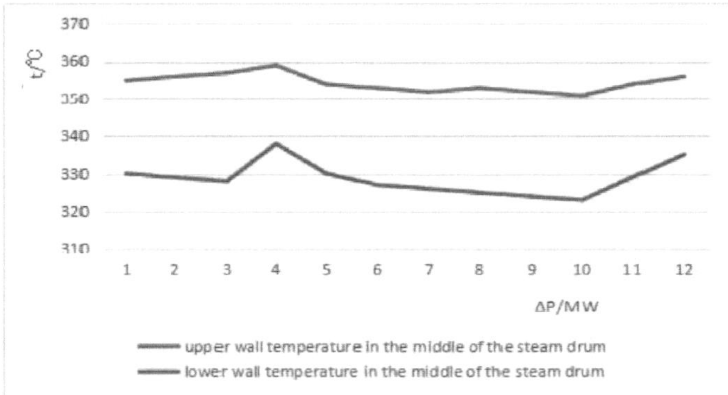

Figure 1. The temperature difference between the inside and outside wall of the drum.

3.3. The safety of steam turbine

The steam turbine was designed with a general life of 30 years. Part of steam turbine life is retained as general life loss distribution is only about 80% and the remaining 20% is for sudden accident. The Mechanical Engineering Handbook gives the team turbine life distribution table. Table 1 shows a life loss rate of 0.01%.

Table 1. The life distribution data recommended Mechanical Engineering Handbook.

Operation model	Loss rate%	The number of years of operation	The cumulative running times	The cumulative loss of life%
Cold start	0.05	4	120	6
Warm start	0.01	1	30	0.3
Hot start	0.01	200	6000	60
Overhaul before shutdown	0.05	1 times in 3 years	10	0.5

Load rejection with auxiliary power	0.1	2 times in 3 years	20	2
Large amplitude variable negative40%	0.005	50	1500	7.5
Small amplitude variable load40%	0.00025	~530	~16500	4
$\sum$				80.3

4. Economic analysis for unconventional peaking regulation

If a single unit stopped running in Shuangliang power plant, it could create a 150MW capacity every hour for wind power in the Jilin power grid. According to that wind turbine generator working with 150MW load operation, each kW·h can save 327.07g of standard coal. Also, during units stop of 33.75 hours, 1655.792t of standard coal can be saved. In other words, if the price of per ton of coal is 650 Yuan, one million Yuan can be saved.

In the process of unconventional peaking regulation, the cost of the power consumption of non-artifical fuel, oil, coal, water, and electricity is about 28.5982 million Yuan. After deducting start-stop loss units, the total economy benefits by 790281 Yuan. The twenty running process can save 158056 Yuan during the five winter months [6].

It can be seen that great loss is caused to power plants with unconventional peak regulation on power unit of peaking shifting start-stop, but it can also greatly improve the entire economy.

References

[1] Yuan Y and Zhang J W 1974 *Test Method for Boiler Combustion Adjustment* (Beijing: Hydraulic and Electric Power Press) pp 111-137
[2] Yuan Z F and Tian Z P 2002 *Principle of Boiler in Power Plant* (Beijing: Chinese Power Press) pp 85-110
[3] Huang X Y 2007 *Running on the Boiler Combustion Adjustment* (Beijing: Chinese Power Press) pp 55-62
[4] Mohan M R, Paranjothi S R and Israel S Prince 2007 Use of pumped-hydro as peak-load management plant in optimal scheduling of power systems *Electric Machines & Power Systems* **25**(10) 1047-1061
[5] Bogachev A F, Kirillina A V and Kozlov Y V 2008 Behavior of phosphates in a high pressure drum boiler with frequent loading *Power Technology and Engineering* **42**(4) 247-251
[6] Zheng H Q and Fang D B 2007 Interconnection electricity prices of peak regulation and frequency regulation on power plants in regional electricity market *East China Electric Power* **35**(2) 107-109

Study on formation of JPC under large standoff distance condition

Y J Wang, W B Li[1], X M Wang, W B Li and Y M Li

ZNDY of Ministerial Key Laboratory, Nanjing University of Science and Technology, Nanjing 210094, China

Email: njustlwb@163.com

Abstract. In order to achieve penetration damage efficiently under larger standoff distance condition, a shaped charge warhead is designed on the basis of jetting penetrator charges (JPC) formation and disruption theory. The paper analyses the formation, elongation and disruption processes of JPC by using LS-DYNA software. The results show that JPC tail begins to break away at 8.5 charge diameters (CD) standoff distance. When the standoff increases, JPC head velocity decrease while JPC tail increase, and JPC length-diameter ratio increased significantly. Based on numerical simulation and theoretical investigation, it was found that the JPC still has good penetration ability at 15.5CD standoff distance, and JPC effective standoff is able to reach 20CD distance. The result provides a reference for the design of JPC warhead at large standoff distance.

1. Introduction

Modern armor protects against the shaped charge jet penetration, of which defeat mechanisms are both breaking the jet up on impact and altering the standoff. The penetration capability and stability are affected, because the jet is easily broken, especially at large standoff distance. Therefore, increasing the effective standoff has great significance on defeating remote target. Jetting Projectile Charge (JPC) [1] has a good penetration capability and a strong adjustability, which has good application prospect for dealing with some warheads aiming to defeat targets in large standoff distance. Aiming at improving penetration capability of warheads in large standoff conditions, Frueh and Heine [2] obtained velocity-distance information about the majority of the particles of two types of small-scale laboratory shaped charge at large standoff distance by using a combined technique of flash X-rays and high speed video cameras. Mayseless et al. [3] presented a new stochastic model to describe adequately the spread in the penetration data at short and long standoffs. Huang et al. [4] analyzed the structure of liner under large standoff distance condition, combining theoretical investigation and experimental verification. Based on JPC design method, Zhang et al. [5] designed three typical liner JPCs, and verified the penetration experiments—the penetration performance of JPC with spherical segment liner is the best. Subsequently, Fu et al. [6] and other researchers studied kinds of typical liner to analyze influence of liner parameters on penetration capability of JPC, and conducted experiments to verify the penetration capability under the condition of large standoff. Nowadays most scholars have focused on the mechanism of liner collapse and distortion and analysis of influencing factors about JPC penetration. However, for the large standoff distance, the influence of standoff distance on the JPC formation has not been reached at home and abroad.

In this paper, a shaped charge warhead is designed, having good penetration capability at large standoff distance, on the basis of JPC formation and disruption theory. The numerical simulation is carried out using dynamic finite element program of LS-DYNA for the influence of standoff distance on the JPC formation. Furthermore, the penetration capability of JPC is verified at the large standoff distance from the comparison of the simulation solution and the theoretical calculation.

2. Structure model and numerical simulation scheme

2.1. Design of structure model

Upon detonation of the shaped charge, the detonation wave, pushed by the rapidly expanding gaseous products along the axis of symmetry, contains higher pressures than the ultimate stress in the material, which allows us to model the material as an inviscid fluid. The jet of material forms, the tip of which moves at high velocity toward the target. And the remaining liner material is formed into a slug which follows the jet at a much lower velocity. The projectile will progressively stretch until it breaks up into small fragments.

The model assumes that equation, based on Zheng's [7] theoretical investigation on the necking and breaking up of a free metal in air produced by shaped charge, applied for describing relation of necking time and liner parameters in this unsteady process. The results are as follows:

$$t_{b1} = C_1 (\frac{\Omega \rho_j}{\sigma})^{1/3} \tag{1}$$

$$t_{b2} = C_2(\phi)(\frac{\Omega \rho_j}{\sigma})^{1/3} \tag{2}$$

where $\Omega = \frac{1}{\pi \rho_j} \cdot \frac{dm}{dv_j} = r_0^2 t^*$, and t_{b1} is jet necking time; t_{b2} is jet breaking time; $1/t^*$ is jet initial velocity gradient; r_0 is jet initial radius; v_j is jet velocity; ρ_j is jet density; ρ_0 is unperturbed density of the air; m is jet quality; C is speed of sound in undisturbed air, an independent normal constant without Ω, ρ_j, σ etc.; α is a constant relating to jet and liner materials; σ is flow stress or yield strength; $C_2(\varphi)$ is reduction of material area; φ is related functions.

Applying the above equation to analysis the method of postponing jet breaking time and improving the stability of jet, it is required that the liner material has good plastic performance and the speed of sound with ρ_j as large as possible and σ as small as possible on process of liner selecting and machining. Secondly, it is required that the jet quality is as much as possible to reduce jet velocity and to improve Ω on process of liner structural designing.

Considering above factors, the configuration of the liner in a shaped charge is sub-hemispherical and the initiation method is single-point initiation with wave-shaper. After analysing effects of shaped charge structure parameters with the help of numerical simulation, the final designed warhead, of which charge diameter was taken as 110mm, has good penetration ability at large standoff distance. A shaped charge is constructed from a robust casing filled with a cylindrical high explosive and a wave-shaper. At one end of the casing there is an axis-symmetric geometric shape precisely indented in the high explosive in which a liner is inlayed. The structure of shaped charge warhead in this paper is shown in Figure 1.

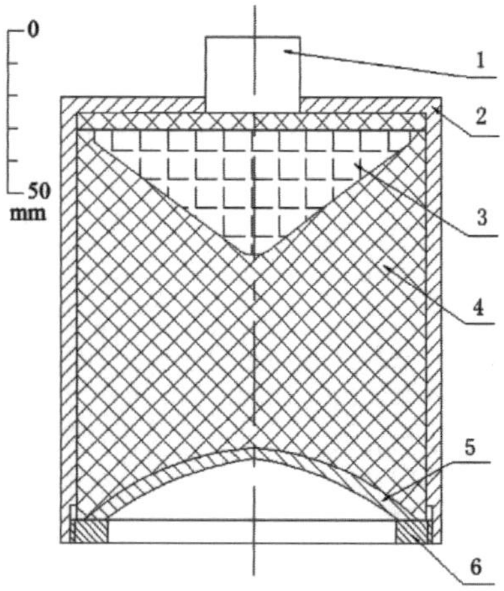

Figure 1. Structure of shaped charge warhead. (where 1 is the detonation device; 2 is the body shell; 3 is the wave-shaper; 4 is the explosive; 5 is the sub-hemisphere liner; 6 is the fixed ring.).

2.2. Establishment of the simulation model

The ALE algorithm provides the coupling mechanism between ALE fluid and Lagrange nodes. To solve the problem of large deformation, ALE algorithm can be used to calculate the JPC formation process, of which the explosive, liner, wave-shaper and air are meshed as ALE fluid, and the shell is meshed as Lagrange nodes. Due to symmetry, the model 1/4 of the geometry is enough. The symmetrical inhibit condition is added to the symmetrical surface of the model to restrict the node's displacement and rotation degrees of freedom. And the outflow boundary conditions of pressure are put in the boundary nodes in order to avoid the reflection of pressure at the boundary. The simulation model is shown in Figure 2.

Figure 2. Simulation model.

The processes of the liner collapse and shell crack generate high temperatures and pressures. The responses of liner material at different pressure levels are shown by both the Johnson-Cook material model and the Gruneisen equation of state. John-Cook's formulation of the flow stress is

$$\sigma_y = (A + B\bar{\varepsilon}^{p^n})\left[1 + C\ln\dot{\varepsilon}^*\right](1 - T^{*m}) \tag{3}$$

where A, B, C, n and M are material constants; $\bar{\varepsilon}^p$ is the effective plastic strain; $\dot{\varepsilon}^*$ is the effective plastic strain rate, $\dot{\varepsilon}_0 = 1s^{-1}$; T^* is the relative temperature given by $T^* = \left(T - T_{room}\right)\left(T_{melt} - T_{room}\right)^{-1}$.

The Gruneisen equation of state takes the pressure of compressible materials as

$$p = \frac{\rho_0 C^2 \mu[1 + (1 - \frac{\gamma_0}{2})\mu - \frac{a}{2}\mu^2]}{[1 - (S_1 - 1)\mu - S_2\frac{\mu^2}{\mu + 1} - S_3\frac{\mu^3}{(\mu + 1)^2}]} + (\gamma_0 + \alpha\mu)E \tag{4}$$

IOP Conf. Series: Materials Science and Engineering **164** (2017) 012020 doi:10.1088/1757-899X/164/1/012020

where s is shock; p is particle; ρ_0 is density of material; E is internal energy of initial unit volume material; C is interception at the vertical axis of the shock velocity versus particle velocity (u_s-u_p) curve; S_1, S_2, S_3 are the coefficients that determine the shape of (u_s-u_p) curve; γ_0 is Gruneisen gamma; a is the first order volume correction to γ_0.

The shaped-charge explosive is 8701, which is described by the "Mat_High_Explosive_Burn" material option in DYNA and the JWL equation of state. The explosive burning equation describes the explosive behavior before its detonation, and controls the release of chemical energy for simulating detonation. The JWL equation describes the pressure, volume and energy of the reaction product in the process of detonation, it defines the pressure as

$$P = A(1-\frac{\omega}{R_1 V})\exp(-R_1 V) + B(1-\frac{\omega}{R_2 V})\exp(-R_2 V) + \frac{\omega E}{V} \tag{5}$$

where A, B, R_1, R_2, ω are all constants, E is the initial internal energy; and V is the relative volume.

The wave-shaper is, which is described by the "Mat_High_Explosive_Burn" material model and the Gruneisen equation of state. The Null material and the Gruneisen equation of state is used for the air as an ideal gas. The parameters of material models used in the calculations are to be found in Li et al. [8].

3. Influence of the standoff distance

The standoff of a shaped charge is the distance from the base of the liner or cavity to the target. It is known that the standoff in shaped charges has an optimum distance for best penetration performance. On one hand, the increase of standoff distance will make the JPC more fully stretched, which is beneficial to the penetration. On the other hand, "too-large" standoff distance will lead to the JPC break and affect the penetration seriously. Calculating numerical simulation models to present the influence of different standoff distance on JPC. Figure 3 is the image of liner collapse and subsequent projectile formation.

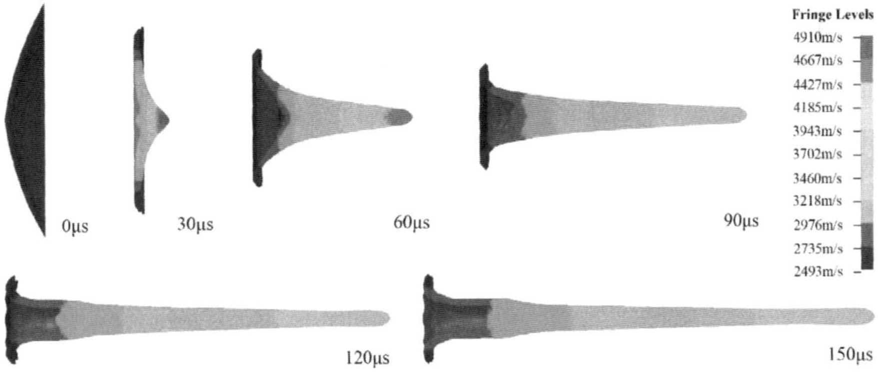

Figure 3. Image of the stages of formation of JPC.

Due to disturbance of wave-shaper, detonation wave is changed with instantaneous angle between the moving walls of the liner increasing. Detonation wave collapses and distorts the liner on its central axis, driving it into a JPC with the tip traveling at around 3~5 km/s. The boundary between jet and slug is not evident as a result of instantaneous angle and liner mass increasing, thus increasing the efficiency of the liner. This is achieved by using the sub-hemisphere liner and the wave-shaper, producing a larger-length-diameter-ratio, slower, uniformer-distributable, more stable projectile.

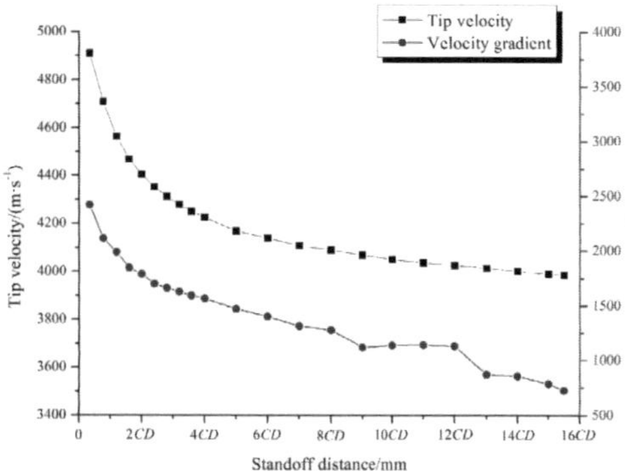

Figure 4. The change curves of velocity at different standoff distance.

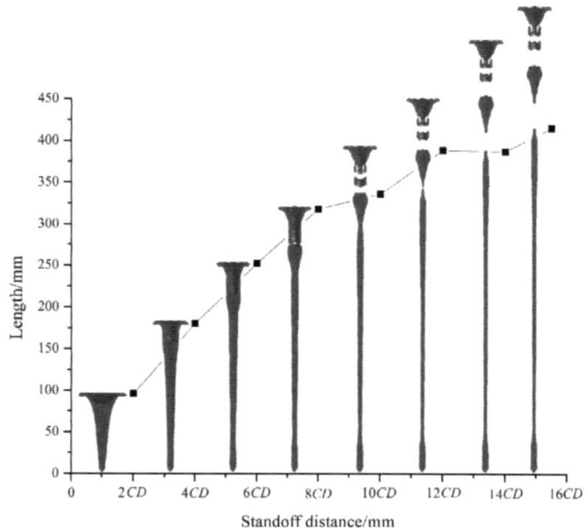

Figure 5. The change curve of length at different standoff distance.

The curve in Figure 4 shows the predicted JPC tip velocity rapidly increases to 4910m/s upon detonation of the shaped charge, and gradually decreases after a brief increase, subsequently stabilize at 4000m/s. But velocity gradient gradually decreases with time increasing. This process is conducive to maintain stable flight of JPC at a large standoff distance. Figure 5 shows that JPC stretches with necking and breaking up similar to the condition of jet at large standoff distance, even if the boundary between jet and slug is not evident. The particulates from broken tail of JPC, whose velocity is gradually decrease, are difficult to defeat the target.

Table 1. Formation parameters of JPC at different standoff distance.

Standoff	Tip velocity ($m{\cdot}s^{-1}$)	Tail velocity ($m{\cdot}s^{-1}$)	Velocity gradient ($m{\cdot}s^{-1}$)	Length (mm)	L/D
2.0CD	4403	2615	1788	96	1.6
4.0CD	4225	2660	1565	180	4.1
6.0CD	4139	2737	1402	252	6.4
8.0CD	4090	2811	1279	318	8.0
10.0CD	4051	2915	1136	336	17.0
12.0CD	4026	2897	1129	389	19.6
14.0CD	4001	3145	856	387	48.9
15.5CD	3986	3261	725	416	52.3

The calculated results of JPC formation can be seen in Table 1 at the different standoff distance, and the standoff distance includes 4CD (Charge Diameter), 6CD, 8CD, 10CD,12CD, 14CD,15.5CD. When the standoff reaches 8.5CD, it is obvious that the standoff influences on JPC formation, and tail of JPC begins to breaking up. With raising standoff, the tip velocities and velocity gradient decrease, but tail velocity increases. Meanwhile, L/D (length-diameter ratio) significantly increases because of radius slimming and length elongating. The JPC has a lot of advantages, including high velocity, uniform velocity gradient, and large L/D, with strong penetration capability. Figure 6 shows that JPC still maintains the continuous body with tail breaking up at 15.5CD standoff distance. At this time the length of continuous partial JPC was 416mm whose tip velocity reached to 3986m/s. The JPC can defeat 4CD target according to experiences. Therefore, it can be concluded that JPC has good penetration capability at 15.5CD standoff distance.

Figure 6. Image of JPC formation at 15.5CD standoff distance.

Based on Chou and Flis assumptions [9], JPC initial strain ratio of each segment is regarded as constant, and then the velocity attenuation can be ignored. With taking the average diameter of JPC as the calculated diameter, JPC breaking time, JPC completely losing penetration capability, on the basis of dimensional analysis and hydrodynamic theory, could be expressed as

$$\bar{t}_b = 3.75 - 0.125\bar{\eta}_0 + \frac{1}{\bar{\eta}_0} \tag{6}$$

where $\bar{t}_b$ is the dimensionless breaking time; $\bar{\eta}_0$ is the dimensionless initial strain ratio.

Introduction of equation of initial strain ratio as

$$\bar{\eta}_0 = \eta_0 r_0 / C_p \tag{7}$$

where η_0 is the initial strain ratio, depended upon the movement velocity (V_0) and initial length (L_0), which $\eta_0 = V_0 / L_0$; r_0 is initial undisturbed radius; $C_p = (Y / \rho_0)^{\frac{1}{2}} = 0.15 km / s$ for copper liner.

Substitution of equation (7) for $\bar{\eta}_0$ and we find

$$t_b = \frac{r_0}{C_p}(3.75 - 0.125\frac{\eta_0 r_0}{C_p} + \frac{C_p}{\eta_0 r_0}) \tag{8}$$

Taking the formation parameters of JPC simulation at 200μs to substitute for related parameters of equation (8), we can get the value of the breaking time. Furthermore, the effective standoff distance can be obtained by multiplying the breaking time with the movement velocity. When JPC flight exceeds effective standoff distance, it is difficult to defeat the target for JPC whose breaking up is serious. The Formation parameters of JPC are shown in Table 2.

Table 2. Formation parameters of JPC at 200μs.

Length (mm)	Tip velocity ($m \cdot s^{-1}$)	Tail velocity ($m \cdot s^{-1}$)	Tip radius (mm)	Tail radius (mm)
291	4107	2790	3.93	19.64

For the breaking time and the effective standoff distance, we take t_b =536μs and H=2201.3mm. The effective standoff distance of the shaped charge warhead reaches 20CD.

4. Conclusions

From the results of the simulation solution and the theoretical calculation studies presented in this paper the following conclusions may be drawn:

1. On the basis of the theory model of jet necking and breaking time as well as other related researches, a shaped charge warhead whose JPC has good penetration capability at large standoff distance was designed.
2. When the standoff reaches 8.5CD, it is obvious that the standoff influences on JPC formation, and tail of JPC begins to break up. With raising standoff, the tip velocities and velocity gradient decrease, but tail velocity increases. Meanwhile, L/D significantly increases because of radius slimming and length elongating.

3. The simulation concludes that JPC of the shaped charge warhead forming has good penetration capability at 15.5CD standoff distance. The effective standoff distance of the shaped charge warhead, on the basis of theoretical calculation, reaches 20CD.

Acknowledgments
The work presented in this paper has been funded by the Graduate Research and Innovation Foundation of Jiangsu province under NO. SJZZ16_0067.

References
[1] Bender D and Carleone J 1993 Explosively formed projectiles *Tact. Miss. Warh.* (Washington DC: The American Institute of Aeronautics and Astronautics) pp 367-86
[2] Frueh P and Heine A 2014 Aiming at quantifying the deceleration of shaped charge JET particles in air at large stand-offs *Prco. 28th Int. Symp. Ballistics (Atlanta)* vol 2 (Lancaster: DEStech Publications) pp 22-26
[3] Mayseless M, Kaufmann H, Misiuk L and Hirsch E 2014 Very long standoff penetration of shaped-charge JETs *Prco. 28th Int. Symp. Ballistics (Atlanta)* vol 1 (Lancaster: DEStech Publications) pp 195-206
[4] Huang Z X, Chen H W and Guan C 2000 The design of liner under large standoff distance condition *J. Proj., Rock., Miss. Guid.* **3** 51-53
[5] Zhang Y H, Fu J P, Li X J, Hou X C and Chen Z G 2013 Preliminary experimental study of jetting penetrator charge penetration at large stand-off *J. N. Univ. Chin. (Nat. Sci. Ed.)* **34** 613-615
[6] Fu L, Li S Q, Wang J W, Yin Z J, Yu G H and Zhou J X 2015 Effect of liner height and curvature radius on jet penetration power under high standoff *J. Proj., Rock., Miss. Guid.* **35** 65-69
[7] Zheng Z M 1981 Stability of jet produced by shaped charge *Expl. Shock Waves* **1** 6-17
[8] Li W B, Wang X M and Li W B 2010 The effect of annular multi-point initiation on the formation and penetration of an explosively formed penetrator *Int. J. of Imp. Eng.* **37** 414-424
[9] Sui S Y and Wang S S 2000 *Terminal Effects* (Beijing: National Defense Industry Press) pp 196-258

Simulation study on combustion of biomass

M L Zhao[1], X Liu[1], J W Cheng[2,3], Y Liu[2] and Y A Jin[2]

[1]Jilin Province Electric Research Institute Co. LTD, Changchun, 130021, China
[2]College of Automotive Engineering, Jilin University, Changchun, 130025, China

Email: 1055236456@qq.com

Abstract. Biomass combustion is the most common energy conversion technology, offering the advantages of low cost, low risk and high efficiency. In this paper, the transformation and transfer of biomass in the process of combustion are discussed in detail. The process of furnace combustion and gas phase formation was analyzed by numerical simulation. The experimental results not only help to optimize boiler operation and realize the efficient combustion of biomass, but also provide theoretical basis for the improvement of burner technology.

Keywords: biomass combustion, numerical simulation analysis, boiler.

1. Introduction

The burning phenomenon is the interaction of complex physical and chemical processes. Combustion is a violent exothermic oxidation reaction, but the related physical processes, specifically the exchange of energy, mass and momentum, also play an important role in the combustion system. The combustion model of the actual processes is determined almost entirely by turbulence. There is a strong interaction between turbulence and the gas phase. Heat released by the reaction can affect turbulence [1]. The turbulence is also influenced by the mixing of the reactant and the product. The $k-\varepsilon$ equation model is most often used to solve the flow problems associated with turbulence [2-3].

2. Simulation calculation conditions

In this paper, the combustion process of a small biomass boiler is simulated by CFD software. In the calculation process, the mass fraction of O_2 is set to 23.3%, the mass fraction of N_2 is 76.7%, the initial temperature of each region is 283.15 K, and the pressure relative to the air pressure is 0 Pa. The boundary conditions are set as follows.

2.1. Boundary condition

Solid wall, velocity slip and no-penetration conditions are assumed. Therefore, relative to the air, solid wall tangential velocity and normal velocity components are zero. The boundary condition for the k−ε equation is $\frac{\partial k}{\partial y}=0$.

2.2. Entrance boundary conditions
Inlet boundary conditions are the combustion chamber and the chamber of a wind and the two winds at the entrance. The turbulence parameters at the entrance are $I=10\%$, $l=0.01m$. I is the turbulent intensity; l is the turbulent mixing length.

The turbulent kinetic energy κ and the dissipation rate of ε are given by the following formula:

$$\kappa=1.5\times(U\times I)^2 \varepsilon=C_u^{0.75}\times\frac{\kappa^{1.5}}{l} C_u=0.09$$

2.3. Exit boundary conditions
The exit boundary is taken as the cross section of the furnace outlet. In the field of export, the distribution of fluid parameters is not generally known in advance. In many cases, the mixture parameter distribution in the outlet section is needed to figure out the exact content [4].

3. Physical model and grid division
Figure 1 is an engineering drawing of the boiler. Figure 2 shows the three-dimensional geometry of the boiler. Figure 3 is a simplified diagram of the combustion chamber network and flue gas flow channel. The construction of these models has greatly improved the accuracy of the experimental results.

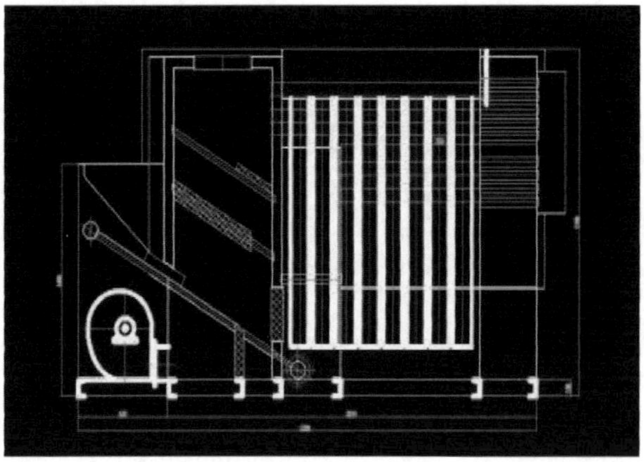

Figure 1. Engineering drawing of the boiler.

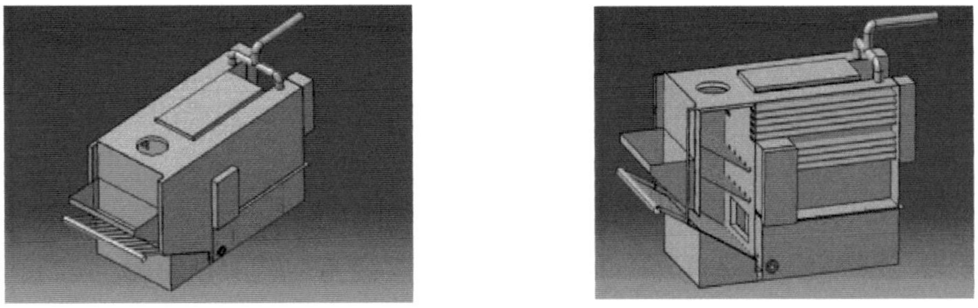

Figure 2. The three-dimensional geometry of the boiler.

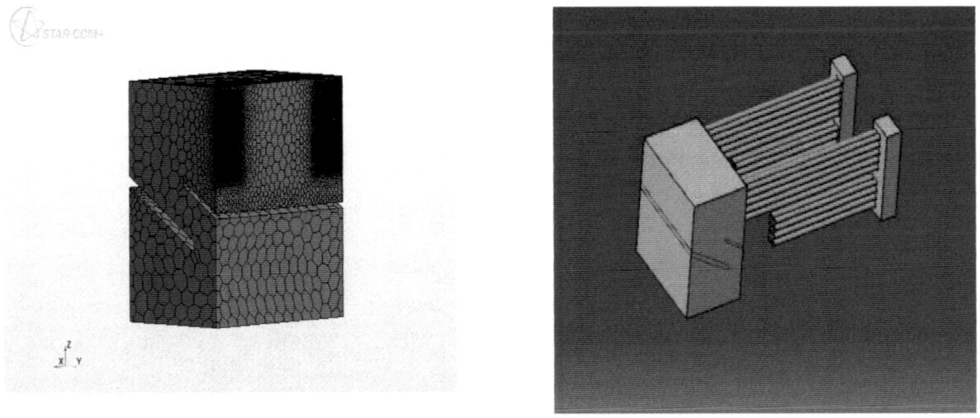

Figure 3. Simplified diagram of combustion chamber network and flue gas flow channel.

3.1. The establishment of the model and the analysis of the operation process

The model shown above was built by three-dimensional CATIA software, based on the design specifications shown in the drawing of the boiler. The operation process is simulated by setting specific inlet and outlet boundary conditions. For the control equation, the simulation equation is K. Due to its demonstrated effectiveness, the two equation model for solving flow problems was used to simulate the actual process.

4. Simulation results and analysis

4.1. Flow characteristics analysis

The flow line trajectory of the flue gas is S-shaped, as shown in Figure 4.

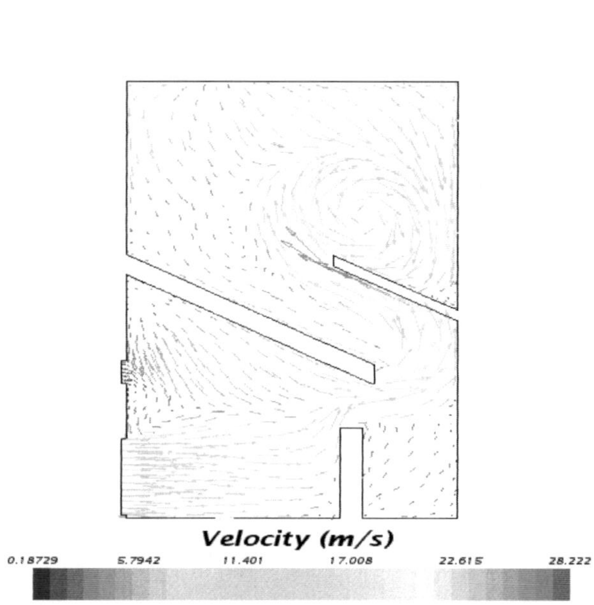

Figure 4. Velocity distribution in the combustion chamber.

It shows that the installation of fire-retardant material on the inclined water discharge in the combustion chamber has played a role in prolonging the distance of smoke. This can increase the heat transfer time and improve the efficiency of heat transfer. Therefore, more of the heat of the flue gas can be transferred to the water in the pipe wall.

4.2. Product distribution analysis of combustion

The product components distribution is shown in Figure 5 and Figure 6.

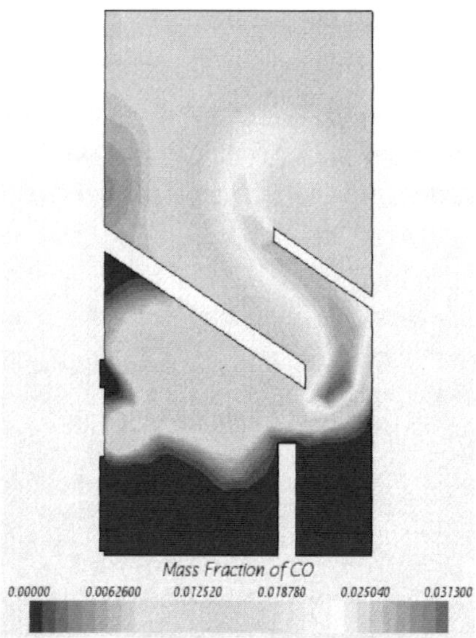

Figure 5. Distribution of CO in combustion chamber.

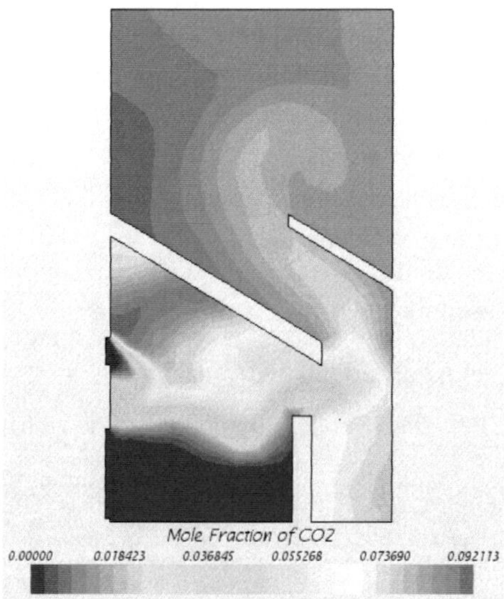

Figure 6. Distribution of CO_2 in combustion chamber.

IOP Conf. Series: Materials Science and Engineering **164** (2017) 012021 doi:10.1088/1757-899X/164/1/012021

By analyzing the distribution of gases in the combustion chamber, combustion can be optimized to burn as much of the fuel in the furnace as possible, reducing the production of carbon monoxide, and improving the thermal efficiency of combustion.

4.3. Analysis of combustion and heat transfer characteristics

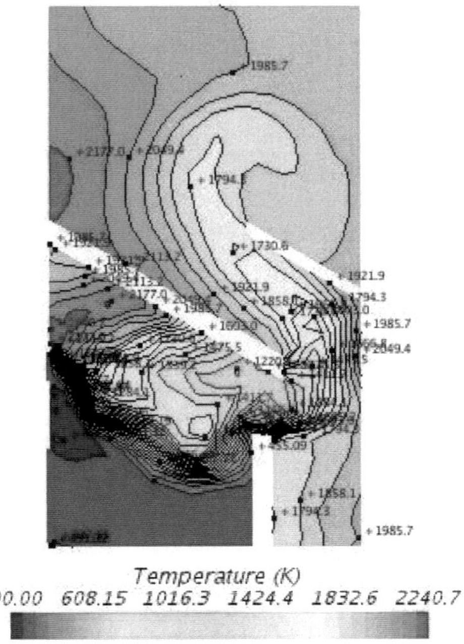

Figure 7. Temperature distribution of the combustion chamber.

From Figure 7 we can see that fuel and air enter the lower left side of the combustion chamber. The first combustion occurs after the maximum temperature has reached more than 1,000 K. The highest temperature in the first combustion chamber is found in the upper left corner. This is due to the application of fire-retardant material to the water channel, so that smoke travels upward along the S-shaped curve, resulting in the accumulation of high-temperature flue gas in the upper left corner of the chamber. As the smoke rises into the large space in the upper part of the chamber, more air can come in contact with the combustion fuel, and the temperature of the flue gas increases. The temperature distribution is S-shaped.

5. Conclusions

In this paper, the design specifications of a boiler were used to create a three-dimensional model of the boiler and set the specific boundary conditions. A numerical simulation using the equations of the turbulence model was conducted to simulate the cloud. Cluster cloud can be analyzed for any type of combustion chamber by changing the boundary conditions and other specifications in the model. This article will help to provide the basis for the research pertaining to boiler combustion and the improvement of the physical structure of the boiler [5].

Acknowledgement
The authors would like to thank the Project by Clean Development Mechanism (CDM) of China (1113050) and the Science and Technology Project of Jilin Province Development and Reform Commission (2013-729).

References

[1] Zhou Z J, Zhuo Y, Wang Z H, Hu X, Zhou J H and Cen K F 2014 Process design and optimization of state-of-the-art carbon capture technologies *Environ Prog. Sustainable Energy* **33**(3) 993-999

[2] Rajika J K A T and Narayana M 2016 Modelling and simulation of wood chip combustion in a hot air generator system *SpringerPlus* **5**(1) 1166

[3] Haseli Y, Oijen J A V and Goey L P H D 2011 A detailed one-dimensional model of combustion of a woody biomass particle *Bioresource Technology* **102**(20) 9772-9782

[4] Rajput P, Sarin M M, Sharma D and Singh D 2014 Atmospheric polycyclic aromatic hydrocarbons and isomer ratios as tracers of biomass burning emissions in Northern India *Environmental Science and Pollution Research* **21**(8) 5724-5729

[5] Wang X, Zhang H, Lin C J, Fu X W, Zhang Y P and Feng X B 2015 Transboundary transport and deposition of Hg emission from springtime biomass burning in the Indo-China Peninsula *J. Geophys. Res. Atmos.* **120**(18) 9758-9771

Low temperature performance prediction model of GAC-20 modified asphalt mixture

Z H Sun[1], H Y Cheng, G Q Zhu and J Ma

The Transportation Engineering School of Shenyang Jianzhu University, Shenyang, 110168, China

Email: happyforevernicety@126.com

Abstract. 25 sets of GAC-20 modified asphalt mixture were designed by means of orthogonal design method. The bending and low temperature creep tests of the GAC-20 were carried out. The related models of the fractal dimension and the road performance evaluation index including low temperature bending failure strain ε_B and bending strength R_B are established by using fractal theory. The model can be used to predict the low temperature performance of GAC-20 modified asphalt mixture according to the design gradation, which can reduce the test workload and improve the working efficiency, so as to provide the reference for engineering design.

1. Introduction

Asphalt mixture low-temperature performance is an important component of road performance, especially for the northeastern region. If the correlation model between asphalt mixture fractal dimension and low temperature performance evaluation index can be established, the low temperature performance of asphalt mixture can be predicted through the gradation fractal dimension to reduce the amount of test work. Based on the correlation analysis between the fractal dimension and the evaluation index of low temperature performance, the low temperature performance prediction model is established and the low temperature performance prediction model of asphalt mixture is recommended through the comparison of multiple models.

2. The raw material performance test

Liaohe petroleum asphalt grade A No.90 [5], which is widely used in the northeast of China and the basic performance test results are shown in Table 1, the basic performance test results of SBS modified asphalt are shown in Table 2 [1].

Table 1. Matrix asphalt technical index.

Detection index	Unit	Test value	Specification requirements
Penetration (25℃, 100g, 5s)	0.1mm	86.3	80-100
Ductility (15℃)	cm	＞100	≥50
Softening Point (R&B)	℃	45.9	≥45

Table 2. SBS Modified asphalt technical index.

Detection index	Unit	Test value	Specification requirements Conclusion
Penetration (25℃, 5s, 100g)	0.1mm	83.6	80~100
Softening Point (R&B)	℃	52.0	≥50
Ductility (5℃, 5cm/min)	cm	＞100	≥40
Kinematic viscosity (135℃)	Pa·s	2.7	≤3
Elastic recovery (25℃)	%	91.2	≥90

The coarse and fine aggregate of GAC-20 modified asphalt mixture use limestone gravel produced by Liaoyang Xiaotun victory quarry. The basic performance test results are shown in Table 3 [3].

Table 3. Technical index of limestone coarse aggregate.

Material specification(mm)		26.5-31.5	19-26.5	16-19	13.2-16	9.5-13.2	4.75-9.5
Technical index	Standard value	Test value					
Crushing value (%)	≤24	15					
Apparent relative density (T/m^3)	≥2.5	2.729	2.726	2.73	2.718	2.729	2.732
Water absorption rate (%)	≤2.0	0.12	1.18	0.26	0.28	0.38	0.62
Adhesion with asphalt (Grade)	≥4	4					
Consistency (%)	≤8	8					
Content of needle and sheet granular (%)	≤12	12					
<0.075Particle content (%)	≤1	0.3		0.3	0.3	0.3	0．3

Grade A No.90 road petroleum asphalt, SBS modified additives and limestone were tested in accordance with the requirements of the road usage.

2. Model building

The experimental results and the corresponding fractal dimensions of the low temperature stability requirements in Northeast China are summarized in Table 4.

It can be seen from Table 4 that the range of fractal dimension satisfying the low-temperature bending strain [4] is D=2.3388~2.5835, Dc=2.3048~2.5171, D_f=2.3734~2.5870.

The ternary linear regression model is established through taking ε_B as the dependent variable, taking D, Dc, D_f as the independent variables, the abnormal point in the data is found by using residual error analysis. The regression residual error chart of the low temperature bending strain [2] and fractal dimension are obtained by regression analysis of low-temperature bending strain and fractal dimension by MATLAB program, as is shown in Figure 1.

Table 4. The fractal dimension of GAC-20 modified asphalt mixture and the low temperature test data.

Gradation number	Average maximum load (N)	Average span deflection (mm)	Bending strain ε_B ($\mu\varepsilon$)	Bending strength Mpa	D	D_c	D_f
2	1150	0.53	2973	9.488	2.5352	2.5046	2.5104
7	667	0.57	3010	5.335	2.3405	2.3767	2.4636
9	778	0.53	2818	6.389	2.5367	2.4487	2.5468
11	824	0.63	3254	7.032	2.5026	2.3988	2.4749
16	701	0.55	2883	5.825	2.3388	2.3048	2.4281
21	883	0.7	3765	7.281	2.4860	2.5171	2.3734
22	1116	0.64	3371	9.199	2.4282	2.4895	2.5870
24	797	0.54	2812	6.571	2.5835	2.5044	2.4632

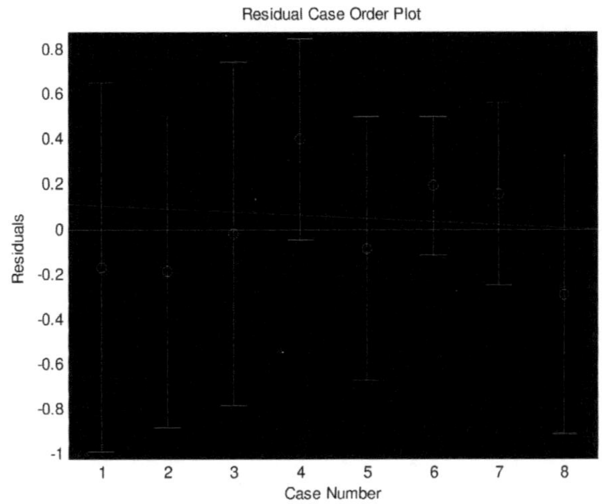

Figure 1. The residual diagram of low temperature bending strain and fractal dimension for GAC-20 modified asphalt mixture.

It can be seen from Figure 1 that there is no abnormal data, the correlation model of the bending strain and the fractal dimension [6-8] is established by the regression analysis, as is shown in formula (1).

$$\varepsilon_B=3884.8-2574.8D+4145.9D_c-1832.3D_f \quad (1)$$

Regression coefficient $R^2=0.5075$.

The regression coefficient is low in formula (1), it can be seen from Figure 1, the data of gradation GAC-20-11 can be considered to eliminate as it is unbalanced and larger deviation, the regression model of bending strain and fractal dimension is obtained by applying software MATLAB, as is shown in formula (2).

$$\varepsilon_B=-524.3-2779.6D+2999.9Dc+1210.9D_f \quad (2)$$

Regression coefficient $R^2=0.9446$.

Similarly, the ternarylinear regression models of bending strength is established, as is shown in the formula (3).

$$R_B=-39.9567-4.0288D+14.8868Dc+8.3331Df \quad (3)$$

Regression coefficient $R^2=0.5784$.

The ternary linear correlation models of bending strength and bending strain with three fractal dimensions are respectively established, the correlations of data in Table 4 are analyzed by using SPSS software to obtain the correlation between the bending strain, bending strength and fractal dimension, as is shown in Table 5.

Table 5. Correlation between low temperature performance evaluation index and fractal dimension of GAC-20 modified asphalt mixture.

	ε_B	R_B	D	D_c	D_f
ε_B	1.000	0.334	-0.072	0.360	-0.307
R_B	0.334	1.000	0.373	0.649	0.476
D	-0.072	0.373	1.000	0.735	0.166
D_c	0.360	0.649	0.735	1.000	0.190
D_f	-0.307	0.476	0.166	0.190	1.000

It can be seen from Table 5, the correlation sequence of low temperature bending strain ε_B and the fractal dimension D, D_C, D_f from large to small is $D_c > D_f > D$, indicating that the relation between the coarse aggregate fractal dimension and bending strain is relatively large, the correlation between ε_B and D is relatively small.

The correlation between R_B and D_c is larger, the correlation between R_B and D is smaller, the correlation of bending strength R_B and fractal dimension D, Dc, D_f is larger than the bending strain ε_B. The correlation model of ε_B, R_B and the larger correlated fractal dimension is established.

The correlation model of ε_B and D_c is established, as is shown in the formula (4).

$$\varepsilon_B=8484.9D_c^2-39460D_c+48828 \quad (4)$$

Regression coefficient $R^2=0.1431$.

The correlation model of R_B and D_c is established, as is shown in the formula (5).

$$R_B=33.656D_c^2-150.03D_c+172.62 \quad (5)$$

Regression coefficient $R^2=0.4309$.

Earlier, the ternary linear correlation model of low temperature index and fractal dimension, and the model of low temperature index and larger correlation fractal dimension are established; then how is the correlation between low temperature performance indexes and two fractal dimension, the correlation model of bending strain, bending strength and two larger correlated fractal dimension are established by using software MATLAB.

The correlation model of ε_B and fractal dimension D, D_c is established, as is shown in formula (6).

$$\varepsilon_B=1086.9-3195.7D+4000.6D_c \quad (6)$$

Regression coefficient $R^2=0.886$.

The correlation model of bending strain ε_B and fractal dimension D_c, D_f is established, as is shown in the formula (7).

$$\varepsilon_B=-2215.38-725.198D_c+2784.562D_f \quad (7)$$

Regression coefficient $R^2=0.432$.

The correlation model of bending strength R_B and fractal dimension D_c, D_f is established, as is shown in the formula (8).

$$R_B=-40.908+11.359D_c+8.181D_f \quad (8)$$

Regression coefficient R^2=0.550.

3. Model selection

As described above, a correlation model of low-temperature bending strain, bending strength and fractal dimension is established, and the results are summarized in Table 6. It can be seen from Table 6 that the prediction accuracy of model 1 and 5, namely, ternary linear predictive model is relatively high, and the model 1 and 5 are recommended as the prediction model of low temperature bending strain and bending strength through multi-model comparison.

Table 6. The prediction model comparison of bending strain and bending strength for GAC-20 modified asphalt mixture.

Model No.	Model expression	Regression coefficient	Advantages and disadvantages
1	$\varepsilon_B=-524.3-2779.6D+2999.9Dc+1210.9D_f$	0.9446	Higher regression coefficient, Factor analysis is more comprehensive
2	$\varepsilon_B=8484.9D_c^2-39460D_c+48828$	0.1431	Low regression coefficient, Factor analysis is single
3	$\varepsilon_B=1086.9-3195.7D+4000.6D_c$	0.8860	Higher regression coefficient
4	$\varepsilon_B=-2215.38-725.198D_c+2784.562D_f$	0.4320	Low regression coefficient
5	$R_B=-39.9567-4.0288D+14.8868Dc+8.3331D_f$	0.5784	Low regression coefficient Factor analysis is more comprehensive
6	$R_B=33.656D_c^2-150.03D_c+172.62$	0.4309	Low regression coefficient Factor analysis is single
7	$R_B=-40.908+11.359D_c+8.181D_f$	0.5500	Low regression coefficient

4. Conclusion

The correlation model recommended between the fractal dimension and the evaluation index of low temperature performance can be used to predict the low temperature performance of GAC-20 modified asphalt mixture according to the design gradation, which can reduce the test workload and improve the working efficiency

Acknowledgement
This research was financially supported by the Natural Science Foundation of China (51178278).

References

[1] Sun Z H, Zhu G Q, Ma J and Bin Q 2014 Comparison Study on Road Performance of Modified Asphalt Mixture for Intermediate Course *Advanced Materials Research* **941-944** 287-290
[2] Occupation Standard of the People's Republic of China JTG E20-2011 2011 Standard Test Methods of Bitumen and Bituminous Mixtures for Highway Engineering
[3] Occupation Standard of the People's Republic of China JTG E42-2005 2005 Test Methods of Aggregate for Highway Engineering
[4] Sun Z H, Yu Q B, Wang T B, Yu B Y, Zhu G Q and Ma J 2014 The Effect of Asphalt and Aggregate Gradation on the Low-temperature Performance of Asphalt Mixtures for Intermediate and Underlying Course *Applied Mechanics and Materials* **505-506** 251-254
[5] Highway Science Research Institute, Ministry of Communications JTG F40-2004 2004 Technical Specifications for Construction of Highway Asphalt Pavements
[6] Sun Z H, Yang G F, Yu B Y and Wang T B 2010 Study on the Gradation Inspection Methold Based on Fractal Theory *Journal of Shenyang Jianzhu University (Natural Science)* **26**(5)

875-880

[7] R K Chakraborti, K H Gardner, J F Atkinson and J E V Benschoten 2003 Changes in Fractal Dimension during Aggregation *Water Research* **37**(4) 873-883

[8] Villani M M 2014 Application of Fractal Analysis for Measuring the Effects of Rubber Polishing on the Friction of Asphalt Concrete Mixtures *Wear* **320**(1) 179-188

3D printing technology as innovative tool for math and geometry teaching applications

M Huleihil[1]

The Arab Academic Institute of Education, The Academic Institute Beit-Berl, Kfar Saba 44905, Israel

Email: cs.berl@gmail.com

Abstract. The industrial revolution and automation of production processes have changed the face of the world. Three dimensional (3D) printing has the potential to revolutionize manufacturing and further change methods of production toward allowing in increasing number of people to produce products at home.
According to a recent OECD (see Backer [1]) publication, "...tapping into the next industrial revolution requires actions on many levels and in many different areas. In particular, unlocking the potential of emerging and enabling technologies requires policy development along a number of fronts, from commercialization to regulation and the supply of skills through education."
In this paper we discuss the role of schools and their responsibility to act as quickly as possible to design a plan of action that will prepare the future citizens to deal with this new reality. This requires planning of action in different directions and on different planes, such as labs, teachers, and curricula.
3D printing requires higher levels of thinking, innovation and creativity. It has the power to develop human imagination and give students the opportunity to visualize numbers, two-dimensional shapes, and three-dimensional objects. The combination of thinking, design, and production has immense power to increase motivation and satisfaction, with a highly probable increase in a student's math and geometry achievements. The CAD system includes a measure tool which enables and alternative way for calculating properties of the objects under consideration and allows development of reflection and critical thinking.
The research method was based on comparison between a reference group and a test group; it was found that intervention significantly improved the reflection abilities of 6th grade students in mathematics.

Keywords: education, mathematical thinking, additive manufacturing, 3D printing.

1. Introduction

In the 21st century, challenges to humanity are diverse. New generations are subject to fast advances in science, technology, and technological systems. These advances expose barriers to human development. The educated citizen should be able to adapt accordingly in order to find a direction to his life.

Perception of the real world depends on the senses. Mathematics and geometry require imagination and abstract thinking, making it very difficult for newcomers to advance in these fields. This (almost pre-ordained) barrier discourages students from pursuing these tracks.

Although, additive manufacturing, such as three-dimensional printing, requires multidisciplinary skills, e.g., critical thinking, problem solving, and computer-based skills, the final product is a physical object that was designed by the student, one that will add another dimension to the learning experience, and increase motivation and curiosity.

What makes 3D printing attractive to teaching mathematics and geometry? What is 3D printing? What are the types and applications of 3D printing? How could 3D printing be implemented in teaching?

In this article, answers to these questions are given by reviewing the literature and by proposing a plan of action for teachers and students.

Intuitively thinking, 3D printing requires skills from different disciplines and skills, mathematics (numbers, operators, and relations), geometry (plane and solid), problem solving, imagination, innovation, design [2], material science, and production processes. Putting these together is a challenge full of motivation and curiosity.

According to Wohlers and Gornet [3], the method of manufacturing by adding layers of material was initiated in 1987 with stereo lithography (SL). A material in the liquid state which is sensitive to light was used and the product was produced by solidifying the layers of the sensitive material to light.

As was described in Ventola [4], the produced object by the 3D printing manufacturing method is obtained by adding (printing) multiple layers of materials.

According to Gross et al. [5], different methods exist in the market. These include several manufacturing methods. The most popular method is fused deposition modelling (FDM) which was developed by Scott Crump of Stratasys. FDM is widely used manufacturing technology for rapid prototyping to fabricate a 3D model by extending (extruding) thermoplastic materials and depositing the semi molten materials onto a stage, layer by layer [6-7].

According to the Leading Edge Forum (LEF) [7], it was found that 3D printing is evolving rapidly, with practical examples in numerous fields and industries including defence, aerospace, automotive, and healthcare.

As was reported by Gross [5], 3D printing technology is suitable for industries such as automotive and aerospace (printing prototypes of car and airplane parts), architecture (printing structural models), medical advances, applications in the food industry, in fashion design, and in other private, and government applications.

As was addressed in [4], the method of 3D manufacturing is expected to make a revolution in medicine and other fields. 3D printing technology is developing continuously and new applications constantly appear in the market. As was reported in Zoran [8], 3D printing technology has the potential to revolutionize the production of acoustic instruments.

Some of the materials used by FDM to fuse or deposit layers to produce 3D objects are plastic, metals, ceramics, powders, liquids, and even living cells. [4].

Although 3D printers were known for many years, they were unaffordable for everyone, which explains why these printers were used mostly for industrial applications. What makes these printers more affordable is their significant growth in the last few years, as a result of which their prices have fallen more than tenfold, which makes them available for almost everyone (personal usage, educational usage, small businesses usage, etc ...) [6].

Berman [9] looked into the properties and fields of applications and compared them with other manufacturing methods. Among many others, 3D printing is superior in low volume production, cheaper products, in the production of replacement parts, dental crowns, and artificial limbs, as well as in bridge manufacturing [9]. Another observation was reported by Szulżyk-Cieplak et al. [6], 3D printing can produce products that cannot be produced by other methods. This is due to the fact that material is added layer by layer which could trace almost any design and any shape.

Another important point was stated by Cambel et al. [10]. As the researchers claimed, the additive manufacturing or the 3D printing technology is about to make new revolution in manufacturing and production of objects.

2. 3D printers in education

Additive manufacturing is an attractive tool for teaching and education. As was reported by Szulżyk-Cieplak et al. [6], 3D printing technology is undoubtedly a new chapter in teaching. 3D printers directly support the teaching process. Computer aided designed (CAD) and printed physical models facilitate better understanding of the creation process. They also enhance students' involvement in the classes. 3D printing enables students to transfer their ideas into reality. Students who can physically examine their projects will more enthusiastically participate in classes, and their spatial imaginations are effectively stimulated. Printed models let students learn about the strengths and weaknesses of their projects [6]. It is important to add that at early ages, teaching starts with physical objects, and only later moves towards abstraction (3D model). 3D printing works the other way around, starting from CAD design and ending with physical objects. Actually, it is possible to link both processes during the learning stage, which aids in developing creativity and imagination.

3D printing in education is expanding rapidly. The following is not intended to be comprehensive, and merely serves to provide a few examples and outline the directions being taken in this developing field.

In her article, Loy [11] addressed the suitability of 3D printing as environment of education and e-learning. The researcher highlights the importance of the virtual and physical worlds in design and education.

Irwin et al. [12] reported and discussed the issue of the 3D printing revolution in the science, technology, engineering, and mathematics (STEM) curriculum, which should have activities with an iterative process that involves statement of the problem, suggesting methods of solution, and optimizing creative solutions and designs. The article discussed the complicated role of educators, who are struggling with how to implement next generation science standards (NGSS) [12].

In his article, Eisenberg [13] addresses several points regarding 3D printing, pointing out that nowadays 3D printers are cheap and affordable, this make them available to young children. This fact raises a challenge of educating the young students to use 3D printers. The education process is a complex of several fields of design and manufacturing, which is not intuitive.

In his study, Segerman [14] discusses the importance of 3D printing for mathematical visualization, comparing 3D printing (additive manufacturing) with "subtractive" manufacturing techniques, in which material must be cut and removed from initially solid objects and objects with complex design can be very hard to manufacture. With the additive procedure of 3D printing, these difficulties are greatly removed. The procedure includes three steps: mathematical concept → computer model → 3D printed object [14].

In a 2013 article, DFE [15] discusses the issue of 3D printers in schools and its uses in a curriculum focused on enhancing the teaching of STEM and design projects. It was pointed out in the study that the 3D printer is greatly suited to project work, where studying stems naturally as part of an analysis or a design project. It was found that the printer encourages cross-curricular thinking [15].

In an article by Rainone et al. [16] entitled "IMAGINARY Math Exhibition using Low-cost 3D Printers," they reported that they attempted to reproduce and construct 17 shapes of the IMAGINARY Open Mathematics Exhibition (www.imaginary.org) by means of cheap, desktop 3D printers.

In Slavkovsky's [17] "Feasibility Study for Teaching Geometry and Other Topics Using Three-Dimensional Printers" thesis, she focused on using 3D printers for teaching mathematics as an illustrating tool. She relied on the work of George Hart (http://www.georgehart.com/sculpture/sculpture.html) regarding some thoughts about design and teaching material. According to her expectations, 3D printers will be needed by students nowadays and in the future in order to enhance their studies and understanding. 3D printers will help students to increase their spatial intuition, evaluate production cost and understand the need of optimization through the process of design and producing (printing) objects [17].

3. Materials and methods

As was discussed in the previous section, the processes of additive manufacturing account for three major steps: Design on paper (mathematical concept), CAD, and printing (manufacturing).

Each step includes several sub-steps, skills, and the existence of a body of knowledge acquired from different disciplines. This might be a barrier to newcomers.

Today, students at early ages are being exposed to computers, almost since the time of their birth. We can easily assume that they learn using technology systems faster than their older peers do, or even their teachers. This might suggest that the teachers and students could learn some subjects and/or disciplines together. In order to break symmetry, teachers are usually pre-prepared for teaching. Due to our fast-changing world, science and technology advances, there might be because of that, some overlapping in the process of teaching of teachers and students. One might suggest in order to minimize overlapping to teach teachers first and together with them develop activities for the students.

In order to be able to teach mathematics (or STEM activities with project-based learning) by using additive manufacturing, several actions should be performed. These include training the teachers to use 3D printing, and then teaching them how to transfer their newly acquired knowledge to their students. The teachers will be able to use this new information to teach mathematics and geometry in middle school. These new techniques can be used to provide instruction in almost any subject of a teacher's specialty. (In fact, it is possible to begin such a program even earlier, with the only barriers being the availability or affordability of 3D printers. The Arab sector in general is poor in budgeting and labs.)

The following subsections summarize the basic concepts and skills required for 3D design and printing.

3.1. Unit 1 – Technical engineering drawings

It is important to teach this unit especially because most of mathematics and science teachers do not study this subject at college or university. The focus in this unit is on projection, front, top, side, isometric view, orthographic view, and perspective view. It should be possible for the learner to work both ways—from the physical body to the projections, and vice versa. Attention must also be given to teach or reinforce the ability to imagine or predict the combination, subtraction, and intersection of bodies.

3.2. Unit 2 – Computer aided design

There are many types of CAD programs available in the market. Some programs are commercial and professional, which are not freeware (such as SolidWork, AutoCAD, Maya, and similar programs), and some are freeware and suitable for school students and children who have computers available at home (such as Tinkercad and 123d design). In these activities, the teaching will be based on Tinker-Cad (a web-based program) for elementary school and 123d design for middle school. These choices are mainly because these platforms are freeware, thereby reducing the cost barrier confronting students.

3.3. 3D printing and methods of manufacturing

This last step or unit gives the student the opportunity to see the printed body and check the result. After testing the result, it is possible to improve. Students will practice saving or exporting the design into different file formats, including STL, the format suitable for 3D printers.

Teaching will be conducted using a hands-on approach and systematic procedures, in order to make the process of teaching as smooth as possible.

3.4. Audience

The action plan will address two populations: teachers and students. Usually, the teachers are specialists in one scientific subject (chemistry, biology, physics, mathematics, and similar fields). The students come from different schools and grade levels, ranging from fourth to ninth grade.

3.5. Location

An experiment is planned to be presented in the sixth grade, to teach the very basics of solid geometry and to calculate volumes and surface areas on paper and by using a CAD system (see appendix).

3.6. Start of action

The aforementioned activities and unit plans should take place during the opening term of the next school year.

3.7. Measuring changes

In order to measure the effect of usage of 3D printing, the following actions should take place.

- Define experimental and reference groups.
- Test both groups before taking action.
- Teach the reference group following the usual curriculum, without any changes.
- Teach one unit of the new curriculum.
- Test both groups using the same test.
- Analyze results and draw some conclusions.

3.8. *Hypothesis and variables*

Hypothesis:
The use of 3D printing to teach math and geometry will improve student's achievement (test score). The use of 3D printing includes the choice of an object to print (cube, sphere, box, and cylinder), choice of the dimensions of the object, representing the object in CAD program, and sending the object to the printer. The CAD program has the measure tool which allows measuring the surface area and the volume of the object. This tool allows an alternative method to find the aforementioned parameters for comparison purposes which in turn enhances reflection and critical thinking.
Variables:
Affecting variable (independent variable) – use of 3D printing (measure tool in CAD system).
Affected Variable (dependent variable) – test score.
These variables will be used to compare between the test group and reference group. Both group should take the same exam pre and post intervention.

4. Data collection and analysis

Data were collected during the last academic year to test the effect of reflection on math achievements in the sixth grade. Two groups took part in the project: a reference group and an intervention group. Table 1 shows results from both groups at three times.

Table 1. Mean test scores in mathematics of 6th grade students at three times.

Term	Reference Group	Intervention Group
A	50.3	56.5
B	54.7	61.5
C	55.2	75.0

At the start of the new academic year, two additional groups (one reference group and one experiment group) will be tested: pre-intervention and post intervention. The scores of the tests will allow a t-test comparison to be made, where the variables can be reduced through the use of the same population.

5. Summary and conclusion

In this article 3D printing or additive manufacturing were considered as tools to enhance teaching math and geometry in the middle school. The model of action described in the text includes three steps or three teaching units: design on paper, CAD design, and manufacturing.

The interdisciplinary features of the manufacturing process make it possible to address math and geometry. Design on paper, and the features and tools of CAD systems, make it possible to teach mathematical concepts and skills through geometrical construction and modeling in both two and three dimensions. Printing the designed object makes it possible to grasp and evaluate the design through checking the product. Thus, 3D printing enables mathematical thinking development.

According to Slavkovsky [17], the initial focus on the pure mathematics of basic geometric shapes was not as interesting as some later experiments. Her findings showed that it did not take very much time for students to learn to design such simple primitive structures as a cube, sphere, or cone, and then print them out. Another observation of her research was that both teachers and students wanted to learn more about 3D printing, which required patience, imagination, and a considerable amount of curiosity [17].

It is important to mention that these 3D basic concepts are taught at the sixth grade level and the students should learn to calculate volumes and surface areas. CAD systems give students the opportunity to perform calculations both on paper and by using the program. This enables them to use both their pencil and paper calculations and their physical printed objects to compare methods and reach the appropriate conclusions.

Acknowledgement
This manuscript was supported by the Center for the study of language, society and Arabic culture which is an important part of the Academic Arab Institute for Education.

Appendix

Solid entities: names and formulae.

Shape #	Name	Volume	Lateral area	Surface area
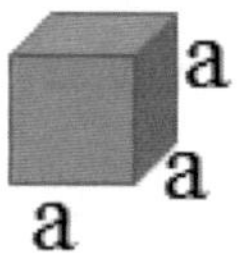	Cube Side a	a*a*a	4*a*a	6*a*a
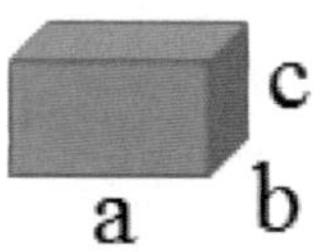	Rectangular prism (Box) Sides a, b, c	a*b*c	2*a*c+2*b*c	2*a*c+2*b*c+2*a*b
	Cylinder Radius r Height h	π*r*r*h	2*π*r*h	2*π*r*h+2*π*r*r

References
[1] Backer K D 2015 Enabling the next production revolution: Issues paper *Shaping the Strategy for Tomorrow's Production Conference* (Copenhagen: Organisation for Economic Co-operation and Development) pp 1-25

[2] Wallace K and Clarkson J 1999 An introduction to the design process University of Cambridge p 170
[3] Wohlers T and Gornet T 2014 History of additive manufacturing Wohlers Report pp 1-34
[4] Ventola C L 2014 Medical applications for 3D printing: current and projected uses *Pharmacy and Therapeutics* **39**(10) 704-711
[5] Gross B C, Erkal J L, Lockwood S Y, Chen C and Spence D M 2014 Evaluation of 3D printing and its potential impact on Biotechnology and the chemical sciences *Anal. Chem.* **86**(7) 3240-3253
[6] Szulżyk-Cieplak J, Duda A and Sidor B 2011 3D printers – new possibilities in education *Advances in Science and Technology Research Journal* **8**(24) 96-101
[7] Leading edge forum (LEF) 2012 3D Printing and the Future of Manufacturing pp 1-36
[8] Zoran A 2011 The 3D printed Flute: Digital fabrication and design of musical instruments *Journal of New Music Research* **40**(4) 379-387
[9] Berman B 2012 3-D printing: The new industrial revolution *Business Horizons* **55** 155-162
[10] Campbell T, Williams C, Ivanova O and Garrett B 2011 *Could 3D Printing Change the World? Technologies, Potential, and Implications of Additive Manufacturing* pp 1-16
[11] Loy J 2014 ELearning and eMaking: 3D printing blurring the digital and the Physical *Educ. Sci.* **4** 108-121
[12] Irwin J L, Pearce J M, Anzalone G and Oppliger D E 2014 The RepRap 3-D printer revolution in STEM education *21st Annual Conference & Exposition* (Indiana Polis: American Society for Engineering Education) pp 1-13
[13] Eisenberg M 2013 3D printing for children: What to build next? *International Journal of Child-Computer Interaction* **1** 7-13
[14] Segerman H 2012 3D printing for mathematical visualization *The Mathematical Intelligencer* **34**(4) 56-62
[15] Department for Education (DFE) 2013 3D Printers in Schools: Uses in the Curriculum Enriching the Teaching of STEM and Design Subjects pp 1-24
[16] Rainone M, Fonda C and Canessa E 2014 Imaginary math exhibition using low-cost 3D printers *ICTP SciFabLab* pp 1-18
[17] Slavkovsky E A 2012 Feasibility study for teaching geometry and other topics using three-dimensional printers Harvard University p 158

IOP Conf. Series: Materials Science and Engineering **164** (2017) 012024 doi:10.1088/1757-899X/164/1/012024

The influence of phase transformation hardening on continuous laser processing of notches for fracture splitting of a C70S6 connecting rod

S Q Kou, Y Gao[1] and Z Shi

Roll Forging Research Institute, Jilin University, Changchun 130022, China

Email: Gaoyan_ly@126.com

Abstract. The dynamic process of local material microstructure and hardness of continuous laser grooving for fracture splitting of a C70S6 connecting rod was studied. According to the phase transformation characteristics of C70S6 steel during laser processing, the coupling calculation between the transient temperature field and phase transformation process of continuous laser grooving was carried out, and then the phase transformation process and phase compositions in the heat affected zone (HAZ) was obtained. The research results showed that the HAZ was composed of martensite and pearlite as well as residual austenite after continuous laser grooving, and the generation of the martensite in the HAZ is beneficial to the subsequent splitting process; meanwhile, the hardening effect of continuous laser grooving is remarkable on the HAZ, and the requirement for the cutting tool and technique used at the subsequent machining process for the fine boring of the big end hole should be higher.

1. Introduction

The advanced fracture splitting technology for the splitting parts such as engine connecting rod has developed rapidly in recent years with high-precision, high-quality, high-efficiency, low-cost and other significant advantages. The principle of fracture splitting process is that the initial fracture source with a certain geometry size, which is the fracture splitting notch, is processed at the predetermined splitting position, then normal stress perpendicular to predetermined fracture surface is applied, and the parts was separated by brittle fracture along the initial fracture source under the condition that there is almost no plastic deformation. The principal means for pre-processing a fracture splitting notch includes laser cutting, wire-electrode cutting and mechanical cutting. Among these cutting techniques, laser grooving has many advantages such as high heating and cooling speed, no tool wear, narrow heat affected zone and others, and it has developed rapidly in grooving for fracture splitting. At present, Nd:YAG pulsed laser grooving for fracture splitting has been widely concerned and applied [1-3], but pulse laser grooving is a repetitive process of laser drilling, fracture splitting notch is actually made up of a series of blind holes, the blind holes distribution is influenced by the laser and processing parameters and then affects the quality of the subsequent splitting process. Therefore, the quality control of pulsed laser grooving is quite complicated. Continuous Nd:YAG laser is one of the lasers used commonly in the industry. The phase transformation hardening generated during laser processing has a great effect on the subsequent splitting and machining process, meanwhile, the sharp and complex changes of the temperature and phase transformation during grooving, which is limited by the material factors and processing factors, are difficult to measure by the experimental method. Therefore, the numerical simulation of continuous laser grooving for fracture splitting of a C70S6 connecting rod was studied in this research, and the coupling calculation between the three-dimensional transient temperature field and phase transformation process was carried out with SYSWELD which has been applied in the other processing fields except for the welding [4, 5]. The influence of the microstructure and hardness in the HAZ around the fracture splitting notch on the quality of the subsequent splitting process was analyzed, and then the feasibility of the new laser source used in the fracture splitting production of connecting rod was explored for evaluating and optimizing the continuous laser grooving technique for fracture splitting.

2. Theoretical basis

Continuous laser grooving is a complexly physical interaction process between laser and materials, and it is a thermal cutting technology for the processing purpose that fracture splitting notch is obtained with required dimension at the appropriate location of the workpiece.

2.1. Heat conduction equation of continuous laser grooving for fracture splitting

Heat conduction differential equation of three dimensional transient temperature field of continuous laser grooving in the coordinate system is shown as follows:

$$\rho c\frac{\partial T}{\partial t}=\frac{\partial}{\partial x}\left(k\frac{\partial T}{\partial x}\right)+\frac{\partial}{\partial y}\left(k\frac{\partial T}{\partial y}\right)+\frac{\partial}{\partial z}\left(k\frac{\partial T}{\partial z}\right)+Q \quad (1)$$

Where T is temperature which is a function of coordinates and time; k is the material thermal conductivity; ρ is the material density; c is the material specific heat; Q is the intensity of internal heat source; t is time for heat transfer.

2.2. Phase transformation theory in continuous laser grooving

Material microstructures and properties around fracture splitting notch could affect the fracture mode, the splitting force and the quality of fracture splitting processing; moreover, they could be predicted by analyzing the phase transformation in the heating stage and cooling stage of continuous laser grooving.

2.2.1. Characteristics of phase transformation in continuous laser grooving and related assumptions. During the continuous laser grooving, phase transformation could inevitably occur due to the rapid increase and decrease of material temperature. The phase transformation process during laser grooving is very similar with the one during quenching; materials would undergo the heating and cooling process accompanied by the transformation from the material microstructures at room temperature into the austenite and martensite. But because the speed of heating and cooling is very high during continuous laser grooving, the evolution of material phase transformation becomes very complex.

C70S6 steel is the typical materials in the fracture splitting connecting rod. Based on the phase transformation characteristic of C70S6 steel and the actual situation of laser grooving, the following hypotheses can be made for the phase transformation research: (1) the original microstructures of C70S6 steel is pearlite at room temperature, and therefore a portion of material microstructures at room temperature could be transformed into austenite; (2) the material temperature must be higher than the critical austenite transformation temperature in the heating stage; (3) the cooling speed of austenite must be over a certain critical value in the cooling stage; (4) the austenite must be cooled below the critical martensite transformation temperature which is namely M_S.

2.2.2. The calculation of phase transformation amount. There is diffusive phase transformation and non-diffusive phase transformation in the whole process of continuous laser grooving for fracture splitting, the transformation from pearlite into austenite belongs to the diffusive type, and the transformation from austenite into martensite belongs to the non-diffusive type.

For the diffusive phase transformation, the relationship between the phase transformation amount and isothermal transformation is described in Avrami equation [6], as is shown in equation (2).

$$V = 1 - \exp(-Bt^n) \quad (2)$$

Where V is transformation amount; t is isothermal time; B and n are the dynamic parameters related to temperature, material composition and austenitizing coefficient.

For non-diffusive phase transformation, Koistinen and Marburger empirical formulas provide the basis for describing transformation from austenite into martensite [7], as is shown in equation (3).

$$V = 1 - \exp[-\alpha(M_S - T)] \quad (3)$$

Where V is the transformation amount; M_S is the martensite transformation temperature; T is the instantaneous temperature; α is the parameter which reflects the speed of martensite phase transformation, and α for steel is set to 0.011 in general.

3. Finite element modeling

The outer contour curve of engine connecting rod is more complex and the fracture splitting notch is located at the inside of the crankshaft hole of the connecting rod big end, as is shown in Figure 1(a). Based on the symmetry of the crankshaft hole, the laser heat source and the fracture splitting notch, the connecting rod big end was simplified to be a rectangle for simplifying the calculation and half of the rectangle along the symmetrical plane of the fracture splitting notch was selected for modeling, as is shown in Figure 1(b).

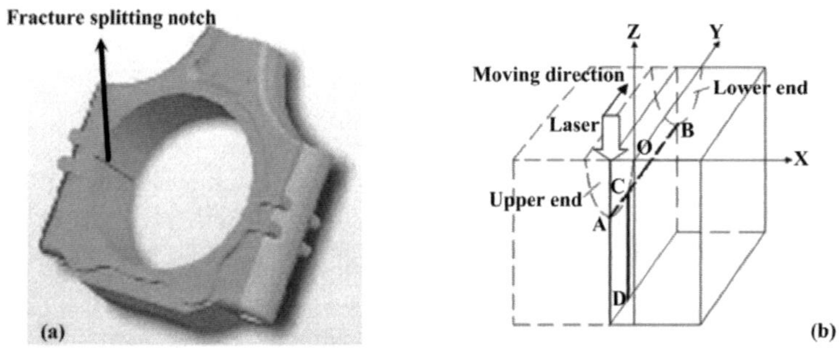

Figure 1. The position of the fracture splitting notch and the simplified model for continuous laser grooving for fracture splitting.

Temperature varies rapidly during continuous laser grooving. Therefore, the changes of material thermophysical parameters with temperature should be considered in the calculation. But for C70S6 steel, its physical properties at high temperature are quite difficult to measure accurately due to the limitation of experimental conditions and the lack of critical data, and the non-linear processing of connecting rod material properties is a difficult issue for the numerical simulation. In this research, the thermophysical parameters of 70 steel whose carbon content is similar to that of C70S6 steel was used for numerical calculation. In addition, martensite transformation temperature of steel C70S6 was 271°C based on the calculation formula of martensite transformation temperature for the pure austenitizing materials [8].

For laser grooving for fracture splitting, the required ratio of depth to width of fracture splitting notch is larger, which shows that heat flow of laser heat source imposes great influence in the thickness direction of the workpiece; therefore, 3D gauss conical heat source model was used in numerical computation. The mathematical expression of 3D gauss conical heat source is shown as follows.

$$q(r,z) = \frac{9\eta P e^3}{\pi H(e^3-1)(r_e^2+r_e r_i+r_i^2)} \exp\left(-\frac{3r^2}{r_0^2}\right) \quad (4)$$

$$r_0(z) = r_e - (r_e - r_i)(z_e - z)/(z_e - z_i) \quad (5)$$

Where r_e and r_i is heat flux distribution radius of the upper and lower surfaces of heat source respectively; z_e and z_i is z coordinate of the upper and lower surfaces of heat source respectively; r_0 is the heat flux distribution radius on the plane of z coordinate; P is laser output power; η, which is affected only by the laser and the material surface status in this research, is effective power coefficient; H is the heat source depth.

In order to guarantee the authenticity and accuracy of the numerical computation, the laser heat source model was checked by means of the heat checking function in SYSWELD software and the experiment result of continuous laser grooving of engine crankcase bearing block [9]. L3020 continuous laser numeric control machine of TRUMP Laser GmbH + Co.KG Company was used in the experiment. The output power (*P*) was 1kW, the cutting speed (*v*) was 25mm/s and the defocusing amount (Δf) was 0mm. The notch width was 0.522mm and the notch depth was 1.06mm. Under the condition of the same parameters, the simulation of continuous laser grooving was carried out, the temperature field result is shown in Figure 2, and the effective absorption rate of materials for laser energy was set to 0.5.

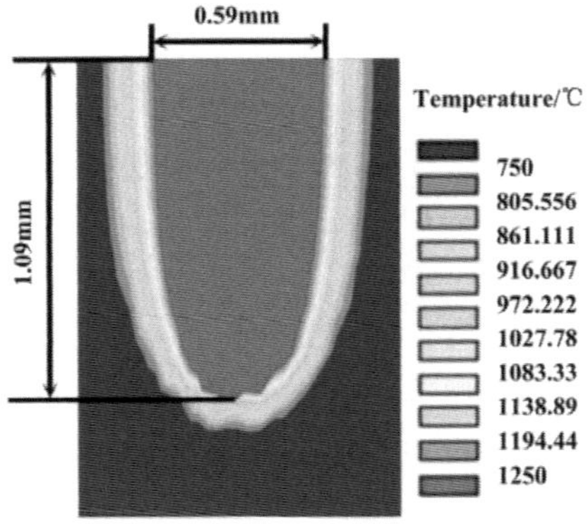

Figure 2. The simulation result of continuous laser grooving.

4. Phase transformation analysis

When the output power (*P*) is 400W, the cutting speed (*v*) is 30mm/s, the spot radius (*r*) is 0.06mm and the Δf is 0mm, the temperature distribution of continuous laser grooving is shown in Figure 3. Continuous laser grooving is a physical process which contains laser melting material or even gasifying material, and high-speed airflow blowing the molten material. Therefore, the isotherm of material melting-point temperature can be as the notch contour line to calculate the dimension of fracture splitting notch, and the notch depth was 0.64mm. The notch depth under the above condition meets the required depth range from 0.45mm-0.7mm in practical production, and then the phase transformation process under the condition of the same parameter combinations was analyzed.

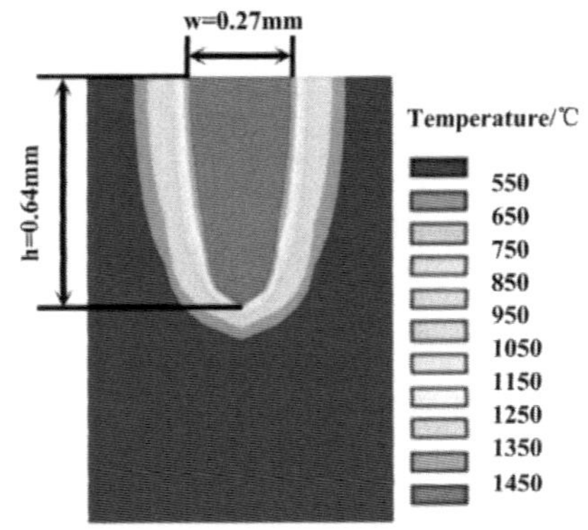

Figure 3. The dimension of the fracture splitting notch.

In the subsequent splitting process, fracture started at the root of fracture splitting notch. Therefore, this research focused on the phase transformation process and the final phase composition in the HAZ of the notch root. Due to the fact that the notch depth was 0.64mm, point E shown in Figure 4 was selected on the cross section where is 1mm away from the fracture splitting notch starting position, the z coordinate of point E is -0.66mm, and the phase transformation process at point E during grooving was analyzed.

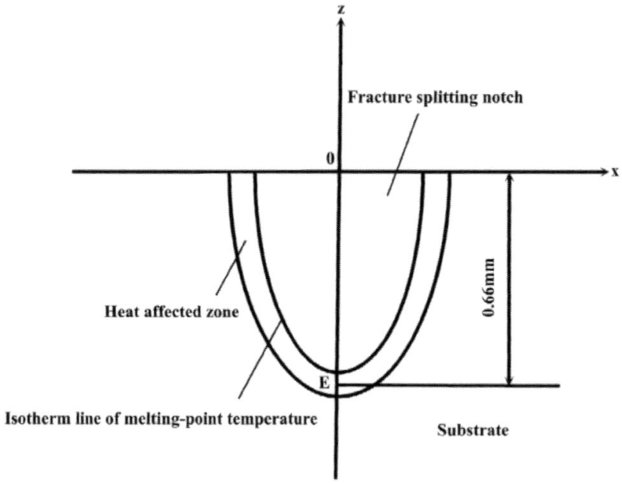

Figure 4. The cross section of fracture splitting notch.

During continuous laser grooving, the relationship curve of the content of each phase composition with temperature at point E is shown in Figure 5. When time is close to 0.02s, temperature begins to rise rapidly. When laser processing time is 0.021s, temperature rises to about 750℃ and reaches the critical austenite transformation temperature (Ac1), then pearlite begins to transform into austenite. When time is close to 0.025s, temperature reaches the highest value.

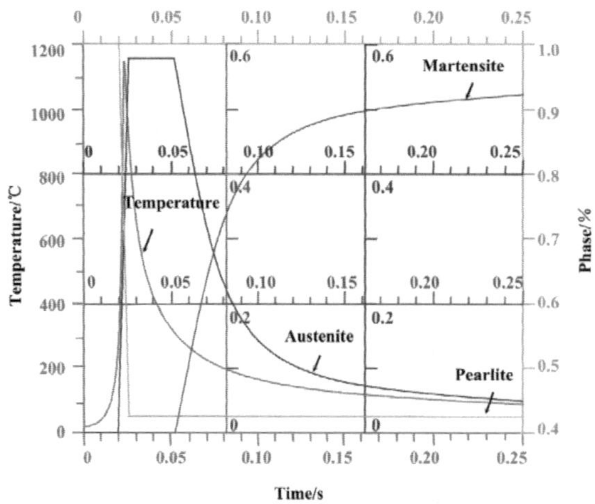

Figure 5. The relationship between the temperature and phase transformation of point E.

Afterwards, temperature at point E begins to decrease rapidly, when temperature is lower than the Ac1, pearlite stops transforming into austenite. Meanwhile, still 42% of pearlite does not transform into austenite, the main reason for which is that temperature changes quickly, the residence time at

IOP Conf. Series: Materials Science and Engineering **164** (2017) 012024 doi:10.1088/1757-899X/164/1/012024

high temperature is too short, and a part of the pearlite cannot transform into austenite in time. Then, temperature at point E continues to decrease, when laser processing time is 0.052s, temperature decreases to 270°C or so, namely Ms. In addition, it can be seen that the cooling speed at point E is very high in the temperature curve. Therefore, the requirements for the transformation temperature and cooling speed of martensite phase transformation are met, and martensite phase transformation would happen at point E. Subsequently, austenite transforms into martensite constantly until time is 0.25s, namely when temperature is below 100°C, the martensite phase transformation is basically completed, and the martensite content is no longer increased. Finally, there is still a small amount of retained austenite in the region of point E. The ultimate phase compositions at point E are described as follows: martensite accounts for 52%, pearlite accounts for 42% and retained austenite accounts for 6%, which show that the solid-state phase transformation zone consists of a large number of pearlite and martensite as well as a small amount of residual austenite after laser grooving.

The above-mentioned microstructures in the HAZ kept consistent with the expected results in application. The generation of martensite, with which material plasticity is decreased and brittleness is increased, is beneficial to brittle fracture in fracture splitting processing. And then splitting force is reduced effectively, plastic deformation during fracture splitting processing is decreased, and fracture defects are reduced, which is also the main reason why laser grooving replaced mechanical broaching and wire cutting. Moreover, microstructures in the HAZ would be removed in the finish-boring of the big end hole of connecting rod subsequently, which can ensure the requirement of microstructures and properties in the final product of connecting rod.

5. Hardness distribution

The hardness and martensite distribution around the notch is shown in Figure 6 after the continuous laser grooving is completed. It can be seen that the material hardness from the notch to the base metal trends to decrease, which shows that the materials hardness in the HAZ has been improved significantly after grooving. The main reason is that the hardness is directly related to the final phase composition when the material chemical composition remains the same. The martensite leads to the material hardness increase, the hardness decreased with the martensite content from the notch surface to the base metal, and the base metal hardness is not affected due to the fact that laser energy has little effect on the matrix microstructure.

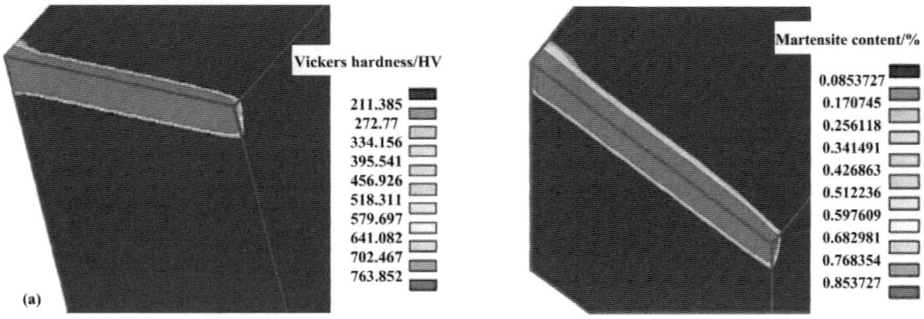

Figure 6. The hardness and martensite distribution.

The hardness distribution along the depth direction of CD line in Figure 1(b) is investigated, and the hardness curve at the CD line is shown in Figure 7. The corresponding material hardness decreases sharply from 680HV to the base metal hardness when the Z coordinate decreases from -0.64mm to -0.70mm, materials hardness at the point E is about 560HV, and the matrix hardness remains the same.

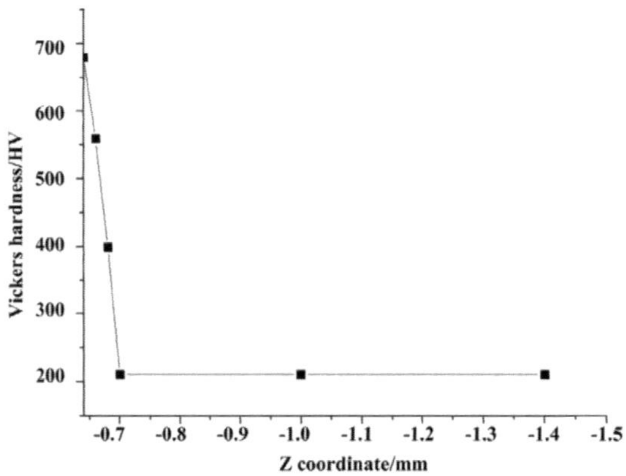

Figure 7. The hardness distribution in the depth direction of CD line.

6. Discussions

Based on the above analysis, the cutting position could be hardened locally by continuous laser, the generation of the martensite in the HAZ increases the material yield strength and reduces the fracture toughness KIC, which makes brittle fracture easier to implement for decreasing the cracking strength and reducing the cracking defects effectively. However, it should be noted that the hardening effect of continuous laser grooving is remarkable for the high carbon micro-alloy non-quenched and tempered forged steel C70S6 of connecting rod material for fracture splitting, materials hardness in the HAZ is larger than the doubled matrix hardness after grooving. After the connecting rod is separated and the bolts are assembled, the fine boring of the big end hole would be done. Therefore, the requirement for the cutting tool and technique used at the subsequent machining process is higher. The tool cuts from the base metal with lower hardness to the narrow high hardness zone at a certain speed, the cutting edge of cutting tool will remain under the impact force, then the "Cutter breakage" or "Cutter relieving" phenomenon could appear easily and the wear of cutting tool becomes severe. The cutting tool with higher hardness should be chosen in subsequent processing, and the problem, such as the size of the feed rate and spindle rotating speed, should be taken into account in the machining process.

7. Conclusions

In this paper, continuous laser was used in grooving for fracture splitting of a C70S6 connecting rod, the phase transformation process during grooving and the hardening effect was studied. Based on the actual physical process of laser processing, the coupling analysis between the three-dimensional transient temperature field and phase transformation process of continuous laser grooving for fracture splitting was carried out, and then the phase transformation process and phase compositions in the HAZ was obtained. The HAZ was constituted by pearlite, martensite and retained austenite after continuous laser grooving; the generation of the martensite in the HAZ is beneficial to the subsequent fracture splitting processing. The hardening effect of continuous laser grooving is remarkable for the HAZ of the fracture splitting notch, and the material hardness in the HAZ can be up to 680HV, which increases the difficulty of the subsequent machining and the requirement for the cutting tool.

Acknowledgements

This work is supported by Special Project of Science and Technology of China (No.2013ZX04002-021) and Key Research and Development Plan of China (2016YFD0701102).

References

[1] Kou S Q, Wang J W, Jian X X and Zheng Q F 2010 *Opt. Precision Eng.* **18** 1340-1346
[2] Kou S Q, Wang J W and Gao Y 2011 *Laser. Eng.* **20** 381-395

[3] Kou S Q, Wang J W and Zheng Q F 2011 *Laser. Eng.* **21** 115-127
[4] Feulvarch E, Fontaine M and Bergheau J M 2013 *Finite Elem. Anal. Des.* **75** 62-70
[5] Ahn D G, Byun K W and Kang M C 2010 *J. Mater. Sci. Technol.* **26** 362-366
[6] Avami M 1940 *J. Chem. Phys.* **8** 212-224
[7] Koistinen D P and Marburger R E 1959 *Acta Metall.* **7** 59-60
[8] Jiang Y, Yin Z D, Zhu J C and Li M W 2003 *Spec. Steel.* **24** 9-12
[9] Yu G 2012 *Ph. D. Dissertation* (Changchun: Jilin University)

IOP Conf. Series: Materials Science and Engineering 164(2017) 012025 doi:10.1088/1757-899X/164/1/01202:

Measurement research for pitting potential of aluminium alloy with chromate passivation layers

Z H Su[1, 3], Y Dai[2], Q Y Ma[2], J N Wang[1] and C Q Zheng[1]

[1]Chongqing Academy of Metrology and quality inspection, Chongqing, 401123, China
[2]College of Materials Science and Engineering, Chongqing University, Chongqing, 400044, China

Email: su_water@163.com

Abstract. This paper describes common technology for measuring corrosion resistant properties of aluminium alloy with chromate passivation layers. It gives a resolution to characterize corrosion resistant properties of pitting potential. The method, including sample preparation, measuring polarization and the determining of pitting corrosion potential, was proposed. Tests of repeatability and reproducibility can meet the demand of measurement precision. In addition, comparison of the positional relationships between polarization curves and nominal pitting corrosion potential was provided to judge products.

1. Introduction

Aluminium alloy is identified by its low specific gravity, good thermal conductivity and electrical conductivity as well as its reflectivity, processability and low cost. Aluminum has also been widely applied in aviation and space, transportation, light industry, building materials and other fields [1-3]. Chromate passivation can greatly improve corrosion resistance, expand the application range and prolong the service life of aluminium alloy. However, in order to meet the requirements of the European Union environmental standards (European Union RoHS directive and WEEE), relevant environmental laws and regulations of China some technologies are being phased out. In China, the hexavalent chromium electroplating technology has been gradually eliinated and environmental trivalent chromium electroplating technology or chromium-free electroplating technology are used instead [4-7]. A gap in the corrosion resistance between trivalent chromium passivation or chromium-free passivation products and hexavalent chromium passivation products can be ascertained via practical test but the quality cannot be easily controlled through this method [8-9]. Further research is necessary in order to standardize aluminum alloy passivation markets as well as guarantee their development. This will also reduce the corrosion waste, expand application of surface treatment technique of aluminium alloy, and enhance the service performance, especially the anti-corrosion capacity. It is also necessary to conduct further research the anti-corrosion capacity of aluminium alloy with chromate passivation layers. This forms the corresponding detection methods which play an important supervising function in the quality control of aluminium alloy with chromate passivation layers.

The current internationally recognized test for corrosion is the "Corrosion Tests in Artificial Atmosphers: Salt Spray Test". This conventional test evaluates the anti-corrosion properties of material surfaces. After hydrochloric acid corrosion, evaluating anti-corrosion properties by assessing weight loss or the degree of corrosion on surface spots. However, simulating a practical environment to assess passivation layers' properties can be time-consuming, costly and fuzzy [10]. Consequently, electrochemical etching, characterised to be sensitive, fast and accurate, is receiving increased attention. The energy of pitting potentials indicates corrosion resistant properties of passivation layers. Pitting corrosion potentials were measured by means of potentiodynamic polarization measurements. This method has received extensive attention in corrosion theory, while in the field of measurement engineering, rare detection methods adopt potentiodynamic polarization measurements with the exception of pitting potential measurement for stainless steels. There is no evaluation criterion in pitting corrosion potential for evaluating corrosion resistant properties so far. This paper aims to seek out such a standard criterion.

2. Experiments

2.1. Sample preparation

Stainless steel has a different sample preparation than that of preparing aluminium with chromate passivation layers for pitting potential. During the process of sample preparation it is necessary to avoid damage to the detecting surface, therefore, this paper excogitated a suit of specific sample preparation methods.

Preparation: a small block with an area equal to 1 cm^2 was cut by linear cutting machine from the chromium passivation aluminum sheet; Polypropylene (PPR) pipe was sectioned into several pieces with each piece being approximately 20 mm long; 0.5 mm^2 conducting wire was cut into several pieces, each approximately 100 mm long; approximately 10 mm of coating was stripped from the two ends of wire. The non-detection surface was cleaned with alcohol or sand paper. One end of the conducting wire was adhered on the non-detection surface with a double face adhesive tape to ensure conductivity between the wire and sample after the stick. The sample was attached with conducting wire on the flat table and with double face tape, then inset into PPR pipe and fixed on the same table. AB glue was mixed in a ratio of 1:1 quickly until smooth, then poured into PPR pipe within three minutes of mixing and along conducting wire. Non-detection surfaces and bare wires should be surrounded and covered completely. Setting of the AB glue started after sealing and ended after approximately six hours. Finally, double face tape was peeled from the table and sample. After the sample was prepared, the surface was delaminated, cleaned and dried using alcohol.

2.2. Polarization detection

Electrochemical workstation type Reference 600, supplied by American Gamry, was introduced and a three-electrode system was used during testing. Electrodes and reference electrodes were platinum pieces and saturated calomel electrode, respectively. According to GB/T 1266, 35g sodium chloride were dissolved into 965ml distilled water of deionized water to obtain sodium chloride with 3.5% concentration. The solution was injected into an electrolysis bath, the ration between solution volume and sample area was over 200mL/cm^2. The solution was then heated to approximately 30±1 degree and held there. Nitrogen with a purity of not less than 99% was bubbled through the solution for over 30 minutes until deaerated and aeration rates equalled 0.5L/min approximately. The sample was immersed in electrolysis bath after being washed and dried, which was below the surface of liquid within 1 cm and after approximately 10 minutes of immersion, the open circuit potential U_0 was measured. Polarized scanned from (U_0-0.1) V to (U_0+0.3) V at a scanning rate of 1mV/s.

3. Results and discussion

3.1. The determined method of pitting potential

GB/T 17899-1999 "The Measurement for Pitting Potential of Stainless Steel" is equivalent to JIS G 0577-1981 and the pitting potential is represented by the most positive, potential value. This corresponds with the 10 μA/cm^2 or 100 μA/cm^2 of the current density on the anodic polarization curve, because there is a continuous self-passivation phenomenon on the surface of stainless steel in the anodic polarization process. Pitting potential is represented by only the most positive potential value when analysed by the current standards, as such, the polarization current shows an increase with this fluctuation.

Stainless steel has a different, obvious curve when compare to the anodic polarization curves of aluminium allows with chromate passivation layers. Figure 1 illustrates a series of curves which are based on the logarithm of aluminium polarization potential and polarization current where a clear inflection point can be identified. Before inflecting, the current is increased slowly then a sharp increase of current is applied. The results show that aluminium alloy chromate passivation layers suffer from pitting corrosion at the inflection point corresponding to the pitting potential. Although there are minute differences between open circuit potential, the potential corresponding to the inflection point are close, therefore, it is more reasonable to use pitting potential to judge corrosion resistance of aluminium alloy with chromate passivation layers.

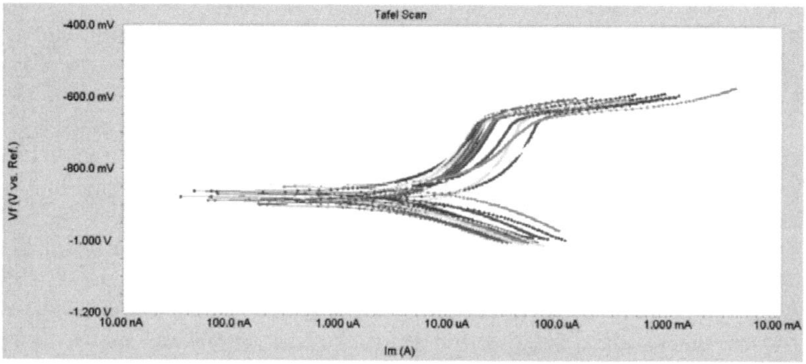

Figure 1. The potentiodynamic polarization curves.

In order for the pitting potential to correspond with the inflection point it is necessary to export dates corresponding to the anodic polarization curves and analyse them independently. By researching several potentiodynamic polarization curves and their corresponding curve date, it can be found that when sample occurred pitting corrosion, every increased 1 mv of voltage, the corresponding current increment will be increased. Based on the results, a method for measuring the pitting corrosion potential is developed. Polarization potential increases by 1mv during the process of potentiodynamic poalarization while polarization current increases by over 1μA.

3.2. Repeatability and reproducibility of analytical methods

The ability of these analytical methods to be reproduced or repeated is essential to determine accuracy of the analytical methods themselves. When it comes to the acceptability of R & R standards for measuring system, the system share common rules. As a result, a value of %R&R was less than 10 % which represents high accuracy, satisfactory testing and a reliable result. Accuracy is mainly effected by testing and analysis methods as well as an influence of repeatability that is larger than that of reproducibility. It has been shown that in the proposed method, reproducibility is within 5%, can ensure the requirements for accuracy in test and analysis methods.

Repeatability can be characterized with relative standard deviation (RSD), standard deviation divided by average value is RSD. Group A and Group B were chosen with each group comprised of 40 samples. At the same time, all samples were consistent. The proposed method were performed to execute the repeatability test with the results listed in Table 1. As can be seen from the table, repeatability of two groups of samples are lower than 5%. Consequently, this method can meet the demand of repeatability and reproducibility.

Table 1. The repeatability of test and analysis methods.

Sample	Pitting Potential / mV										Average Value / mV	Standard Deviation / mV	RSD (%)
A	634	599	637	645	619	620	649	647	636	637	628.8	14.5	2.31
	619	646	634	627	647	629	628	636	596	624			
	614	630	646	643	627	625	620	639		632			
	595	605	610	625	629	642	639	634	651	613			
B	654	653	644	658	628	634	649	632	622	619	635.4	14.9	2.35
	646	657	655	628	620	646	657	649	643	631			
	627	649	634	600	628	637	636	628	668	615			
	618	628	627	630	622	645	639	618	613	630			

3.3. Criterion for quality

A determination method of pitting potential has direct significance to the research for pitting potential of aluminium alloy with chromate passivation layers, but also demands to analyse the polarisation curves which is relatively sophisticated. A new and simple method for determining pitting potential is required in order to maintain quality control during analytical testing.

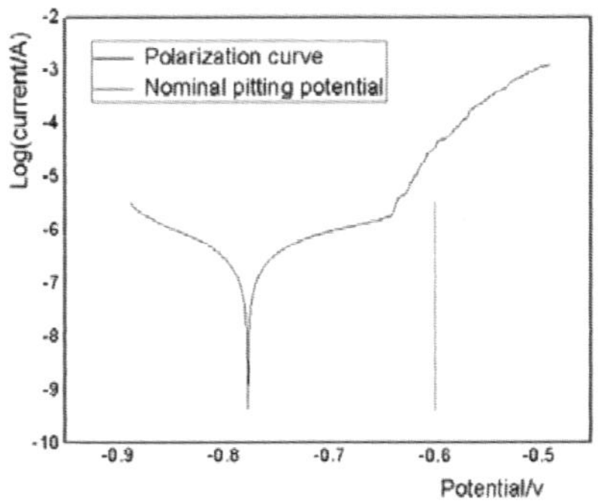

Figure 2. The judgement of product quality.

Based on pitting corrosion potential in a graph showing polarization potentiodynamic measurements, a line called pitting corrosion potential was drawn. The products' quality of chromate passivation layers determine the position of this nominal straight line. When the affecting point of polarization curves is located at the left side of the line, we conclude that the quality does not meet quality control or performance requirements. When the affecting point of polarization curves is located at the right side of the nominal pitting corrosion potential, we conclude that the quality does not meet quality control or performance requirements. When two point nearly coincide, we can tell the relationship by doubling tests. Only if all tests tell relationship between inflection point and the line considerably, can we judge them as qualified products. A criterion for quality is represented in Figure 2.

4. Conclusion

There are negatives associated with conducting salt-mist corrosion tests including the time involved, the high cost and the indeterminate results. A kind of evaluation of pitting potential of aluminium alloy with chromate passivation layers was proposed through a great deal of experimentation. This analytical method introduced the preparation process of the polarization corrosion sample and testing process of the dynamic potential polarization corrosion in detail. A new method for measuring the putting potential was developed through analysis and processing of this large data set resulting in the following conclusions. During the process of potentiodynamic polarization, while polarization potential increases by 1 mv, polarization current increases by over 1 μA, it is confirmed that the polarization potential is pitting corrosion potential. Repeatability and reproducibility of this analytical methods is helpful in determining how accurately the method is. This method can meet the demand of repeatability and reproducibility. In order to simplify the process of the criterion, comparison of the positional relationship between polarization curves and nominal pitting corrosion potential were put forward.

References

[1] Teng S L, Yin Y S, Du J B, Cheng S and Yan L N 2007 Research Progress of Rare Earth Passivation on Metal Surface *Shandong Metallurgy* **29** 7-9

[2] Yu H C, Chen B Z, Shi X C, Li B and Wu H Y 2008 Preparation and Electrochemical Properties of Trivalent Chromium Coating on 6063 Aluminium Alloy *Acta Phys Chim Sin* **24**(8) 1465-1470

[3] Kang K, Huang S H, Feng Y, Dong L Q and Zhu L A Y 2008 Fabrication of Colloidal Photonic Crystals by a Micro-cell Based on AL Thin Film *Acta Chimica Sinica* **66**(14) 1615-1619

[4] Peng J D, Deng C Y, Shi Y and Xiong Y 2011 Nature Color Passivation Process of Trivalent Chromium for Aluminum Alloy Castings *Foundry* **60**(2) 147-149

[5] Zhao Y C 2010 Trivalent Chromium Passivation of Aluminum Alloy *Electroplating & Pollution Control* **30**(2) 47

[6] Feng Y, Mcguire G, Shenderova O, Ke H and Burkett S 2016 Fabrication of Copper/Carbon Nanotube Composite Thin Films by Periodic Pulse Reverse Electroplating Using Nanodiamond as a Dispersing Agent *Thin Solid Films* **615** 116-121

[7] Jordan M, Feng Y and Burkett S 2015 Development of Seed Layer for Electrodeposition of Copper on Carbon Nanotube Bundles *Journal of Vacuum Science & Technology B* **33**(2) 021202

[8] Li J, Sun J and An C Q 2008 Research Development of Aluminium Alloy with Chromate-free Passivation *Surface Technology* **137**(4) 60-62

[9] Fang Z 2009 Development Trends of Chemical Conversion Films *Electroplating & Finishing* **28**(9) 30-34

[10] Ge S S, Yang Y X and Shao Q 2006 Morphology and Anticorrosion Property of A Black Conversion Film with Non-chromate on Casting Aluminum Alloy *Corrosion Science and Protection Technology* **18**(3) 228-230

IOP Conf. Series: Materials Science and Engineering **164** (2017) 012026 doi:10.1088/1757-899X/164/1/012026

A moderate method for preparation DMSA coated Fe_3O_4 nanoparticles

L N Song, N Gu and Y Zhang[1]

State Key Laboratory of Bioelectronics, Jiangsu Key Laboratory for Biomaterials and Devices, School of Biological Science and Medical Engineering, Southeast University, Nanjing 210096, PR China

Email: zhangyu@seu.edu.cn

Abstract. A moderate way to prepare water soluble magnetic Fe_3O_4 nanoparticles has been developed. Firstly, oleic acid coated Fe_3O_4 is prepared by coprecipitation. Second, oleic acid were replaced by 2,3-dimercaptosuccinnic acid (DMSA) to prepare DMSA/Fe_3O_4 in the mixed solution of n-hexane and acetone. After dialysis and filtration the DMSA/Fe_3O_4 can be transferred into distilled water to form stable Fe_3O_4 nanoparticle solutions. The TEM images indicated that the particles had spherical shape and the nanoparticles were found to be 12 nm with a relatively narrow size distribution with the hydrodynamic size of 30 nm. And the result of VSM shows that DMSA/Fe_3O_4 nanoparticles have a saturation magnetization of 31 emu/g. The IR spectra indicated that the iron oxide was located by carboxyl matrix.

1. Introduction

Magnetic Oxide (Fe_3O_4, γ-Fe_2O_3) nanoparticales has been widely applied in the biomedical field such as magnetic separation [1, 2], magnetic resonance contrast agent [3, 4], DNA detection [5], drug delivery [6, 7] and magnetic hyperthermia [8, 9].

The chemical method of preparing magnetic iron oxide nanoparticles can be divided into homogeneous method and Non-homogeneous method. Homogeneous method includes chemical coprecipitation, hydrothermal reactions and thermolysis of precursors. Non-homogeneous method contains microemulsions, sol-gel syntheses and so on [1, 10]. The reasons for choosing DMSA are as the follows: Firstly, DMSA can form strong complexes with Fe_3O_4 nanoparticles. Second, the SH group and COOH group can offer the coupling group for biomolecule.

Among all of them, chemical coprecipitation is a simple, convenient and widely used method. A two-step exchange method was used in the previous report [11]. In our paper, a one-step exchange method was used to synthesize DMSA coated magnetic iron oxide nanoparticles. One-step exchange method was simple, high efficiency, easy to operate and low cost. Oleic acid coated Fe_3O_4 is prepared firstly, and it was replaced by ligand exchange for DMSA/Fe_3O_4. The Size, morphology, microstructure, magnetic behavior and element of these two kinds of nanoparticles were characterized by infrared spectroscopic analysis (IR), transmission electron microscopy (TEM), dynamic light scattering (DLS), power X ray diffraction (XRD) and vibrating sample magnetometer (VSM). The nanoparticles surface was modified by DMSA, which made it contain the carboxyl group.

2. Materials

All reagents and solvents were obtained from commercial suppliers. They were applied without additional purification. Ferric chloride hexahydrate ($FeCl_3 \cdot 6H_2O$), ammonium hydroxide ($NH_3 \cdot H_2O$, 25%), ferrous sulfate heptahydrate ($FeSO_4 \cdot 7H_2O$), and oleic acid (90%) were purchased from Shanghai Chemical Reagents Company, China. Acetone, n-hexane and absolute alcohol were reagent from Sinopharm Chemical Reagent Co., Ltd. 2,3-dimercaptosuccinic acid (DMSA) was supplied by shanghai zhourui Chemical Reagent Co., Ltd.). The distilled water was deoxygenated by passing inert gas for 1 h before use.

3. Experimental

The OA coated magnetic Fe_3O_4 nanoparticles were synthesized by the method of chemical co-precipitation. A mixed solution was obtained by adding 13 g $FeCl_3 \cdot 6H_2O$ and 9.05 g $FeSO_4 \cdot 7H_2O$ into 25 mL deoxygenated water. 15 mL of 25% $NH_3 \cdot H_2O$ was rapidly poured to the mixture under vigorous stirring at 70°C. After 5 min, 2.5 mL of OA was dropped to the reaction liquid. The reaction was continued for 30min and the temperature was rose to 85°C for 1h under vigorous stirring. Then the precipitate was separated by magnetic separation. A powerful magnet was placed under the beaker with magnetic nanoparticles solution and continues until the supernatant clarified. The solution was washed with anhydrous ethanol for several times. The obtained OA coated Fe_3O_4 (OA/Fe_3O_4) were dispersed in n-hexane.

One hundred milligrames of OA/Fe_3O_4 nanoparticles was dissolved in 100 mL n-hexane. 50 mg of DMSA was dissolved in 100 mL acetone. Mixed these two together into a reaction bottle and agitate the mixture at 60°C for 4h. The black precipitate, which was DMSA/Fe_3O_4 nanoparticles, was observed by magnetic separation. Half of the black precipitate was washed with distilled water and transferred into water (Figure 1A). The other half was dried into power for sample characterization (Figure 1B).

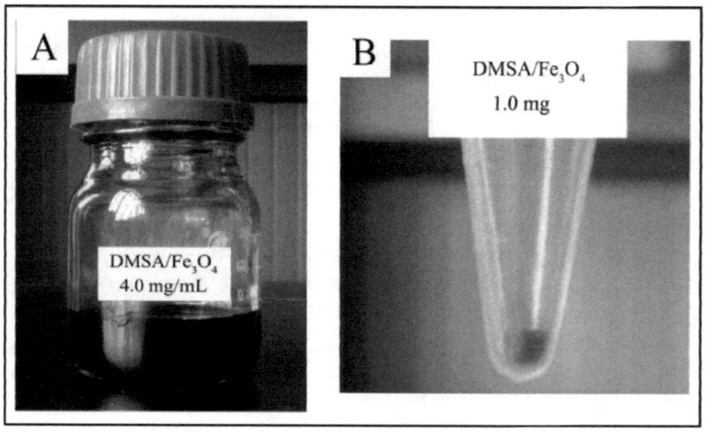

Figure 1. The aquous solution of DMSA/Fe_3O_4 nanoparticles (A) and DMSA/Fe_3O_4 nanoparticles powder (B).

4. Results and discussion

The core size and morphology of OA/Fe_3O_4 nanoparticles and DMSA/Fe_3O_4 nanoparticles were observed by TEM. The TEM micrograph of OA/Fe_3O_4 nanoparticles and DMSA/Fe_3O_4 nanoparticles show in Figure 2. Result shows that the morphology of OA/Fe_3O_4 and DMSA/Fe_3O_4 nanoparticles are quasi-spherical. The image indicates that the DMSA/Fe_3O_4 nanoparticles are not aggregated and the mean diameter is 12 nm. The value is obtained from the measurement of more than 100 particles.

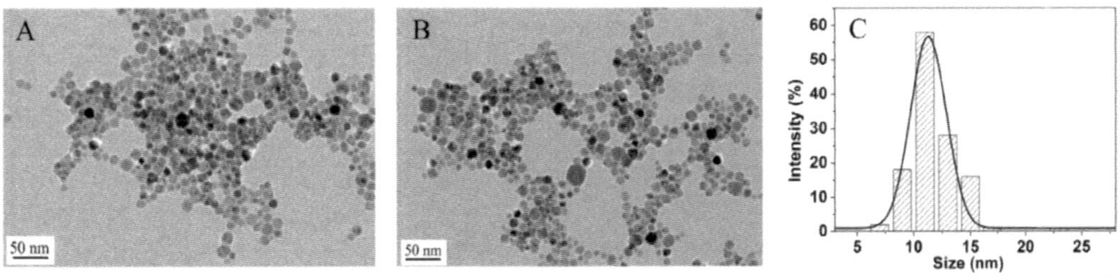

Figure 2. TEM image of OA/Fe_3O_4 (A) DMSA/Fe_3O_4 (B) and the histogram of DMSA/Fe_3O_4 size distribution (C).

The hydrodynamic size distribution of DMSA/Fe_3O_4 nanoparticles is shown in Figure 3. The hydrodynamic size of DMSA/Fe_3O_4 nanoparticle is about 25 nm. They are larger than that measured by TEM. Because the dynamic light scattering measures the hydrodynamic size of nanoparticles in the hydrated state, which includes the DMSA molecular from surface of DMSA/Fe_3O_4 nanoparticles. However, the DMSA molecular are not viewed in the TEM images because of their poor contrast to the background [12].

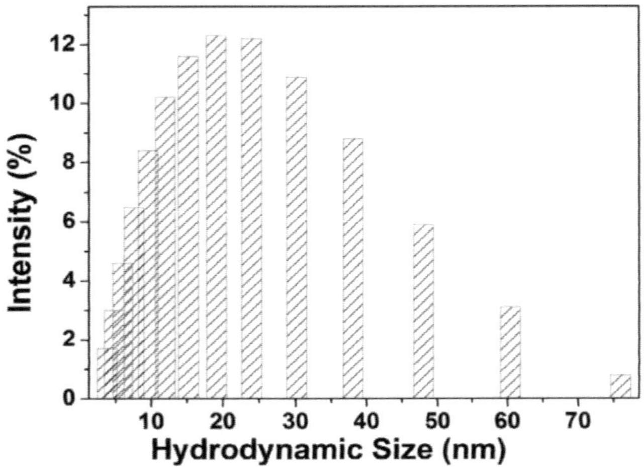

Figure 3. The hydrodynamic size distribution of DMSA/Fe_3O_4.

Magnetic measurement indicates the superparamagnetic behavior of DMSA/Fe_3O_4 at 25°C. As it is shown in Figure 4, zero coercivity and remanence are observed from the hysteresis loop. The saturation magnetization value (Ms) for DMSA/Fe_3O_4 is 31 emu/g, which is much less than the bulk ferromagnetic value of 90 emu/g [13]. This reduction of magnetization value is results from the existence of noncollinear spins at the surface of nanoparticles [14, 15].

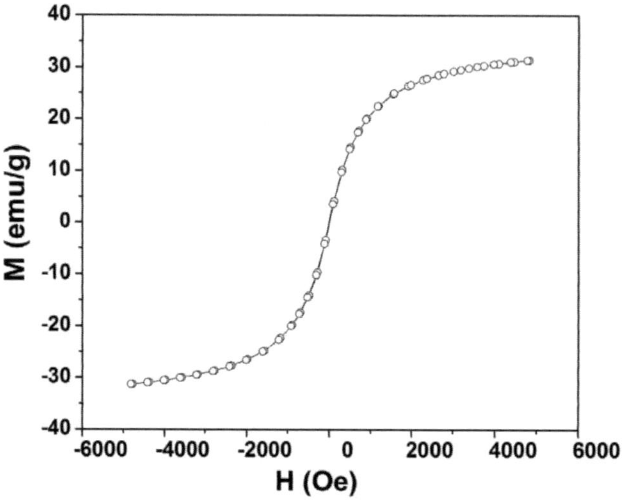

Figure 4. The hysteresis loops of DMSA/Fe_3O_4 measured by VSM at 25°C.

In order to investigate the role of DMSA in the nanoparticles, IR was used to detect the synthetic magnetic nanoparticles. As shown in Figure 5, the wide absorption band at 3410 cm^{-1} refers to O-H vibrations for the remains of water in the sample. The DMSA/Fe_3O_4 nanoparticles presents absorption peaks in 1620 cm^{-1} and 1390 cm^{-1}, which are attributed to the asymmetric and symmetric carboxylate radical stretches. And the characteristic absorption peak of Fe-O is observed at 569 cm^{-1}. Based on the IR spectra, DMSA is believed to coat on the surface of the magnetite particles. The crystalline structure of DMSA/ Fe_3O_4 nanoparticles was characterized by XRD as shown in Figure 6. The diffraction peaks of nanoparticles match well with the JCPDS card (19-0629) of magnetite.

The potential further bio-applications of Fe_3O_4 nanoparticles is to be MR contrast agent and neotype iron supplements. So it is essential to exchange oil phase to biocompatible water phase for next in vitro and vivo application.

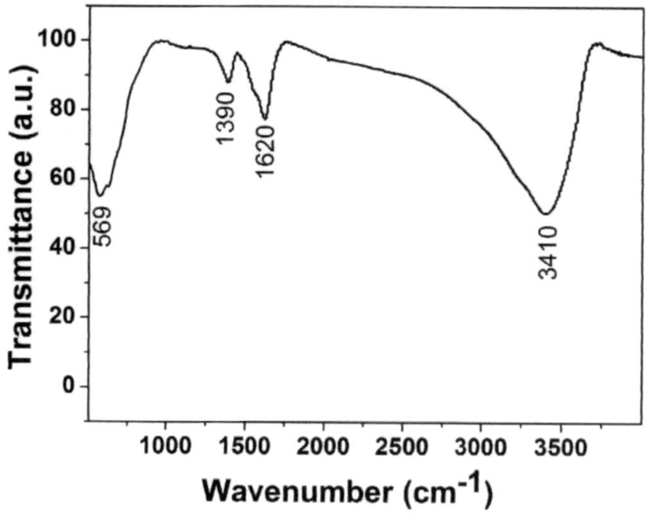

Figure 5. IR spectra of DMSA/Fe_3O_4.

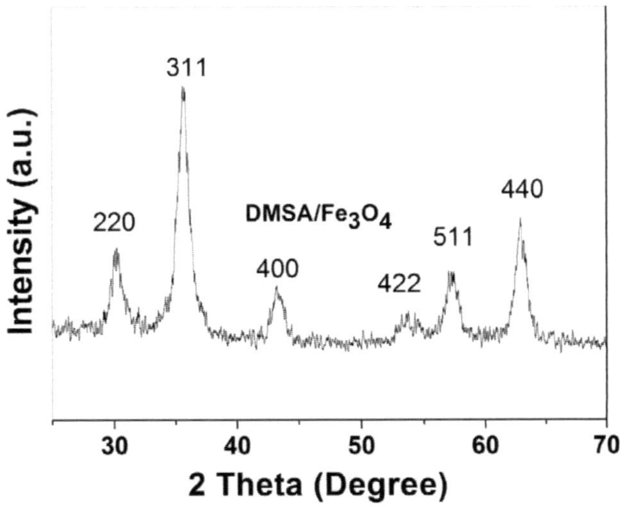

Figure 6. XRD pattern of DMSA/Fe_3O_4.

5. Summary

In a word, we described a moderate method to transfer OA coated Fe_3O_4 nanoparticles into DMSA capped nanoparticles. The DMSA/Fe_3O_4 nanoparticles have quasi-spherical morphology and core size of 12 nm. For DMSA has no toxicities and good biocompatibility, the DMSA/Fe_3O_4 nanoparticles may have potential in biologically relevant applications.

Acknowledgement

This work was supported by the function of National Key Basic Research Program of China (2011CB933503).

References

[1] Fan Z, Senapati D, Khan S A, Singh A K, Hamme A, Yust B, Sardar D and Ray P C 2013 *Chemistry–A European Journal* **19**(8) 2839-2847
[2] Shao M, Ning F, Zhao J, Wei M, Evans D G and Duan X 2012 *J Am Chem Soc* **134**(2) 1071-1077
[3] Laurent S, Forge D, Port M, Roch A, Robic C, Elst L V and Muller R N 2008 *ChemInform* **108**(35) 2064-2110
[4] Corot C and Warlin D 2013 *Wiley Interdisciplinary Reviews: Nanomed and Nanobiotec* **5**(5) 411-422
[5] Ruiz-Hernandez E, Baeza A and Vallet-Regí M 2011 *ACS Nano* **5**(2) 1259-1266
[6] Veiseh O, Gunn J W and Zhang M 2009 *Adv Drug Deliver Rev* **62**(3) 284-304
[7] Wang C, Ravi S, Garapati U S, Das M, Howell M, Mallela J, Alwarappan S, Mohapatra S S and Mohapatra S 2011 *J Mater Chem B* **1**(35) 4396-4405
[8] Mehdaoui B, Tan R P, Meffre A, Carrey J, Lachaize S, Chaudret B and Respaud M 2013 *Phys Rev B* **87**(17) 69-75
[9] Martirosyan K S 2012 *J Nanomed Nanotech* **03**(6) 325-341
[10] Babes L, Denizot B, Tanguy G, Le Jeune J J and Jallet P 1999 *J Colloid Interf Sci* **212**(2) 474-482
[11] Chen Z P, Zhang Y, Zhang S, Xia J G, Liu J W, Xu K and Gu N 2008 *Colloids Surf A* **316**(1-3) 210
[12] Tan H, Xue J M, Shuter B, Li X and Wang J 2010 *Adv Funct Mater* **20**(5) 722-731

[13] Han D H, Wang J P and Luo H L 1994 *J Magn Magn Mater* **136**(136) 176-182
[14] Morales M P, Andres-Vergés M, Veintemillas-Verdaguer S, Montero M I and Serna C J 1999 *J Magn Magn Mater* **203**(1) 146-148
[15] Coey J M D 1971 *Phys Rev Lett* **27**(27) 1140-1142

IOP Conf. Series: Materials Science and Engineering **164** (2017) 012027 doi:10.1088/1757-899X/164/1/012027

The toughening mechanism of rubber particles in polypropylene composite

L Shi and J M Xiao[1]

Department of Applied Mechanics, University of Science and Technology Beijing, Beijing 100083, China

Email: jiujiu@sas.ustb.edu.cn

Abstract. Filling polypropylene materials with rubber particles can effectively increase the toughness of PP material and improve its cushioning properties. In this paper, we used the two kinds of method of the finite element analysis and experiment to study the rubber particles toughening mechanism, got the deformation process of particles when polypropylene material compressed and the yield stress of polypropylene after compression with particles filled or not.

1. Introduction

Polypropylene is commonly used as a functional material with good comprehensive mechanical properties, but its brittleness is big, which limit its application in engineering practice. So modifying the PP to improve its ability of absorbing energy is of great importance in engineering practice. There are many kinds of modification methods such as Blending modification, filling modification and composite modification, chemistry modification, surface modification. The widely used method among them is particles-filled [1-3]. Common filler particle is mainly divided into rigid filler particles and the flexible filler particles. The previous research shown that rigid filler particles like glass beads have no obvious effect on improving the performance of polymer [4]. But powdered acrylonitrile-butadiene rubber is an ideal filling material which often used in the study of polymer modification. It uses the excellent resin modifying agent and can be directly blended with a variety of resin, it also has good compatibility. Compared with block nitrile rubber, the advantage is that the small particle size and good dispersion when blended with resin and can be directly put in the extruder, making process continuous, automation, convenient, energy conservation and reducing the cost. Rubber particles as a filler particles can through own large elastic deformation and the interaction with substrate to improve material absorption ability, so we can achieve the purpose of increasing the toughness of pp matrix [5-7].

Studies show that using natural rubber and epdm rubber to toughen polypropylene respectively, with the increase of mass fraction, the impact strength and elongation at break are dramatically increased, but the tensile strength and young's modulus decreases [8, 9]. Ravi ayyer researched the influence of rubber particle size and surface area for the structure and mechanical properties of pp matrix. AJ Kinloch used rubber to toughen epoxy resin, studied the micro structure and fracture of materials [10].

In the present article, we used powdered NBR to fill polypropylene matrix, compared to the matrix filled with glass beads of particles, combined with the finite element method and experimental methods to analyze the toughening mechanism of flexible particles.

2. Preparation and measurements

2.1. Materials

Polypropylene, Guangdong Maoming Petrochemical Factory, brand for T46F - H;
Hollow glass beads, Sinosteel Maanshan Mining Research Institute Co., LTD., types of H40;
Nitrile rubber powder, Jiangsu Jingjiang Guangsheng Pixels Materials Plant, models for GM50.

2.2. Experiment method

This experiment with micron grade nitrile rubber powder filled polypropylene, production of rubber powder/pp composites. Nitrile rubber powder and polypropylene blend with the quality ratio of 18:82. Hybrid system as a set of rubber powder and polypropylene, pure polypropylene system as the control group, two groups of materials respectively by the double screw extruder with mixing granule. Squeeze in turn set to a temperature of 205 °C, 215 °C, 215 °C, 215 °C, 215 °C, 205 °C. Speed is set to 250 r/min, melt pressure of 26 bar, melting temperature is 205 °C.

Mixing after all material system of particles with injection molding, injection molding processing of compressed sample size to 10 mm x 10 mm x 4 mm, made after the completion of the sample after cooling to finalize the design for subsequent experiments.

Normally, uniaxial compression tests were performed at room temperature by an electronic universal testing machine in accordance with GB T1041-2008. Measurement range of the machine is 10kN. The measurement accuracy is about 1%. The size of Specimens which were made for uniaxial compression tests were 10mm×10mm×4mm. The speed of crosshead was 5 mm / min [11].

The morphologies of fractured surfaces of the samples broken during impact measurement were observed through Scanning Electron Microscopy (SEM). The fractured surfaces of the samples were sputtered with gold or carbon (improve the electical conductivity)before observation by SEM [11].

2.3. Ansys method

In the process of finite element modeling, it is necessary to accurately describe the mechanical behavior of materials and establish the constitutive relation consistent with the experimental data. Powdered butadiene-acrylonitrile rubber chose the Mooney - Rivlin model that elastic materials often used in large deformation, modeling material parameters are shown in Table 1, polypropylene materials used the elastoplastic model and the elasticity parameters of the material are shown in Table 2. All the data obtained from compression experiment results.

Table 1. Modeling material parameters of Mooney-Rivlin model.

C10	C01	D1
0.769	0.012	0.001

Polypropylene's plastic section parameters in the elastic-plastic material model need to set the yield stress and plastic strain. Due to the compression test results is engineering stress and strain curve, in the finite element simulation, we need to convert it into true stress and strain, conversion relations as follows:

$$\varepsilon_t = In\left(1 + \varepsilon_n\right)$$
$$\sigma_t = \sigma_n \cdot \left(1 + \varepsilon_n\right)$$

True stress and strain for finite element calculation [12, 13].

For accurately analyzing the distribution of particle and the surrounding stress in the deformation process, a single particle model is established first. This article selects the two-dimensional model, simplify the rubber powder into the spherical, model for the spherical particles embedded in the pp matrix of cuboid. The model size should be consistent with the actual material filling ratio and particle size. The actual quality of material particles and matrix than for 18:82, we can know the particle volume fraction in general by density function conversion. The processing method of interface between the filler particles and matrix is two phase idealized adhesive, namely interface nodes implement fully binding. In order to study the interaction of filler particles and their effects for yield and deformation of the matrix, this paper established a two-dimensional plane stress model of five particles.

And in order to study the shape of filler particles on the substrate and the influence on the result of the deformation, this paper established two two-dimensional models of filling particles in different shapes. Nitrile rubber as buffer energy absorption mechanism of the flexible and rigid particles, thus established the matrix particles and glass bead filled rubber particles (rigid particles) model is used in the comparison.

Table 2. The material parameters of PP.

Material	Modulus of elasticity(MPa)	Poisson's ratio
PP	375	0.43

3. Result

3.1. Deformation process of rubber particles with different shapes in the PP composite filled with rubber particles

To study the particle shape on the particles of different energy absorption mechanism, the influence of two irregular shape particles were taken under the condition of same compression simulation, get their compression stress nephogram, as shown in Figure 1, the particles began to deformation matrix are around the edge position is the largest tensile stress, the upper and lower edge position is one of the biggest compressive stress, similar to round particles. But uniform stress distribution around a circular particles, and irregular shape of the particles at the edges and will produce a certain degree of stress concentration, but relatively little impact on the whole collective.

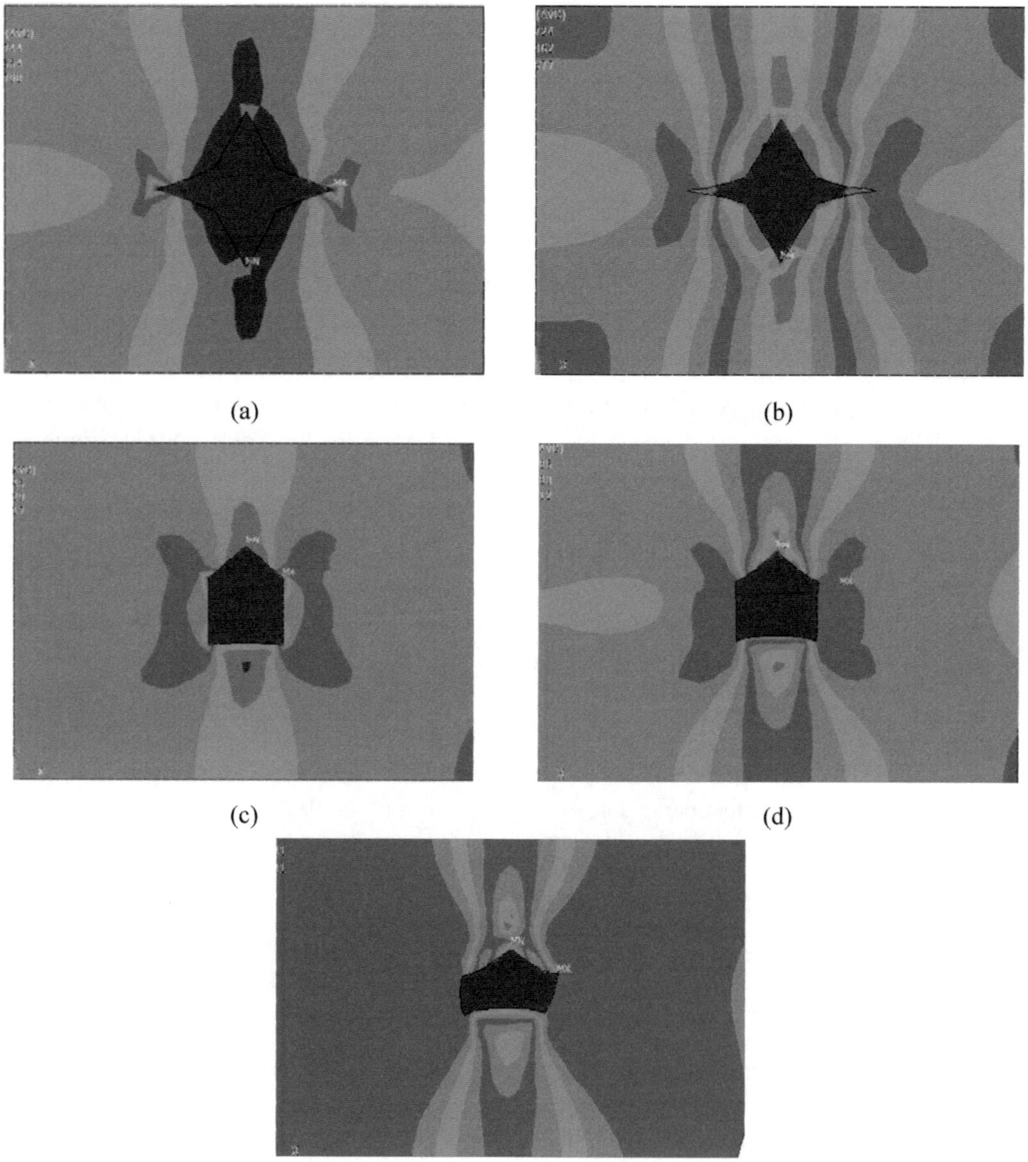

(a) (b) (c) (d) (e)

Figure 1. Deformation process of rubber particle in the PP composite filled with rubber particles (pentagon).

Table 3. The absorption of energy filled with different rubber particles.

Different shapes	circle	Shape 1	Shape 2
Absorption of energy	33	32	31

3.2. Deformation process of rubber particle in the PP composite filled with rubber particles

The particles of two-dimensional plane stress model is presented in this paper, the rubber particles is set to round, pressure on the simulation process in the vertical direction to load, the stress cloud of the compression process. As shown in Figure 2, during the compression process, the left and right sides of the rubber particles produced by the tensile stress is greater than the rest of the matrix stress, and the closer distance of rubber particles produced by the stress, the greater the first stress that rubber particles to produce large deformation, has a tendency to like horizontal expansion. Rubber ball and has been a compressive stress near the loading direction, poisson is bigger, because the rubber particles under compression load deformation speed is greater than the pp matrix [14, 15]. Particles was crushed, as shown in Table 3, we find that in the event of large elastic deformation in the process of the absorbed energy, delayed the matrix of the yield process, so as to achieve the effect of buffer energy absorption. Contrast experiments of particle microstructure before and after compression, and finite element simulation results conform to the basic, rubber particles are compressed in a plane perpendicular to the direction of loading the elongated, matrix near the particles reach the yield, first occurred plastic deformation, the mechanical properties of weakened.

The compression process of rubber particles as shown in Figure 3, the rubber particles by spherical gradually became elongated elliptic, large deformation occurred. Rubber particles formed the yield zone, near the plastic deformation of the formation of yield with led matrix, and combined with the large deformation of rubber particles, and become the main cause of buffer endergonic.

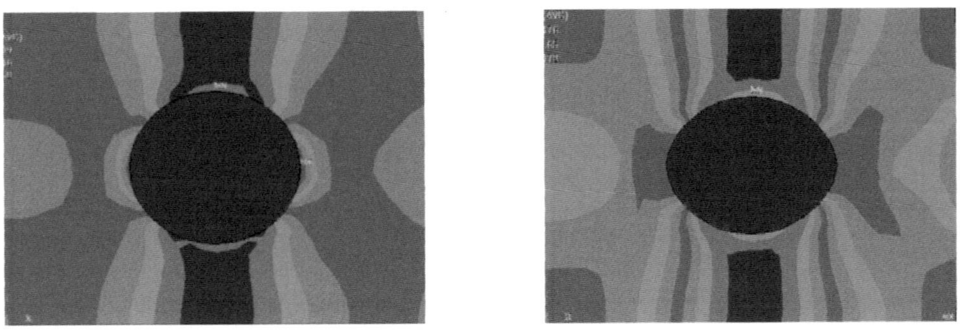

(a) (b)

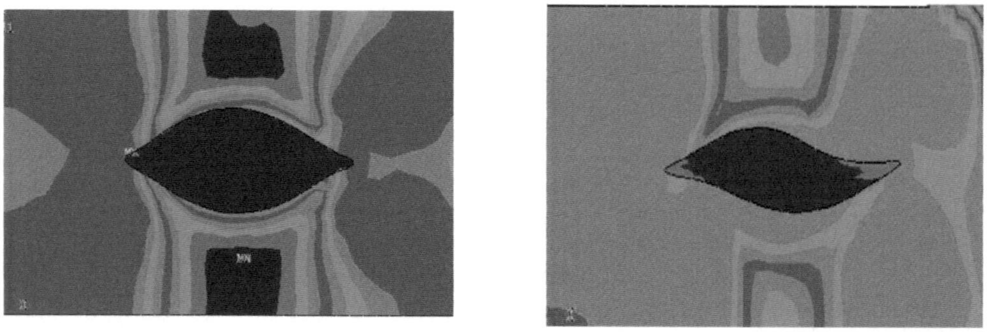

(c) (d)

Figure 2. Deformation process of rubber particle in the PP composite filled with rubber particles (circle).

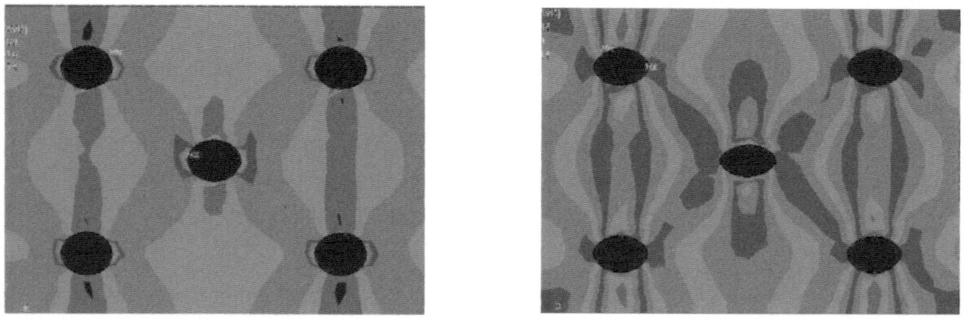

(a) (b)

Figure 3. Deformation process of rubber particles in the PP composite filled with rubber particles (five circles).

As shown in Figure 4, observation of scanning electron microscopy (SEM), sample before compression can clearly see the rubber particle and polypropylene matrix, after compression materials reach the yield limit, the rubber particle and matrix together elongated, deformation is severe, and in line with the results of finite element simulation.

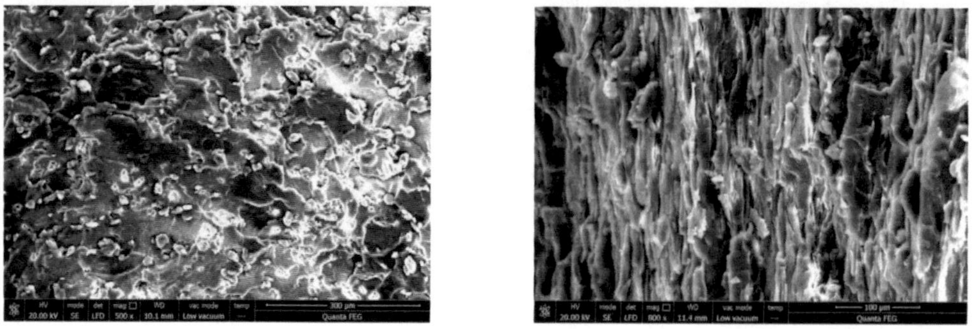

(a) (b)

Figure 4. Scanning electron microscopy (SEM) profile of the PP composite filled with rubber particles.

3.3. Effect of rubber particles on morphology and impact property of the PP composite including rubber particles

In pp matrix along the horizontal direction to take three nodes on ABC, get energy time plot as shown in Figure 5. Can be seen from the diagram, at the beginning of the curve levels did not absorb energy, over time, began to absorb energy, distance of rubber particles far point of energy absorption rate faster, that forced deformation of rubber particles absorb energy has effect on the substrate, forces acting on the matrix of smaller. Three points at the same time into the platform area, and then began to absorb energy. Beginning level relative only polypropylene matrix for a long time, the compression process at the beginning of the rubber particles absorb energy first, and the substrate to maintain current state.

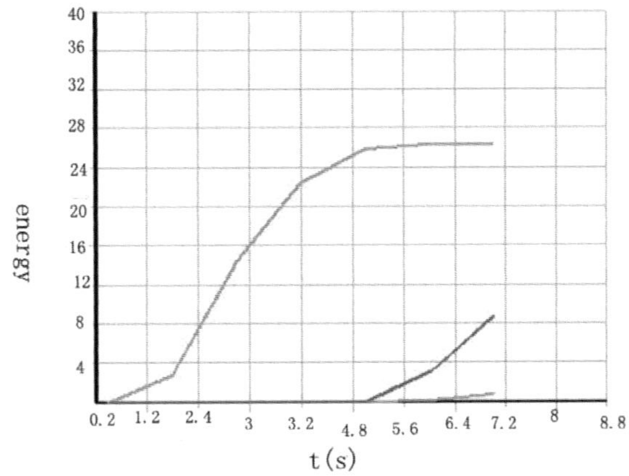

Figure 5. Curve of energy-time of the PP composite filled with rubber particles.

It can also be seen from Figure 6 that the PP composite including BNRPs could absorb more energy than that of the PP under the same stress. It's proved that filling of the rubber particles can effectively improve the toughness of pp matrix and its cushioning feature.

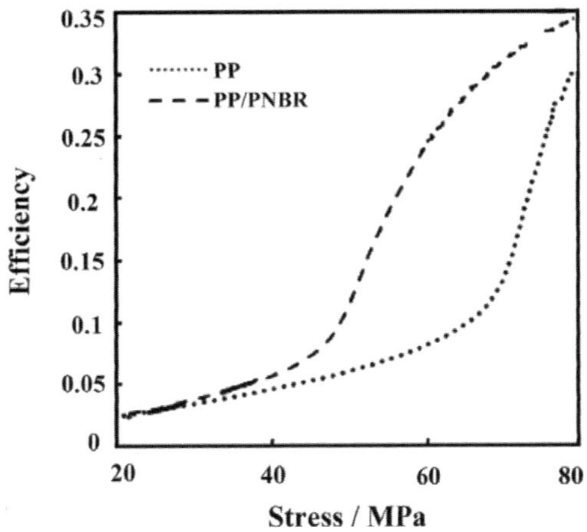

Figure 6. Efficiency-stress curves for the studied materials: the PP composite including BNRPs and PP.

4. Conclusions

Based on the finite element analysis and experimental results, the rubber particles filled polypropylene can through own deformation to absorb some of the impact energy, so as to improve the toughness of pp material, to achieve the effect of buffer energy absorption. When enough hours filling particles, the shape of the rubber particles on the energy absorption capacity is not big, the influence of the distribution of density can produce certain effect. Compared with glass beads filling particles, rubber particles mainly in the prophase work of polypropylene, and toughening effect than that of glass beads.

References

[1] Zebarjad S M and Golmakaniyoon S 2013 Influence of strain rate on the toughening effect of $caco_3$ in polypropylene/$caco_3$ composites *Journal of Vinyl and Additive Technology* **19**(4) 271-275

[2] Liu G D and Xiang M A 2004 Study on sliding wear of ultrahigh molecular weight polyethylene/polypropylene blends *Polymer Engineering and Science* **44**(1) 197-208

[3] Delhaye V, Clausen A H and Moussy F *et al* 2010 Mechanical response and micro structure investigation of a mineral and rubber modified polypropylene *Polym Test* **29**(7) 793

[4] Wu X F, Li J W and Xiao F J *et al* 2013 Fabrication and properties of hollow glass beads loaded carbon nanotubes/epoxy composites *J Macromol Sci Part B* **52**(2) 355

[5] Chan C M, Wu J S and Li J X *et al* 2002 Polypropylene/calcium carbonate nanocomposites *Polymer* **43**(9) 2981-2992

[6] Zebarjad S M, Sajjadi S A and Tahani M 2006 Modification of fracture toughness of isotactic polypropylene with a combination of EPR and $CaCO_3$ particles *Journal of Materials Processing Technology* **175**(1-3) 446-451

[7] Leong Y W, Ishak Z A M and Ariffin A 2004 Mechanical and ther-mal properties of talc and calcium carbonate-filled polypropylene hybrid composites *Journal of Applied Polymer Science* **91**(5) 3327-3336

[8] Sun L Y, Gibson R F and Gordaninejad F 2011 Multiscale analysis of stiffness and fracture of nanoparticle-reinforced composites using micromechanics and global-local finite element models *Engineering Fracture Mechanics* **78**(15) 2645-2662

[9] Boycea M C, Socrate S and Kear K *et al* 2001 Micromechanisms of deformation and recovery in thermoplastic vulcanizates *Journal of the Mechanics and Physics of Solids* **49**(6) 1323-1342

[10] Boesl B P, Bourne G R and Sankar B V 2011 Insitu multiscale analysis of fracture mechanisms in nanocompositions *Composites: Part B Engineering* **42**(5) 1157-1163

[11] Xiao J M and Chen Y N 2015 New microstructure designs of a polypropylene (PP) composite with improved impact property *Materials Letters* **152** 210-212

[12] Xu C H, Cao L M and Chen Y K 2013 Ternary blends based on poly(ethylene-naphthalate)/glass fibers/nitrile rubber: Preparation, properties and effect of dynamic vulcanization *Polymer Testing* **32**(8) 1529-1537

[13] Ayyer R, Rosenmayer T and Schreiber W *et al* 2013 Effects of micronized rubber powders on structure and properties of polypropylene composites *Waste and Biomass Valorization* **4**(1) 65

[14] Ku H, Wang H and Pattarachaiyakoop N 2011 A review on the tensile properties of natural fiber reinforced polymer composites *Composites: Part B Engineering* **42**(4) 856-873

[15] Williams J R 2010 Particle toughening of polymers by plastic void growth *Composites Science and Technology* **70**(6) 885-891

Effect of different filtration materials on oil and water separation efficiency

D Zhang[1], W X Wu and L Zhang

Key Laboratory on EOR of Education of China, Northeast Petroleum University, Daqing, Heilongjiang, 163318, China

Email: zhangdong084@163.com

Abstract. The key to remove oily wastewater by coarse graining is the selection of filtration materials. In this paper, the four kinds of filtration materials were tested the efficiency of oil and water separation, which are walnut shell, ceramsite, oil sand and polyethylene. The results of laboratory experiments showed that walnut shell and oil sands as oil-soluble media had higher separation efficiency. With the increase of sewage flow velocity, the separation efficiency of porous media decreases. With the increase of the thickness of the porous medium, the separation efficiency increases, and the separation efficiency can reach 100% when the thickness reaches a certain value.

1. Introduction

In China, most of the oil field has entered the middle and late stages of oil development, the amount of water that needs to be injected into the reservoir increases year by year, so the amount of sewage produced also increases [1-3]. Oil-water separation produces a large number of oily wastewater that can cause pollution to the environment hazards if it is not disposed of directly. It will ultimately threaten the lives of people and cause economic losses. The primary treatment of oily wastewater by using the filtration material can greatly reduce the oil content, and then a two stage treatment and advanced treatment can make the sewage indicators reach the quality requirements of the reinjection water [4-5]. In this paper, the filtering materials are selected for on-site wastewater treatment through the use of indoor experiment. The filter material can make small oil droplets coalesce into large beads and easy gravity separation, so as to improve the processing efficiency [6-7].

2. Experimental condition

2.1. Physicochemical properties of oily wastewater

Oilfield wastewater is usually oil-in-water dispersion system. Oil in oil wastewater is mainly present in four states which were free oil, dispersed oil, emulsified oil and dissolved oil, the results shown in Table 1.

Table 1. Oil status in oily waste water.

Status	Particle size	Stability	Removal method
Free oil	>100μm	Instable	Gravity separation
Dispersed oil	25~100μm	Instable	
Emulsified oil	0.1~25μm	Stable	Filtering material
Dissolved oil	<0.1μm	Extremely stable	Chemical method

The data in the table show that the stability of free oil and dispersed oil is poor and crude oil can be removed by using gravity degreasing facilities. The dissolved oil particle size is very small which needs to be removed by chemical methods.

In this paper, emulsified oil with particle size between 0.1 and 25μm was used as the object of study which is made by using the screw pump mixing white oil and water in accordance with the ratio of 1: 9 mass ratio. The average droplet size of the oil droplets in the three parallel experiments was 18.4, 17.8 and 17.5μm, which satisfied the experimental requirement. The oily wastewater obtained in the experiment is shown in Figure 1.

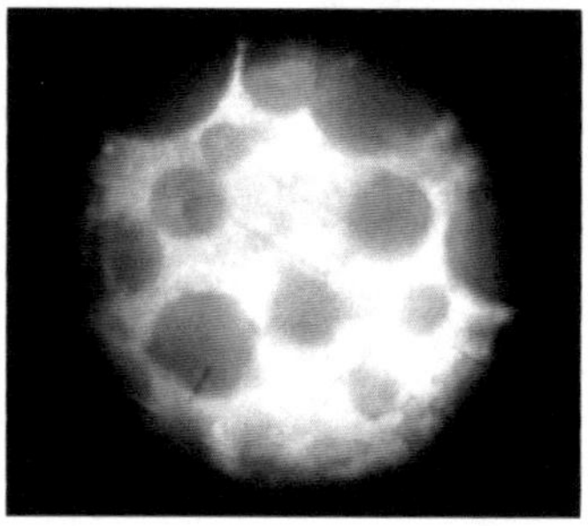

Figure 1. Oil droplets in sewage.

2.2. Oil-water separation device and filtration materials

- Experimental equipment: advection pump, sand filling tube, test tube;
- Filtration materials: walnut shell, ceramsite, polyethylene, oil sand;
- Experimental temperature: room temperature.

2.3. Experimental procedure and parameters

First of all, t is loaded into the sand filled pipe to the specified thickness, and then the oil-water mixture is filled above the filter material. The displacement of simulated wastewater to the material adsorption saturation at a speed of 2.44×10^{-4}m/s. Displacement of sewage into the adsorbed saturated media at different speeds, the amount of injected water and the oil content at the outlet end were calculated. 10% of the injected water quality is defined as the amount of oil entering and the cumulative oil level at the outlet end is defined as the oil output. Separation efficiency can be calculated. The experimental procedure is shown in Figure 2.

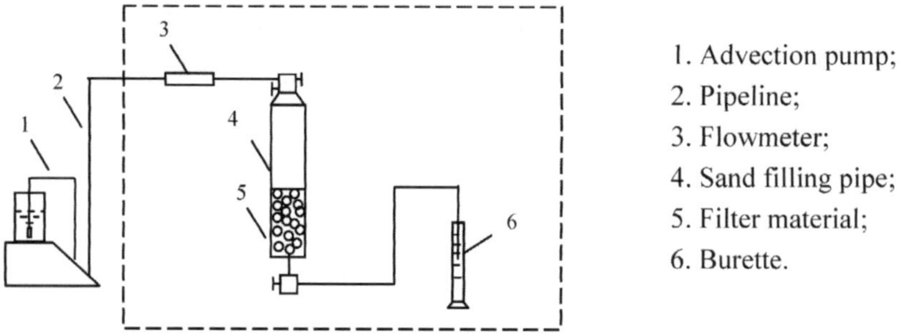

Figure 2. Experimental setup.

3. Experimental results and analysis

3.1. Seepage capacity of sewage in different filtration materials

The suction force of the oil droplets in the porous media formed by the filter material is equal to the seepage resistance. According to Darcy's formula:

$$Q = \frac{kA\Delta p}{\mu L} \tag{1}$$

The formula is deformed to:

$$\frac{\Delta p}{L} = \frac{Q\mu}{KA} \tag{2}$$

The inner diameter of the sand filling pipe is 3.148cm and the length is 6.884cm, the pressure gradient can be calculated as shown in Table 2.

Table 2. Pressure gradients of different filtration materials at different velocities.

T(s)	V(10^{-4}m/s)	ΔP/L(MPa/m)			
		walnut shell	ceramsite	polyethylene	oil sand
50	4.88	0.00359065	0.009245321	0.002273043	0.010111547
100	2.44	0.001795325	0.004622661	0.001136521	0.005055774
200	1.22	0.000897662	0.00231133	0.000568261	0.002527887
500	0.488	0.000359065	0.000924532	0.000227304	0.001011155

When the fluid and porous media are determined, the flow rate, viscosity, and cross-sectional area are all known, and the pressure gradient is inversely proportional to the permeability. The particle diameters of the four kinds of filtration materials from large to small order are polyethylene, walnuts, pro oil sands and ceramsite. The porous media with large particle diameter has large permeability and less resistance under the same velocity condition, so the pressure gradient is smaller. The porous medium with smaller particle diameter is just opposite.

3.2. Separation efficiency of sewage in different filtration materials

The oil and water separation efficiency of four kinds of filtration materials with the thickness of 1cm are shown in Figure 3.

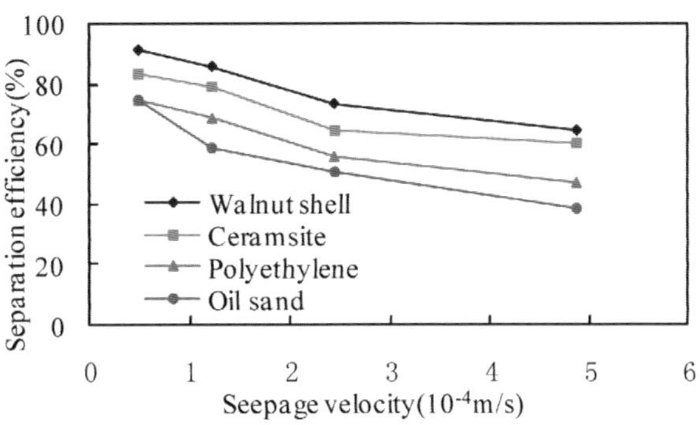

Figure 3. Separation efficiency of different filtration materials.

Under the same flow rate, the separation efficiency of walnut shell was significantly higher than the other three kinds of filter media, the separation efficiency of ceramsite was the second, and the separation efficiency of polyethylene and sand was the lowest. This is because the lipophilic of the walnut shell is the strongest, the adsorption of oil droplets in the water so that the adhesion of oil beads coalescence, from small oil beads into large oil beads, thus increasing the separation efficiency. When the sewage flows through the channel, it increases the collision chance between the oil beads. On the other hand, the oil bead is squeezed and deformed to release the oil film, which improves the separation efficiency. The oil-wettability of polyethylene and sand is poor and the separation efficiency is low.

At this time, the thickness of filter material is low, the short residence time of oil droplets in the filter media leads to a certain amount of oil droplets which have no chance to contact with the pore wall. The coalescence of the ceramisite is weaker than that of the walnut shell. Its coalescence principle is mainly collision and squeezing, forming many curved channels. It is necessary to increase the thickness of the filter material to conduct experiments. The separation efficiency of different thickness of the same media is shown in Table 3.

Table 3. Separation efficiency of walnut shell with different thickness.

$V(10^{-4}m/s)$	Separation efficiency		
	H=1cm	H=2cm	H=3cm
0.488	91.3	100	100
1.22	85.9	89.7	100
2.44	73.7	82.5	94.7
4.88	64.9	78	86.8

As can be seen from the figure, with the increase in the thickness of the filter material, the separation efficiency is also increased. This is because the increase in the thickness of the filter material to extend the oil droplets and filter media and pipe wall contact time, which is more conducive to floating up the oil droplets and increase the amount of oil. Walnut shell filter thickness of up to 3cm, the walnut shell as a filter material in different sewage flow separation efficiency of 100%. At this time, the oil output is 0ml, into the amount of oil is equal to floating oil. So the increase in the thickness of filter media is conducive to the oil droplets in the full contact with the filter, thereby enhancing the separation efficiency.

4. Conclusion

Through the experiment of sand-filling pipe, the separation efficiency of oil and water of four kinds of filter materials decreased with the increase of sewage flow rate. Under the same flow rate, the separation efficiency of walnut shell was significantly higher than the other three kinds of filter media. With the increase in the thickness of the filter material, the separation efficiency is also increased. When the thickness of walnut shell is up to 3cm, the separation efficiency can reach to 100%.

Acknowledgments

The project was supported by National Science and Technology Major Project of the Ministry of Science and Technology of China (2011ZX05009).

References

[1] Thew M T and Smyth I C 1998 *Proceedings of IMM Conference on Innovation in Physical Separation Technologies* (London:The Institution of Mining and Metallurgy) pp 77-83

[2] Li J and Zhang M 2016 *Research Progress of the Materials with Special Wettability for Oil-Water Separation* (Fuzhou: Fujian Forestry College Press) pp 37-38

[3] Zhao J F and Fan Y G 2016 Study on Oil Water Separation Test and Determination of Oil Content *Guangdong Chemical Industry* **9** 257-265

[4] Wu Y X and Xu J 2015 Oil and Water Separation Technology *Advances in Mechanics* **45**(1) 179-214

[5] Luo Y L and Xie Q H 2015 Research Progress of New Membrane Materials for Oil/Water Separation *New Chemical Materials* **43**(11) 221-223

[6] Rubio J and Souza M L 2002 *Overview of Flotation as a Wastewater Treatment Technique(Minerals Engineering)* (London: Elsevier Publishing) chapter 15 pp 139-155

[7] Svoboda J 2001 *A Realistic Description of the Processof High-Gradient Magnetic Separation (Minerals Engineering)* (London: Elsevier Publishing) chapter 14 pp 1493-1503

Stripping of titanium from TBP-decanol phase

X H Mao[1]

College of Biological and Chemical Engineering, Panzhihua University, Panzhihua 617000, Sichuan, China

Email: chdmaoxh@sina.com

Abstract. The stripping of titanium from TBP-decanol phase with hydrochloric acid has been investigated. The results showed that the stripping rate of titanium increased with decreased hydrochloric acid concentration in the aqueous phase, decreased organic and aqueous phase ratio and increased stripping temperature. Extracted titanium was easy to strip and 0.5 mol/L hydrochloric acid was suitable stripping agent. The kinetics of the stripping process was fast, since the equilibrium was reached in 10 minutes. The extraction and stripping isotherms showed that through extraction and stripping titanium (IV) was separated and enriched as about five fold. The simulated extraction and stripping of titanium (IV) in leaching solution of blast furnace slag were proceeded.

1. Introduction

Hydrochloric acid method was reported by Altair for the manufacture of nano-TiO_2 [1]. The advantage of hydrochloric acid method is the low cost titanium raw material and less waste. In China the content of TiO_2 reaches 20-22% in the blast furnace slag smelted from vanadium-titanium magnetite, which comprehensive utilization is an urgent problem. The blast furnace slag has poor leaching performance and high impurity content. Hydrochloric acid method is with strong leaching and process flexibility of extraction operation. Therefore hydrochloric acid method is suitable for the extraction of titanium in blast furnace slag.

The extraction separation is the key process of hydrochloric acid method. In Altair process titanium was separated by extraction with TRPO as the extraction agent. However TRPO is expensive and poor to phase disengagement [2, 3].

A lot examination has been made for the extraction of titanium. The main extractants are acidic organophosphorus extractants and neutral organophosphorus extractants. But the extraction equilibrium time of titanium by acidic organophosphorus extractants is long (30-120 minutes) [4-6]. Neutral organophosphorus extractants including of TRPO, TOPO and TBP have been studied respectively for the extraction of titanium [7-13]. Among these studies the extraction of titanium (IV) with mixed extractants of TBP and decanol, TOPO and decanol was reported [7]. As the addition of decanol the extraction yield of titanium increases largely. At the same time the synergistic extraction of metals has been investigated [14-15]. However the stripping behavior of titanium form TBP-decanol phase haven't been studied.

In the present study the stripping of titanium from TBP-decanol phase was studied. The potential of TBP-decanol for the separation and recovery of titanium from hydrochloric leaching solutions of titanium slag has been assessed.

2. Experimental

2.1. Materials

Titanium (IV) solutions were prepared from $TiCl_4$. TBP and decanol were provided from AODA Chemical Co., Ltd. Other chemicals were of analytical reagent grade.

2.2. Methods

Solvent extraction experiments were conducted by mixing aqueous and organic phases under the certain conditions. After phase separation, the following methods were used to determine the concentration of the metal ions in the aqueous phase. Titanium (IV) was analyzed by spectrophotometry with hydrogen peroxide. Titanium (IV), calcium (II), magnesium (II) and aluminum (III) in the hydrochloric leaching solutions of titanium slag were analyzed by ICP. The concentration of the metal ions in the organic phase was obtained by mass balance.

3. Results and discussion

3.1. Extraction of titanium (IV) by TBP and decanol

The extraction of titanium in the acid chloride solutions was studied with TBP and decanol. And the extraction conditions were optimized on the basis of the relevant report [7]. The results showed that with the increasing of extractants total concentration the extraction rate of titanium (IV) first increased and then unchanged (Figure 1), with the increasing of the extraction temperature the extraction percentage of titanium (IV) decreased, with the increasing of the O/A the extraction rate of titanium (IV) increased and the concentration of titanium (IV) in the organic phase decreased, titanium (IV) in the acid chloride solutions could be extracted mostly by TBP and decanol within 10 minutes. And that the suitable extraction conditions was that using 1.0 mol/L TBP and 0.67 mol/L decanol as the extractant, with O/A equal to 1, 25℃ extrantion temperature and 10 minutes extraction time.

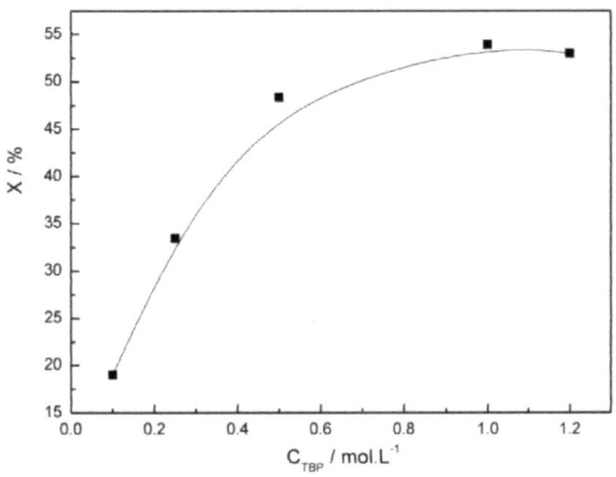

Figure 1. Effect of TBP-decanol concentration on the extraction of titanium (IV).

X-Extraction rate of titanium (IV)

Based on the above results extraction isotherm has been obtained for a solution containing 0.15 mol/L titanium (IV) in 8 mol/L hydrochloric acid with 1.0 mol/L TBP and 0.67 mol/L decanol in kerosene as extraction agent (Figure 2). The result showed that the concentration of titanium (IV) in aqueous phase was below 0.02 mol/L in three or four stages at O/A equal to 1.

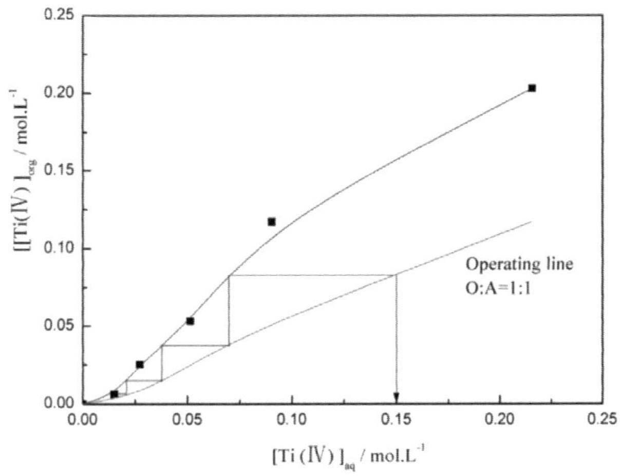

Figure 2. McCabe-Thiele plot for titanium (IV) extraction.

3.2. Stripping of titanium

3.2.1. Choice of stripping agent. To realize the separation and recovery of titanium (IV) the stripping behavior of titanium from TBP-decanol phase has been studied with hydrochloric acid as the stripping agent (Figure 3).

The results showed that with the increase of hydrochloric acid concentration the stripping rate of titanium decreased. When using 0.5 mol/L hydrochloric acid as the stripping agent, the stripping

IOP Conf. Series: Materials Science and Engineering **164** (2017) 012029 doi:10.1088/1757-899X/164/1/012029

percentage of titanium was above 90%.

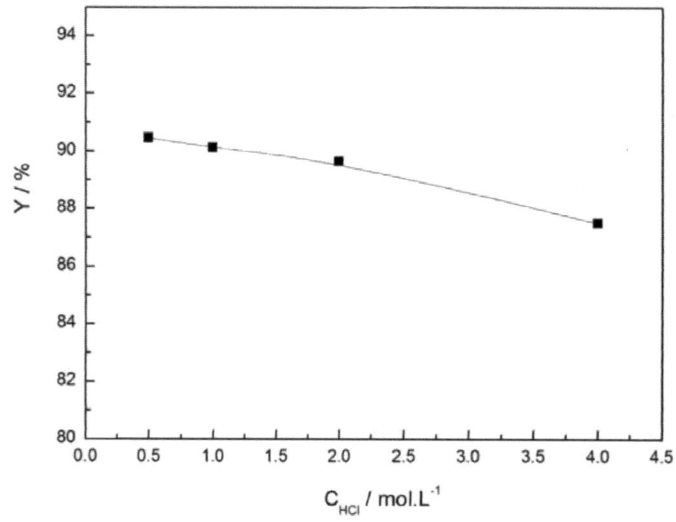

Figure 3. Stripping of titanium from TBP-decanol phase with HCl.

Y-Stripping rate of titanium

3.2.2. Effect of O/A. In the case of stripping by 0.5 mol/L hydrochloric acid the influence of stripping O/A was studied (Figure 4). The results showed that an increase in the O/A led to an increase in the stripping percentage of titanium and a decrease in the titanium concentration in stripping solutions. The suitable stripping O/A was 5.

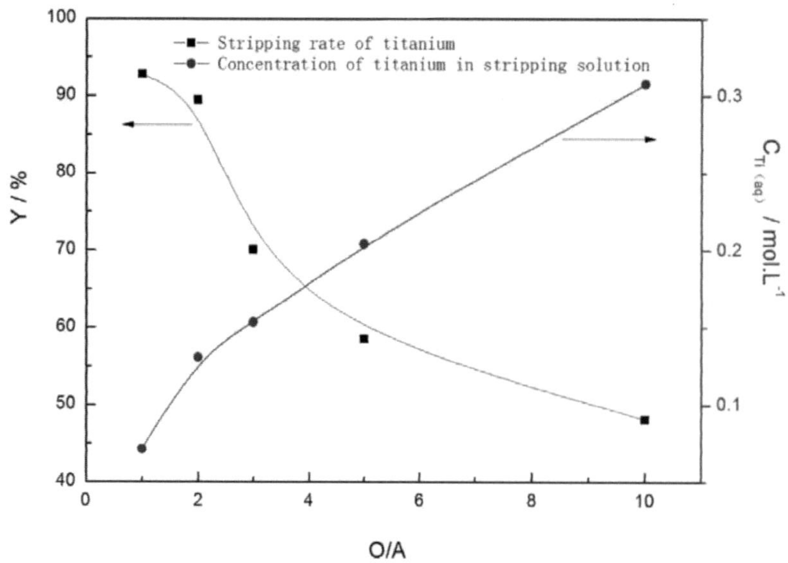

Figure 4. Effect of the phase radio on the stripping of titanium.

3.2.3. Effect of stripping temperature. The impact of stripping temperature on the stripping of titanium was examined. The result showed that with the increase of the stripping temperature the stripping percentage of titanium increased slightly (Figure 5). The related stripping reaction is then endothermic. And the suitable stripping temperature was room temperature.

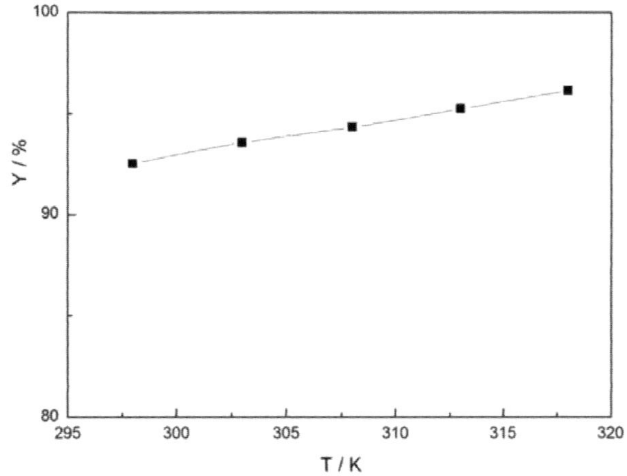

Figure 5. Effect of temperature on the stripping of titanium.

3.2.4. Effect of stripping time. Titanium in TBP-decanol phase was mixed with 0.5 mol/L hydrochloric acid for 1-30 minutes. The stripping percentage of titanium was above 90% within 10 minutes (Figure 6). From the result it was found that the stripping of titanium with 0.5 mol/L hydrochloric acid was a quick process, suitable to industrial application.

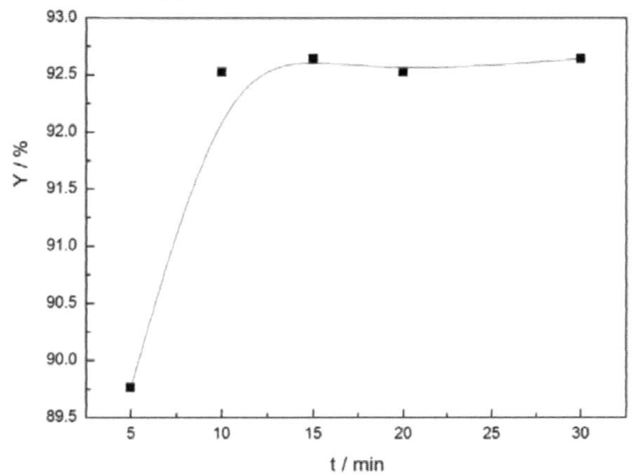

Figure 6. Effect of stripping time on the stripping of titanium.

3.2.5. Stripping isotherm. Based on these results, a stripping isotherm for titanium (IV) was obtained from a organic phase containing 0.15 mol/L titanium with 0.5 mol/L hydrochloric acid as stripping agent. A McCabe-Thiele plot (Figure 7) of titanium stripping showed that the concentration of titanium in stripping phase was below 0.01 mol/L in four stages with A/O equal to 1:5.

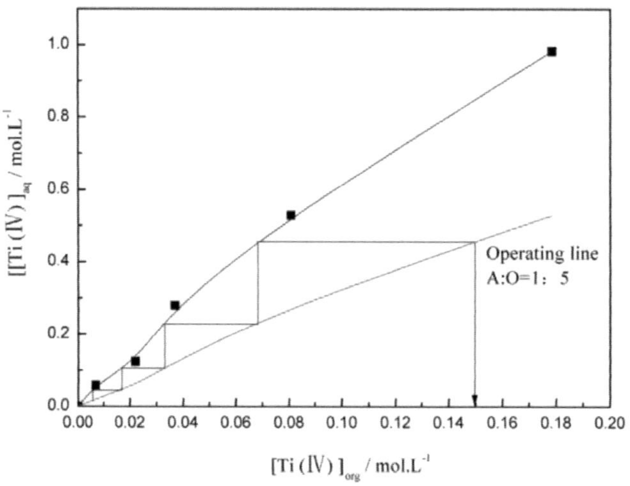

Figure 7. McCabe-Thiele plot for titanium stripping

3.3. Extraction and separation of titanium (IV) from leaching solutions of titanium slag

The developed method was applied for the separation and recovery of titanium (IV) from the hydrochloric leaching solutions of titanium slag with 1.0 mol/L TBP and 0.67 mol/L decanol as extraction agent. As it is well known that ferric iron can be extracted selectively with TOA, in the present study the hydrochloric leaching solutions of titanium slag firstly was extracted to separate ferric iron by TOA.

Simulated counter-current extraction and stripping were made at laboratory scale at 298 K. The process comprised the steps of (1) extraction separation of ferric iron by TOA, (2) selective extraction of titanium (IV) from the iron raffinate in three stages of counter-current extraction (O/A = 1/1) with the mixed solvent system and (3) stripping of titanium in organic phase with 0.5 mol/L HCl in four stages (A/O = 1/5) to recover titanium. Through the concentration determination by ICP the results of the process were given in Table 1. The present investigations revealed that the mixed solvent system can be used as a selective extractant for the separation and recovery of titanium (IV) in the hydrochloric leaching solutions of titanium slag. Titanium (IV) was separated and concentrated through the process.

Table 1. Extraction and separation of titanium (IV) in the leaching solutions of titanium slag.

	Fe (mol/L)	Ti (mol/L)	Mg (mol/L)	Al (mol/L)	Ca (mol/L)
Feed	0.02697	0.151857	0.1986	0.2716	0.4868
Raffinate of Fe	0.00161	0.152079	0.1988	0.2712	0.4858
Raffinate of Ti	——	0.008732	0.2128	0.3169	0.5535
Strip solution of Ti	——	0.568839	0.0027	0.0009	0.0313

4. Conclusion

The mixed extraction agent of TBP and decanol is effective and selective for the extraction of titanium (IV) in acidic chloride solutions. The extracted titanium (IV) is easy to strip and the stripping process is fast. The suitable stripping condition was that using 0.5 mol/L hydrochloric acid as stripping agent, with A/O equal to 1:5, 25℃ stripping temperature and 10 minutes stripping time. Extraction and stripping isotherms were generated. The concentration of titanium (IV) in aqueous phase was below 0.02 mol/L in three or four stages with O/A equal to 1. The concentration of titanium (IV) in stripping phase was below 0.01 mol/L in four stages with A/O equal to 1:5. The extraction and stripping of titanium (IV) in leaching solution of blast furnace slag were proceeded.

References

[1] Zhang W, Zhu Z and Cheng C Y 2011 *Hydrometallurgy* **108**(3) 177-188
[2] Reddy M L P and Saji J 2002 *Miner. Process Extr. M.* **23**(3-4) 199-227
[3] Zhu Z W, Zhang W S and Cheng C Y 2011 *Hydrometallurgy* **105**(3) 304-313
[4] Da Silva G C, Da Cunha J W D, Dweck J and Afonso J C 2008 *Miner. Eng.* **21**(5) 416-419
[5] Saji J, John K S and Reddy M L P 2000 *Solvent. Extr. Ion. Exc* **18**(5) 877-894
[6] Awwad N S and Ibrahium H A 2013 *J. Environ. Chem. Eng.* **1**(1-2) 65-72
[7] Allal K M, Hauchard D, Stambouli M, Pareau D and Durand G 1997 *Hydrometallurgy* **45**(1) 113-128
[8] El Dessouky S I, El Nadi Y A, Ahmed I M, Saad E A and Daoud J A 2008 *Chem. Eng. Process* **47**(2) 177-183
[9] Narayanan R P and Lakshmipathy R M 2004 *J. Chem. Technol. Biotechnol.* **79**(7) 734-741
[10] Saji J and Reddy M L P 2001 *Hydrometallurgy* **61**(2) 81-87
[11] Saji J K, Saji J, Reddy M L P, Ramamohan T R and Rao T P 1999 *Hydrometallurgy* **51**(1) 9-18
[12] Seyfi S and Abdi M 2009 *Miner. Eng.* **22**(2) 116-118
[13] Hao X L, Lv L, Liang B, Li C, Wu P and Wang J 2012 *Hydrometallurgy* **113-114**(3) 185-191
[14] Zhang Y, Zhang T A, Lv G Z, Zhang G Q, Liu Y and Zhang W G 2016 *Hydrometallurgy* **166** 87-93
[15] Zhao Z G, Baba Y, Kubota F, Kamiya N and Goto M 2014 *J. Chem. Eng. Japan* **47**(8) 656-662